CHRONIC OBSTRUCTIVE PULMONARY DISEASE:
PATHOGENESIS TO TREATMENT

The Novartis Foundation is an international scientific and educational charity (UK Registered Charity No. 313574). Known until September 1997 as the Ciba Foundation, it was established in 1947 by the CIBA company of Basle, which merged with Sandoz in 1996, to form Novartis. The Foundation operates independently in London under English trust law. It was formally opened on 22 June 1949.

The Foundation promotes the study and general knowledge of science and in particular encourages international co-operation in scientific research. To this end, it organizes internationally acclaimed meetings (typically eight symposia and allied open meetings and 15–20 discussion meetings each year) and publishes eight books per year featuring the presented papers and discussions from the symposia. Although primarily an operational rather than a grant-making foundation, it awards bursaries to young scientists to attend the symposia and afterwards work with one of the other participants.

The Foundation's headquarters at 41 Portland Place, London W1N 4BN, provide library facilities, open to graduates in science and allied disciplines. Media relations are fostered by regular press conferences and by articles prepared by the Foundation's Science Writer in Residence. The Foundation offers accommodation and meeting facilities to visiting scientists and their societies.

Information on all Foundation activities can be found at http://www.novartisfound.org.uk

Novartis Foundation Symposium 234

CHRONIC OBSTRUCTIVE PULMONARY DISEASE: PATHOGENESIS TO TREATMENT

2001

JOHN WILEY & SONS, LTD

Chichester · New York · Weinheim · Brisbane · Singapore · Toronto

Other Wiley Editorial Offices

John Wiley & Sons, Inc., 605 Third Avenue,
New York, NY 10158-0012, USA

WILEY-VCH Verlag GmbH, Pappelallee 3,
D-69469 Weinheim, Germany

Jacaranda Wiley Ltd, 33 Park Road, Milton,
Queensland 4064, Australia

John Wiley & Sons (Asia) Pte Ltd, 2 Clementi Loop #02-01,
Jin Xing Distripark, Singapore 129809

John Wiley & Sons (Canada) Ltd, 22 Worcester Road,
Rexdale, Ontario M9W 1L1, Canada

Novartis Foundation Symposium 234
ix+286 pages, 34 figures, 12 tables

Library of Congress Cataloging-in-Publication Data

Chronic obstructive pulmonary disease : pathogenesis to treatment
⠀⠀⠀⠀⠀p. ; cm. – (Novartis Foundation symposium ; 234)
⠀⠀⠀Editors, Derek Chadwick and Jamie A. Goode.
⠀⠀⠀Includes bibliographical references and index.
⠀⠀⠀ISBN 0-471-49437-2 (hbk.)
⠀⠀⠀1. Lungs – Diseases, Obstructive. I. Chadwick, Derek. II. Goode, Jamie. III. Series.
⠀⠀⠀[DNLM: 1. Lung Diseases, Obstructive – physiopathology – Congresses.⠀⠀2. Lung
Diseases, Obstructive – therapy – Congresses. WF 600 C5527 2000]
⠀⠀⠀RC776.O3.C478 2000
⠀⠀⠀616.2′4–dc21⠀⠀⠀⠀⠀⠀⠀⠀⠀⠀⠀⠀⠀⠀⠀⠀⠀⠀⠀⠀⠀⠀⠀⠀⠀00-063291

British Library Cataloguing in Publication Data

A catalogue record for this book is available from the British Library

ISBN 0 471 49437 2

Contents

Participants

A. G. N. Agustí   Institut de Medicina Respiratoria, Hospital Universitari Son Dureta, Andrea Doria 55, 07014 Palma de Mallorca, Spain

P. J. Barnes   National Heart and Lung Institute, Imperial College School of Medicine, Dovehouse Street, London SW3 6LY, UK

P. M. A. Calverley   University Clinical Departments, University Hospital Aintree, Longmoor Lane, Liverpool L9 7AL, UK

E. J. Campbell   Department of Internal Medicine, University of Utah School of Medicine, 410 Chipeta Way, Room 108, Salt Lake City, UT 84108, USA

T. E. Cawston   Department of Rheumatology, Department of Medicine, University of Newcastle, Framlington Place, Newcastle upon Tyne NE2 4HH, UK

M. Dunnill   Merton College, Oxford OX1 4JD, UK

J. C. Hogg   St Paul's Hospital, The University of British Columbia, McDonald Research Laboratories, 1081 Burrard Street, Vancouver, Canada V6Z 1Y6

A. Jackson   Novartis Horsham Research Centre, Wimblehurst Road, Horsham, West Sussex RH12 4AB, UK

P. K. Jeffery   Lung Pathology Unit (Department of Gene Therapy), Imperial College School of Medicine at The Royal Brompton Hospital, Sydney Street, London SW3 6NP, UK

D. A. Lomas   Department of Medicine, University of Cambridge, Cambridge Institute for Medical Research, Wellcome Trust/Medical Research Council Building, Hills Road, Cambridge CB2 2XY, UK

W. MacNee (*chair*)   ELEGI/Colt Laboratories, Department of Medical and
Radiological Sciences, Wilkie Building, The University of Edinburgh, Medical
School, Teviot Place, Edinburgh EH8 9AG, UK

A. Mantovani   Cattedra di Patologia Generale, Università di Brescia e Istituto
Mario Negri, via Eritrea 62, 20157 Milan, Italy

D. Massaro   Lung Biology Laboratory, Georgetown University School of
Medicine, Washington, DC 20007-2197, USA

J. A. Nadel   Cardiovascular Research Institute, Department of Medicine,
University of California, San Francisco, CA 94143-0130, USA

P. Nicklin   Novartis Horsham Research Centre, Wimblehurst Road, Horsham,
West Sussex RH12 5AB, UK

P. D. Paré   University of British Columbia Pulmonary Research Laboratory,
St Paul's Hospital, Vancouver, Canada, V6Z 1Y6

C. Poll   Novartis Horsham Research Centre, Wimblehurst Road, Horsham,
West Sussex RH12 5AB, UK

S. I. Rennard   Pulmonary and Critical Care Medicine Section, Department of
Internal Medicine, University of Nebraska Medical Center, Omaha,
NE 68198-5300, USA

D. F. Rogers   Department of Thoracic Medicine, National Heart and Lung
Institute (Imperial College), Dovehouse Street, London SW3 6LY, UK

R. M. Senior   Respiratory and Critical Care Medicine, Barnes-Jewish Hospital
(North Campus), 216 S. Kingshighway Boulevard, St Louis, MO 63110, USA

S. D. Shapiro   Department of Pediatrics, Medicine and Cell Biology,
Washington University School of Medicine at St Louis Children's Hospital,
St Louis, MO 63110, USA

E. K. Silverman   Channing Laboratory/Pulmonary and Critical Care Division,
Brigham and Women's Hospital, 181 Longwood Avenue, Boston, MA 02115,
USA

R. A. Stockley   Department of Medicine, Queen Elizabeth Hospital, Edgbaston, Birmingham B15 2TH, UK

J. Stolk   Department of Pulmonology (C3-P), Leiden University Medical Center, PO Box 9600, 2300 RC, Leiden, The Netherlands

J. A. Wedzicha   Academic Respiratory Medicine, St Bartholomew's and Royal London School of Medicine and Dentistry, St Bartholomew's Hospital, West Smithfield, London EC1A 7BE, UK

T. J. Williams   Leukocyte Biology Section, Biomedical Sciences Division, Sir Alexander Fleming Building, Imperial College School of Medicine, London SW7 2AZ, UK

E. F. Wouters   Department of Pulmonology and Department of Human Biology, Maastricht University, Maastricht, The Netherlands

Introduction

W. MacNee

ELEGI/Colt Laboratories, Department of Medical and Radiological Sciences, Wilkie Building, The University of Edinburgh, Medical School, Teviot Place, Edinburgh EH8 9AG, UK

The purpose of this meeting is to discuss pathological processes in chronic obstructive pulmonary disease (COPD), with the aim of providing a view on the role of future treatments for this condition based on pathogenesis. There have been a number of previous Ciba/Novartis Foundation symposia on respiratory topics. In 1962 there was a meeting on lung structure and function, at which COPD (or chronic bronchitis and emphysema, as it was then known) was discussed. However, the meeting that was most influential was not actually one of the major symposia, but was a Guest Symposium held in 1958 and published in 1959 in *Thorax*. This is what the late Professor David Flenley told me to read when I first became interested in COPD. The participants of the Guest Symposium were exclusively British and included some famous names such as Charles Fletcher, Philip Hugh-Jones and Guy Scadding. The aim of the 1958 meeting was to discuss terminology, definitions and classification of this condition; as you are all aware, we haven't moved on very much further in these particular areas over the last 40 years! The main conclusion of this meeting was to propose a revised terminology for this group of conditions. It is interesting to note that asthma does not feature in the new terminology and neither of course does 'small airways disease', since the relevance of this condition had not been described.

Fortunately at this meeting we are not focusing our discussions on definitions and terminology, or else we would be unlikely in three days to come to any firm conclusions. One of the problems with COPD — and this is likely to be a recurring theme in this meeting — is that it is a group of conditions produced by different pathological processes involving different sites in the lungs — chronic bronchitis in the large airways, 'small airways disease' or bronchiolitis, and emphysema. Clearly, one pathogenic process does not produce these diverse pathologies. Furthermore, different pathologies are more prominent in different individuals. Thus in future it may be necessary to provide a more detailed classification of COPD.

It may be useful to pose some questions that we may wish to address in this symposium. *First, which pathologies should we target for treatment?* More knowledge

1

of the pathogenic processes at the different lung sites may provide new targets for future treatment. In the Ciba Guest Symposium of 1958 considerable attention was given to the different pathologies, for example differences between the two major types of emphysema, central lobular and panlobular emphysema. There may be some key findings in these different pathological conditions which relate to the mechanisms of disease.

What can we learn from studies of present therapies? We now have much more information on the treatment of COPD, and new information from large clinical trials are becoming available. Among the more important new information are recent data on the effects of corticosteroid therapy in COPD.

Another critical question concerns what we can learn from studies of the susceptibility to COPD. Studies such as that by Burrows and colleagues show that although the degree of airways obstruction increases with increasing pack years of smoking there is huge variability in the effects of smoking, so that only 15–20% of smokers develop clinically significant airways obstruction (Sherrill et al 1994). This susceptibility to the effects of smoking can also be shown in the pathological changes of COPD. In studies by David Lamb on post mortem lungs, morphometric measurements of distal airspace size shows a clear relationship between AWUV (airspace wall surface area per unit volume of lung tissue) and age in non-smokers, with increasing age airspace size increases (Gillooly & Lamb 1993). Interestingly most cigarette smokers had values of distal airspace size within the normal non-smoking range and only 20% of smokers had an increase in airspace size for their age. This again suggests that there is susceptibility in some smokers to the development of distal airspace enlargement, a defining feature of emphysema.

There are two further key questions that will be addressed at this meeting. *These are, which mechanism(s) and which cell(s) should we target for treatment?* I hope that we don't fall into the trap of believing that our own pet mechanism or cell is the key to this disease. Instead, the idea behind this meeting is to discuss the different cells and mechanisms involved in the pathogenesis of COPD and put this into a meaningful perspective, with a view to future therapy for this condition. We will be discussing neutrophils, macrophages, T cells, protease/antiprotease balance, oxidant/antioxidant balance, and mucus hypersecretion. I hope we will be able to consider these topics individually, discuss their interactions and their relative importance in the pathogenesis of this condition.

Looking to the future are there new mechanisms which we should target for treatment? We will be hearing of the effect of retinoids on the regeneration of alveoli and of other new potential therapies for COPD. We will also hear of other new hypotheses related to exacerbations and to the systemic effects of COPD. I hope that as part of our discussion we might also develop ideas for key future studies in COPD.

I will end with a quote from an article on chronic airways disease published in 1923. 'The trite observation that "familiarity breeds contempt" is essentially true for the outlook for chronic bronchitis. Those afflicted are inclined to accept the complaint, and those called upon to treat it don't really find it sufficiently interesting enough to study it closely.' I hope that we will help to change this rather negative view of COPD by our efforts in this symposium.

References

Ciba Foundation 1962 Pulmonary structure and function. J. A. Churchill, London (Ciba Found Symp 69)

Ciba Guest Symposium 1959 Terminology, definitions and classifications of chronic pulmonary emphysema and related conditions. Thorax 14:286–299

Gillooly M, Lamb D 1993 Microscopic emphysema in relation to age and smoking habit. Thorax 48:491–495

Sherrill DL, Holberg CJ, Enright PL, Lebowitz MD, Burrows B 1994 Longitudinal analysis of the effects of smoking onset and cessation on pulmonary function. Am J Respir Crit Care Med 149:591–597

Chronic obstructive pulmonary disease: an overview of pathology and pathogenesis

James C. Hogg

St Paul's Hospital, The University of British Columbia, McDonald Research Laboratories, 1081 Burrard Street, Vancouver, Canada V6Z 1Y6

Abstract. A cigarette smoke-induced inflammatory process underlies the pathogenesis of the majority of pathologic lesions associated with chronic obstructive pulmonary disease (COPD). In chronic bronchitis, this process is located in the mucosa, gland ducts and glands of intermediate sized bronchi with an internal diameter of 2–4 mm. The mucus-containing exudate produced by the inflammatory response overpowers the normal clearance mechanisms, resulting in the cough and expectoration that characterize chronic bronchitis. In some cases of chronic bronchitis, the inflammatory process extends to smaller bronchi and bronchioles less than 2 mm in internal diameter. In this location, the inflammatory process thickens the wall, narrows the lumen and destroys the parenchymal support of the airways. These changes progressively increase peripheral airways resistance and eventually reduce the patient's ability to empty their lungs to a degree that can be measured by a reduction in FEV_1 (forced expiratory volume in one second). The reduction in lung surface area produced by parenchymal inflammation contributes to the decline in FEV_1 by reducing lung elastic recoil, which is the major force driving air out of the lung. It also contributes to the reduction in diffusing capacity by reducing the lung capillary bed. The purpose of this presentation is to review the quantitative aspects of these pathological changes and attempt to provide insight into factors which result in progression of these lesions.

2001 Chronic obstructive pulmonary disease: pathogenesis to treatment. Wiley, Chichester (Novartis Foundation Symposium 234) p 4–26

The inflammatory process is responsible for the lesions that cause chronic obstructive pulmonary disease (COPD). This includes the lesions that cause chronic cough and sputum production (Mullen et al 1985), peripheral airways obstruction (Cosio et al 1978), emphysematous destruction of the lung surface (Janoff 1983) and the changes in the pulmonary vasculature that contribute to right heart failure (Wright et al 1983, Peinado et al 1999). It also contributes to recurrent exacerbations of COPD where acute changes are superimposed on chronic disease (Smith et al 1980, Saetta et al 1994). In the interest of time, this presentation will concentrate on the pathology and pathogenesis of the lesions responsible for the bronchitis, airways obstruction and emphysema leaving the other features of COPD to be discussed elsewhere.

The tobacco smoking habit is the major aetiological factor for the inflammatory lesions in the airways and parenchyma. There is now good evidence that everyone who smokes develops lung inflammation and one of the most interesting features of this observation is that only a minority of these smokers (approximately 15–20%) develop COPD (Speizer & Tager 1979, Fletcher et al 1976). This strongly suggests that the cigarette smoke-induced lung inflammation must be amplified by other host and environmental risk factors to cause disease and, although many additional risk factors have been identified, it is unclear how this amplification takes place.

Chronic bronchiti

Figure 1 shows a histologic section of a normal bronchus where a duct connects a mucous gland to the surface epithelial lining of the bronchial lumen. Reid and her associates used the size of the mucous glands as a yardstick for measuring chronic bronchitis but did not implicate inflammation as a cause for the gland enlargement and excess mucus production (Reid 1960). In a study of a large number of cases, Thurlbeck and colleagues were able to show that although the size of the bronchial mucous glands was increased in established bronchitis, there was no clear separation in airway mucous gland size between patients with chronic bronchitis and controls (Thurlbeck & Angus 1964). A re-evaluation of this problem some years later (Fig. 2) showed that chronic bronchitis was associated with inflammation of the airway mucus secreting apparatus including the mucosal surface, the submucosal glands and gland ducts particularly in the smaller bronchi between 2 and 4 mm in diameter (Mullen et al 1985).

Chronic inflammation of the bronchial wall is also associated with connective tissue changes that include increased amounts of smooth muscle and degenerative changes in the airway cartilage as well as a metaplastic process in the epithelium that results in greater numbers of goblet and squamous cells (Thurlbeck & Angus 1964, Jamal et al 1984, Carlile & Edwards 1983, Mackenzie et al 1969, Haraguchi et al 1999, Thurlbeck et al 1974). The nature of the inflammatory cells present in the airway tissue and lumen in chronic bronchitis and how inflammatory cells interact with the mucus secreting apparatus is currently under active investigation and will be discussed in later presentations at this symposium.

The site of airways obstruction

The smaller bronchi and bronchioles less than 2 mm in diameter are the major site of airways obstruction in COPD (Hogg et al 1968). These small airways offer very

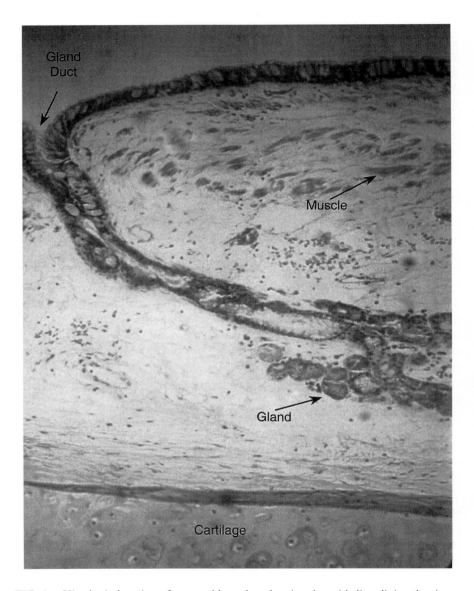

FIG. 1.   Histological section of a normal bronchus showing the epithelium lining the airway lumen and a duct leading to a bronchial mucus gland. The opening of the duct, the airway smooth muscle, the gland and the cartilage are labelled for orientation. (Original photograph courtesy of the late Dr W. M. Thurlbeck.)

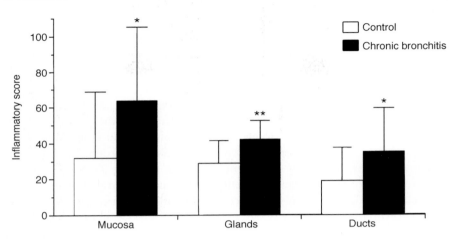

FIG. 2.   Data obtained using a scoring system to grade inflammation in various components of the bronchial wall. This scoring system documented the increased inflammatory response in the mucosal surface of the airway lumen, glands and gland ducts in patients with chronic bronchitis compared to age and smoking matched controls. From Mullen et al (1985).

little resistance in the normal lung suggesting that a considerable amount of disease can accumulate in these airways before symptoms of airways obstruction appear (Hogg et al 1968, Macklem & Mead 1967). Although the proportion of total resistance attributed to the peripheral airways of the normal lung has been challenged (van Braband et al 1983), there is agreement that these smaller airways become the major site of obstruction in COPD (Hogg et al 1968, van Braband et al 1983).

The proposed mechanisms for the increase in peripheral airways resistance include destruction of their alveolar support (Dayman 1951), loss of elastic recoil in the parenchyma that provide this support (Butler et al 1960), a decrease in the elastic force available to drive flow out of the lung (Mead et al 1967) and structural narrowing of the airway lumen by the inflammatory process (Cosio et al 1978, Hogg et al 1968, Matsuba & Thurlbeck 1972). Direct measurements of peripheral airways resistance in postmortem lungs from cases of COPD with severe airways obstruction suggest that the major cause of the increase in resistance is due to a chronic inflammatory process in the airway wall (see Fig. 2) that eventually narrows the lumen of these small airways (Matsuba & Thurlbeck 1972) (Fig. 4). These measurements did not support the argument that a loss in elastic recoil and/or destruction of the alveolar attachments supporting the airway wall account for the increase in peripheral airways resistance (Hogg et al 1968).

Pulmonary emphysema

Pulmonary emphysema was first described by Laennec in 1834 from observations of the cut surface of postmortem human lungs that had been air-dried in inflation (Laennec 1834). He postulated that the lesions were due to over-inflation of the lung which compressed capillaries leading to atrophy of lung tissue, and this hypothesis appeared in a major textbook of pathology as late as 1940 (McCallum 1940). Reports implicating infection and inflammation in the pathogenesis of emphysema were summarized in an important contribution from McLean in 1958 (McLean 1958). The observation that emphysema could be produced experimentally by depositing the enzyme papain in the lung (Gross et al 1964), the description of emphysema in patients with α_1-antitrypsin deficiency (Laurel & Erickson 1963) led naturally to the current hypothesis that emphysema results from a proteolytic imbalance caused by cigarette smoking (Janoff 1983, Gadek et al 1979). The exact sources of the responsible proteolytic enzymes and their inhibitors remain controversial and will be discussed later in this symposium.

Terminology

Pulmonary emphysema has been defined as 'abnormal permanent enlargement of airspaces distal to terminal bronchioles, accompanied by destruction of their walls without obvious fibrosis' (Ciba Guest Symposium 1959, Snider et al 1985). The terms used to describe the lesions are based on the anatomy of the normal lung where the 'secondary lobule' has been defined as that part of the lung surrounded by connective tissue septae which contains several terminal bronchioles (Fig. 3A). The acinus is the portion of the lung parenchyma supplied by a single terminal bronchiole (Fig. 3B). As each secondary lobule contains several terminal bronchioles (Fig. 3A) it follows that there are several acini in each secondary lobule.

Centrilobular/centriacinar emphysema

Figure 3C shows a line drawing from Leopold and Gough's original descriptions of centrilobular emphysema as well as a postmortem radiograph illustrating the effect of dilatation and destruction of the respiratory bronchioles. This type of lesion was briefly described by Gough (1952), by McLean in Australia (McLean 1956), and in a more detailed report by Leopold & Gough (1957). It affects the upper regions of the lung more commonly than the lower (Heard 1958) and the lesions are also larger and more numerous in the upper lung (Thurlbeck 1963). Heppleston & Leopold (1961) emphasized that the parent airways supplying the centriacinar lesions were often narrowed due to an inflammatory reaction that was peribronchiolar in location and involved both a polymorphonuclear and mononuclear leukocyte infiltration. They used the term focal emphysema to

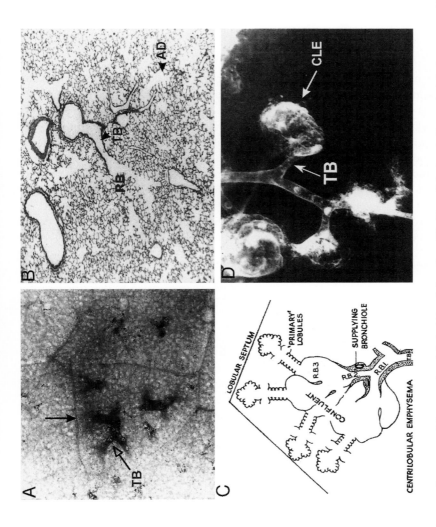

FIG. 3. The normal anatomy of the secondary lobule (A) and the acinus (B). The solid arrow indicates the terminal bronchiole supplying the acinus in both A and B. The diagram (C) of the lobule containing the centrilobular emphysematous lesion is reproduced from Leopold & Gough (1957) and the postmortem bronchogram (D) illustrates a terminal bronchiole (TB) supplying a centrilobular space (CLE).

describe a less severe form of the disease, but Dunnill's excellent review of the subject indicates that this separation is based largely on semantic arguments (Dunnill 1982). He suggests that the two conditions probably have a similar origin with focal emphysema being more widely distributed in the lung and less severe than the classic centrilobular form (Leopold & Gough 1957). He also prefers the term centricinar to centrilobular, which is logical, based on the fact that the secondary lobule contains several acini (Fig. 3).

Panacinar/panlobular emphysema

In panacinar emphysema, there is more uniform destruction of the acinus and all of the acini within the secondary lobule are involved. Wyatt, Fisher and Sweet (Wyatt et al 1962) provided a detailed account of this lesion in 1962 and Thurlbeck (1963) pointed out that in its mildest forms, it is difficult to discern from normal lung unless fixed inflated lung slices are impregnated with barium sulfate and examined under water using low power magnification. In contrast to centrilobular emphysema, the panacinar form tends to be more severe in the lower compared to the upper lobe but this difference only becomes statistically significant in severe disease (Thurlbeck 1963). Panacinar emphysema is commonly associated with α_1-antitrypsin deficiency but is also found in cases where no clear-cut genetic abnormality has been identified.

Miscellaneous forms of emphysema

Distal acinar, mantle or paraseptal emphysema are terms used to describe lesions that occur in the periphery of the lobule. This type of lesion can be commonly found along the lobular septae particularly in the subpleural region. It can occur in isolation where it has been associated with spontaneous pneumothorax in young adults and bullous lung disease in older individuals where individual cysts can become large enough to interfere with lung function. Unilateral emphysema or McLeod's syndrome occurs as a complication of severe childhood infections caused by rubella or adenovirus, and congenital lobar emphysema is a developmental abnormality affecting newborn children. The emphysema that forms around scars lacks any special distribution in the lobule and is referred to as irregular emphysema. These conditions are more fully discussed in Dunnill (1982).

Quantitation of emphysema

The method for quantitating normal lung structure in precise terms was developed in the Cardiopulmonary Laboratory of the Department of Medicine in Bellevue Hospital in New York by Ewald Weibel (Weibel 1963). The English pathologist, Michael Dunnill who worked at Bellevue just after Weibel, was among the first to apply these quantitative principles to the study of lung pathology (Dunnill 1962). Both acknowledge the assistance they received from Dr Domingo Gomez, a biological scientist and talented mathematician who helped them develop and apply quantitative methods to the structure of normal and diseased lungs.

Histology provides two-dimensional representations of three-dimensional objects where lines appear as points, surfaces as lines, and volumes as areas. For example, a section through a sphere yields a circle where the parameters of the sphere and circle are interrelated by the following formulae:

$$2\pi r = \text{circumference of a circle}$$

$$\pi r^2 = \text{area of a circle}$$

$$4\pi r^2 = \text{surface area of a sphere}$$

$$\frac{4}{3}\pi r^3 = \text{volume of a sphere}$$

The constant π is defined by the ratio of the circumference of a circle to its diameter, and the dimension that links the volume of the sphere to its two-dimensional representation as the area of a circle is referred to as the mean chord length (or L_M), which works out to be 2/3 of the diameter of the sphere.

$$L_M = \frac{\text{sphere volume}}{\text{circle area}} = \frac{\frac{4}{3}\pi r^3}{\pi r^2} = \frac{4}{3}r \text{ or } \frac{2}{3} \text{ diameter} \qquad (1)$$

The area of a circle is linked to the surface area of a sphere by the relationship:

$$\frac{\text{circle area}}{\text{surface area of sphere}} = \frac{\pi r^2}{4\pi r^2} = \frac{1}{4} \qquad (2)$$

Rearranging 1 and 2 to solve for circle area and setting the results equal to each other, yields a useful relationship for calculating surface area.

$$\frac{\text{sphere volume}}{L_m} = \frac{\text{surface area}}{4}$$

$$\text{or surface area} = \frac{4 \times \text{volume}}{L_m} \qquad (3)$$

This relationship can be used to calculate lung surface area by measuring the total volume of the fixed lung (tissue + air) by water displacement and determining the proportion of the volume taken up by alveoli by point-counting the cut surface of the lung slices. The alveolar mean linear intercept (L_M) is calculated by projecting test lines on histological sections of the lung and dividing the total length of the lines by the number of times an alveolar wall intersects them (using two intersections of the surface for each alveolar wall crossing). The shrinkage that occurs when fresh tissue is fixed and processed into paraffin is determined by comparing the area of the template used to cut the histological block to the area of the histological section after it is mounted and stained (Weibel 1963, Dunnill 1962).

Duguid et al (1964) were among the first to estimate the internal surface area of the lung in emphysema, but Thurlbeck's extensive study of postmortem lungs suggested that this measurement was too variable to be a useful way of comparing the amount of emphysema present in different subjects (Thurlbeck 1967a,b). Recent reports of both an animal model of emphysema (Massaro & Massaro 1997) and the human disease (Coxson et al 1999) confirmed that measurements of total surface area fail to pick up the early lesions. However, both studies showed that the early changes are reflected in measurements of the surface area:volume ratio (SA:V) which can be measured using the same morphometric approach.

The SA:V ratio can be estimated with the same test line system by keeping track of the number of times the points formed by the end of the lines fall on tissue (P) as well as the number of times alveolar walls intersect the lines. The ratio between the

$$SD = \frac{4\Sigma I}{l \Sigma P} \tag{4}$$

parameters used to determine the surface area $4 \times \Sigma I/l$ and the tissue volume ΣP is referred to as surface density (SD) which becomes the surface to volume ratio in three dimensions: where $l =$ the length of the test lines, ΣI the sum of the intercepts, ΣP is the sum of the number of times the points fall on tissue. We have recently used the pre-operative CT scan in combination with histological measurements of SD made on resected lung tissue to calculate the SA:V ratio and total surface area of the lung (Coxson et al 1999).

Computed tomography

The definition of emphysema presupposes knowledge of normal airspace size. As this varies considerably with lung volume, it is difficult to separate fully inflated normal from emphysematous lung. This problem restricted the diagnosis of the milder forms of emphysema to pathologists who were willing to examine fixed inflated postmortem specimens until the introduction of the computed tomography (CT) scan.

The Edinburgh group led by the late Professor David Flenley was the first to use CT to provide a quantitative diagnosis of emphysema in living patients. They

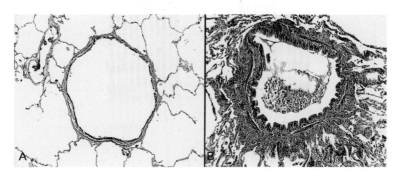

FIG. 4. Compares a normal peripheral airway to an airway that is involved by a chronic inflammatory process in the airway wall and lumen. These inflammatory changes are associated with abnormal peripheral airway function and increased peripheral airways resistance (see text for further discussion).

based their analysis on the frequency distribution of the CT measurement of Hounsfield units (HU) to separate normal lung from permanent distal airspace enlargement (Hayhurst et al 1984). The HU measures the degree to which X-rays are attenuated by tissue. It varies between −1000 (air) and +1000 (water), and can be converted to lung density by adding 1000 to the HU in each voxel and dividing that number by 1000. Müller and colleagues developed a density mask to separate the CT voxels below a fixed density (0.910 HU) and used it to separate normal from emphysematous lung (Müller et al 1988). The same group subsequently showed that this method identified emphysematous lesions down to those that were approximately 5 mm in diameter but did not identify the milder forms of the disease (Miller et al 1989).

Measurements of lung density (weight/volume) are readily converted to specific volumes (volume/weight) and have been used to calculate regional lung volumes in canine lungs where the thorax had been frozen intact (Hogg & Nepaszy 1969). This approach was later applied to calculate the regional lung volume of subjects with normal lung function using CT measurements of lung density. The results provided useful measurements of regional differences in the expansion of the normal lung that were used to predict the pleural pressure gradient (Coxson et al 1995). The same group also found that the summed weight of each voxel (density × volume) measured from the preoperative CT scan of a lung or lobe compared favourably with the weight of the same lung or lobe measured after it was resected from patients undergoing surgical treatment for cancer (Coxson et al 1999). The upper limits of normal lung expansion were determined by measuring total lung capacity using whole body plethysmography and the weight of both

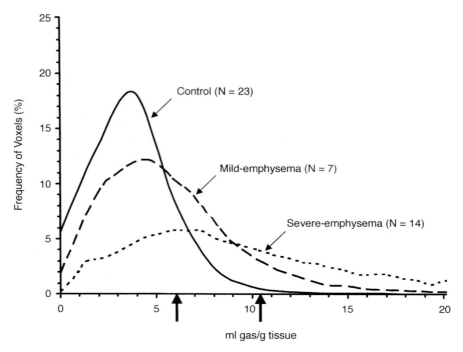

FIG. 5.   Data from Coxson et al (1999) showing the frequency distribution of the regional lung
volume of all of the lung voxels in CT scans from 23 control cases with normal lung function, 7
cases with mild emphysema, and 14 cases with severe emphysema. The arrows indicate the upper
limits of normal lung volume (6.0±1.1 ml gas/gram tissue) and the lung volume determined by
the cut-off defined by the density mask used by Muller et al (1988) (10.2 ml/gram). Lesions less
than 0.5 mm in diameter are not found using the density mask but are identified between 6 and
10 ml/gram.

lungs using the CT scan. This showed that the maximally expanded normal human
lung contains 6.0±1.1 ml air/gram of lung tissue at total lung capacity (Coxson et
al 1995).

Analysis of the frequency distribution of the volume of gas/gram tissue for all
of the voxels in the CT scan of lungs from patients with normal lung function
showed that these values were normally distributed. In mild emphysema, this
distribution was shifted to the right with a further shift to the right occurring
in lungs with severe emphysema (Fig. 5). The value added by analysing the CT
scan in terms of volume (Coxson et al 1999) rather than density (Miller et al 1989)
is that it separates fully inflated normal lung (6.0±1.1 ml/gram) from the milder
increases in lung volume that define emphysema (Ciba Guest Symposium 1959,
Snider et al 1985).

Combining CT measurements with quantitative histology

The robust cascade sampling design for quantitative histology introduced by Cruz-Orive and Weibel uses the fixed inflated lung as its reference volume (Cruz-Orive & Weibel 1981). The cascade nature of the design allows quantitation to be extended from the gross to the electron microscopic level. Coxson and associates have applied this to living patients by substituting the CT-determined lung volume as the reference in the cascade sampling design and used biopsies, resected lobes and more recently tissue obtained from lung volume reduction surgery for the histological analysis (Coxson et al 1995, 1997, 1999). They found that an equation (surface area/volume $= /\,6.84-0.32\,\text{ml gas/gram tissue}$) could be used to link the CT-determined volume to histologically measured surface area/volume ratio and calculate the surface area to volume ratio and total surface area of the lung from the CT scan. These estimates of total surface area correlated with measurements of diffusing capacity made in the same subjects (Coxson et al 1999). This type of analysis provides the quantitative estimate of the extent of lung destruction, and localizes it within the lung in a way that will be useful in following both the natural history of the disease and the effects of treatment (Coxson et al 1999).

Quantitation of the inflammatory response in emphysema

We have recently extended the quantitative methods described above to estimate the number of inflammatory cells present/unit surface area of the lung parenchyma. This showed that emphysematous lung destruction is associated with a marked amplification of the inflammatory response in patients with emphysema compared to control subjects with equal smoking histories (Elliott et a 1999). We believe this finding is relevant to the observation that everyone who smokes develops lung inflammation and the evidence that only a minority of heavy smokers develop chronic airways obstruction (Fletcher et al 1976). The exact mechanisms by which the cigarette smoke-induced inflammation is amplified to cause emphysematous lung destruction remain to be elucidated but our work suggests that latent adenoviral infection is one of the factors that contributes to this amplification process (Elliott et al 1999, Hogg et al 1989, Matuse et al 1992, Keicho et al 1997a,b, Matsuba et al 1998, Hogg 1997).

Summary

A cigarette smoke-induced inflammatory process underlies the pathogenesis of the major lesions associated with COPD. In chronic bronchitis, this process is located in the mucosa, gland ducts and glands of intermediate sized bronchi between 2–4 mm internal diameter. The cellular and molecular contents of this mucus containing inflammatory exudate are responsible for the cough and expectoration that characterize chronic bronchitis. The increase in airway obstruction in COPD is due to the presence of a local inflammatory process in the smaller bronchi and bronchioles less than 2 mm in internal diameter. These small airway lesions may occur either with or without those responsible for chronic bronchitis and result in a narrowing of the small airway lumen, as well as destruction and decline in elasticity of the parenchymal support of the smaller airways. These changes progressively increase peripheral airways resistance and eventually reduce the patient's ability to forcibly empty their lungs to a degree that can be measured by a reduction in FEV_1 (forced expiratory volume in one second). Emphysematous destruction of the lung parenchyma is caused by an inflammatory process which first reduces the surface area/volume ratio and then the total surface area of the lung. The reduction in the capillary surface area that results from this destruction process decreases the gas exchanging capability of the lung and further reduces the FEV_1 by lowering the elastic forces that drive air out of the lung.

The introduction of CT scanning and its use in combination with quantitative histology has made it possible to move the diagnosis of emphysema out of the postmortem room into the clinical setting. The analysis of the electronic record of the CT scan can provide accurate estimates of lung weight and volume as well as valuable estimates of surface/volume ratio and total surface area of the lung. These new methods should provide insight into the natural history of emphysematous lung destruction as well as an objective way of evaluating new treatments.

References

Butler J, Caro C, Alkaler R, Dubois AB 1960 Physiological factors affecting airway resistance in normal subjects and in patients with obstructive airways disease. J Clin Invest 39:584–591

Carlile A, Edwards C 1983 Structural variations in the main bronchi of the left lung. A morphometric study. Br J Dis Chest 77:344–348

Ciba Guest Symposium 1959 Terminology, definitions and classifications of chronic pulmonary emphysema and related conditions. Thorax 14:286–299

Cosio M, Ghezzo M, Hogg JC et al 1978 The relation between structural changes in small airways and pulmonary-function tests. N Eng J Med 298:1277–1281

Coxson HO, Mayo JR, Behzad H et al 1995 The measurement of lung expansion with computed tomography and comparison with quantitative histology. J Appl Physiol 79:1525–1530

Coxson HO, Hogg JC, Mayo JR et al 1997 Quantification of idiopathic pulmonary fibrosis using computed tomography and histology. Am J Respir Crit Care Med 155:1649–1656

Coxson HO, Rogers RM, Whittall P et al 1999 Quantification of the lung surface area in emphysema using computed tomography. Am J Resp Crit Care Med 159:851–856

Cruz-Orive LM, Weibel ER 1981 Sampling designs for stereology. J Microsc 122:235–257

Dayman H 1951 Mechanics of airflow in health and emphysema. J Clin Invest 3031:1175–1190

Duguid JR, Young A, Cauna D, Lambert MU 1964 The internal surface area of the lung in emphysema. J Clin Pathol 88:405–421

Dunnill MS 1962 Quantitative methods in the study of pulmonary pathology. Thorax 17:320–328

Dunnill MS 1982 Emphysema. In: Pulmonary pathology. Churchill Livingstone, London, p 81–112

Elliott WM, Retamales I, Meshi B et al 1999 Inflammatory cell recruitment in emphysema. Am J Respir Crit Care Med 159:810 (abstr)

Fletcher C, Peto R, Tinker C, Speizer FE 1976 The natural history of chronic bronchitis and emphysema. Oxford University Press, Oxford

Gadek JE, Fells JA, Crystal RG 1979 Cigarette smoking induces functional antiprotease deficiency in the lower respiratory tract. Science 206:1315–1316

Gough J 1952 Discussion of diagnosis of pulmonary emphysema. Proc R Soc Med 45:576–577

Gross P, Babyuk MA, Toller E, Kashak M 1964 Enzymatically produced pulmonary emphysema. J Occup Med 6:481–484

Haraguchi, M, Shemura S, Shirata K 1999 Morphometric analysis of bronchial cartilage in chronic obstructive pulmonary disease and bronchial asthma. Am J Respir Crit Care Med 159:1005–1013

Hayhurst MD, MacNee W, Flenley DC et al 1984 Diagnosis of pulmonary emphysema by computerized tomography. Lancet 2:320–322

Heard BE 1958 Further observations on the pathology of pulmonary emphysema in chronic bronchitis. Thorax 14:58–90

Heppleston AG, Leopold JG 1961 Chronic pulmonary emphysema: anatomy and pathogenesis. Am J Med 31:279–291

Hogg J 1997 Latent adenoviral infections in the pathogenesis of COPD. Eur Respir Rev 7:216–220

Hogg JC, Nepaszy S 1969 Regional lung volume and pleural pressure gradient estimated from lung density in dogs. J Appl Physiol 27:198–203

Hogg JC, Macklem PT, Thurlbeck WM 1968 Site and nature of airway obstruction in chronic obstructive lung disease. N Engl J Med 278:1355–1360

Hogg J, Irving WL, Porter H, Evans M, Dunnill MS, Fleming K 1989 In situ hybridization studies of adenoviral infections of the lung and their relationship to follicular bronchiectasis. Am Rev Respir Dis 139:1531–1535

Jamal K, Cooney TP, Fleetham JA, Thurlbeck WM 1984 Chronic bronchitis. Correlation of morphological findings and sputum production and flow rates. Am J Respir Dis 129:719–722

Janoff A 1983 Biochemical links between cigarette smoking and pulmonary emphysema. J Appl Physiol 55:285–293

Keicho N, Elliott WM, Hogg JC, Hayashi S 1997a Adenovirus E1A gene dysregulates ICAM-1 expression in transformed pulmonary epithelial cells. Am J Respir Cell Mol Biol 16:23–30

Keicho N, Elliott WM, Hogg JC, Hayashi S 1997b Adenovirus E1A upregulates interleukin-8 expression induced by endotoxin in pulmonary epithelial cells. Am J Physiol 272:L1046–L1052

Laennec RTH 1834 A treatise on diseases of the chest and on mediate auscultations, 4th edition. (transl. J Forbes) Longman, London

Laurel CB, Erickson S 1963 The electrophoretic alpha-1 globulin pattern of serum α-1 anti-trypsin deficiency. Scand J Clin Lab Invest 15:132–140

Leopold JG, Gough J 1957 Centrilobular form of hypertrophic emphysema and its relation to chronic bronchitis. Thorax 12:219–235

Mackenzie HI, Glick M, Outhred KG 1969 Chronic bronchitis in coal miners: ante-mortem–post-mortem comparisons. Thorax 24:527–535

Macklem PT, Mead J 1967 Resistance of central and peripheral airways measured by the retrograde catheter. J Appl Physiol 22:395–401

Massaro GD, Massaro D 1997 Retinoic acid treatment abrogates elastase-induced emphysema in rats. Nat Med 3:675–677

Matsuba K, Thurlbeck WM 1972 The number and dimensions of small airways in emphysematous lungs. Am J Pathol 67:265–275

Matsuba T, Keicho N, Higashimoto Y et al 1998 Identification of glucocorticoid- and adenovirus E1A-regulated genes in lung epithelial cells by differential display. Am J Respir Cell Mol Biol 18:243–254

Matsuse T, Hayashi S, Kuwano K, Keunecke H, Jefferies WA, Hogg JC 1992 Latent adenoviral infections in the pathogenesis of chronic airways obstruction. Am Rev Respir Dis 146:177–184

McCallum WG 1940 Types of injury — destruction of the respiratory tract. In: A textbook of pathology, 7th edition. Saunders, Philadelphia, PA, p 419–428

McLean KA 1958 Pathogenesis of pulmonary emphysema. Am J Med 25:62–74

McLean KH 1956 Microscopic anatomy of pulmonary emphysema. Aust Ann Intern Med 5:73–88

Mead J, Turner JM, Macklem PT, Little J 1967 Significance of the relationship between lung recoil and maximum expiratory flow. J Appl Physiol 22:95–108

Miller RR, Müller NL, Vedal S, Morrison NJ, Staples CA 1989 Limitations of computed tomography in the assessment of emphysema. Am Rev Resp Dis 139:980–983

Mullen JBM, Wright JL, Wiggs B, Paré PD, Hogg JC 1985 Reassessment of inflammation in the airways of chronic bronchitis. Br Med J (Clin Res Ed) 291:1235–1239

Müller NL, Staples CA, Miller RR, Abboud RT 1988 'Density mask'. An objective method to quantitate emphysema using computed tomography. Chest 94:782–787

Peinado VI, Barbéra JA, Abate P et al 1999 Inflammatory reaction in pulmonary muscular arteries of patients with mild chronic obstructive pulmonary disease. Am J Respir Crit Care Med 159:1605–1611

Reid L 1960 Measurement of the bronchial mucous gland layer: a diagnostic yardstick in chronic bronchitis. Thorax 15:132–141

Saetta M, Di Stefano A, Maestrelli P et al 1994 Airway eosinophilia in chronic bronchitis during exacerbations. Am J Respir Crit Care Med 150:1646–1652

Smith CB, Golden CA, Kanner RE, Renzetti AD Jr 1980 Association of viral and *Mycoplasma pneumoniae* infections with acute respiratory illness in patients with COPD. Am Rev Respir Dis 121:225–232

Snider GL, Kleinerman JL, Thurlbeck WM, Bengally ZH 1985 Definition of emphysema. Report of a National Heart, Lung and Blood Institute, Division of Lung Diseases. Am Rev Respir Dis 132:182–185

Speizer FE, Tager IB 1979 Epidemiology of chronic mucus hypersecretion and obstructive airways disease. Epidemiol Rev 1:124–142

Thurlbeck WM 1963 The incidence of pulmonary emphysema with observations on the relative incidence and spatial distribution of various types of emphysema. Am Rev Respir Dis 87:207–215

Thurlbeck WM 1967a The internal surface area of nonemphysematous lung. Am Rev Respir Dis 95:765–773

Thurlbeck WM 1967b Internal surface area and other measurements in emphysema. Thorax 22:483–496

Thurlbeck WM, Angus GE 1964 The distribution curve for chronic bronchitis. Thorax 19:436–442

Thurlbeck WM, Pun R, Toth J, Fraser RG 1974 Bronchial cartilage in chronic obstructive lung disease. Am Rev Respir Dis 109:73–80

van Braband T, Cauberghs M, Verbeken E et al 1983 Partitioning of pulmonary impedance in excised human and canine lungs. J Appl Physiol 55:1733–1742

Weibel ER 1963 Morphometry of the human lung. Springer-Verlag, Heidelberg

Wright JL, Lawson L, Paré PD et al 1983 The structure and function of the pulmonary vasculature in mild chronic obstructive pulmonary disease. The effect of oxygen and exercise. Am Rev Respir Dis 128:702–707

Wyatt JP, Fischer VW, Sweet AC 1962 Panlobular emphysema: anatomy and pathogenesis. Dis Chest 41:239–259

DISCUSSION

Jeffery: I appreciated your overview. I certainly remember the days of Lynne Reid, my mentor, when the original experimental models that were developed for bronchitis specifically aimed to exclude inflammation. We were about the business of producing the changes in mucus hypersecretion without inflammation. Since then, the whole scene has shifted: you mentioned inflammation several times. Why do we now not reconsider the definition of COPD, and include the word 'inflammation', as we now do with asthma?

Hogg: The term 'inflammation' describes the pathology quite well. There are a number of lines of evidence to suggest that the inflammatory response is different in chronic bronchitis than it is in asthma, for instance. Not least among these is the response to treatment. We need more carefully thought out ideas about how these mechanisms might work.

Calverley: The new GOLD (Global Obstructive Lung Disease initiative) definition of COPD includes the terms 'produced by inflammation'. Inflammation is now going to be recognized as part of the definition of this disease.

Nadel: I think inflammation was an acceptable term in 1890, but now in the year 2000 it doesn't help much with regard to the actual cell biology: it is too broad a term. I would encourage people to use terms that have specific functional meaning, e.g. neutrophil activation and release of elastase, which degranulates goblet cells, etc.

Dunnill: You raised an interesting and fundamental point that is often ignored: why is it that only 25% of people get into trouble from smoking? This is also true, of course, for smoking-related coronary disease and carcinoma of the lung: not everyone gets them. Why is this? In any other disease group, we recognise that there are two factors in causing disease. One is the environmental factor (e.g. pollution or infection), and the other is the genetic factor. Would you care to speculate on what is being done about this?

Hogg: We believe that one of the factors that amplifies this inflammatory process is latent adenoviral infection, and we have growing data to support this view. When I had a sabbatical year with you, we were able to show that it targeted the epithelial cells that were relevant. Since then, we have shown that adenovirus remains latent to a greater degree in the subjects who have airway obstruction. We have also shown that adenovirus expresses a very interesting protein in airway epithelial cells, the E1A protein. We have experimental evidence, both in cell lines and in animals, to show that latent adenoviral infection is capable of amplifying the inflammatory process. This suggests that one of the additional risk factors for COPD is childhood bronchitis and bronchiolitis. We think that viruses (our work is done just on adenovirus, but there isn't any reason that others might not play an equally significant role) that persist in the respiratory tract and integrate into the DNA of the host cells are capable of amplifying the inflammatory process. The interrelation of viral infection with genetics is very interesting. There is good evidence that the ability of the virus to persist in a latent form depends on the host's HLA type. The HLA is critical in presenting the virus to the cell surface in a way that allows the cytotoxic T cell to identify and destroy the infected cell. The adenoviral E3 protein prevents surface expression by binding to the host HLA proteins and this binding efficiency is HLA dependent.

Silverman: One reason that only a minority of smokers develop COPD is because of competing risks. A patient dying of early-onset lung cancer may not live long enough to get severe emphysema.

The aetiology of the variable susceptibility to develop COPD remains unresolved. I recently reviewed some of the early autopsy studies of emphysema by Auerbach et al (1972) and Petty et al (1967) to determine whether there was compelling evidence that only a small minority of smokers ever developed emphysema. However, a clear majority of smokers at autopsy did have at least some evidence for emphysema. How much of the variable development of airflow obstruction among smokers is related to a low correlation between emphysema and airflow obstruction as opposed to the existence of a subset of individuals who never develop significant smoking-related lung damage?

Hogg: Our data are from individuals who have had lung resections. Even in these, which are a selected sample of the population, there is a clear number that never develop emphysema. Now that we can diagnose emphysema accurately with the CT scan, we need a population-based study to see how prevalent it really is.

MacNee: Didn't you tell me a long time ago that all cigarette smokers get small airways disease to some extent, and that all of them get some degree of centrilobular emphysema?

Hogg: I don't think I ever said that! I have said that all smokers get lung inflammation while only some get emphysema and airways obstruction. I think that this statement is supported by work from several laboratories using different techniques ranging from bronchoalveolar lavage (Hunninghake & Crystal 1983) to biopsy (Ollenshaw & Woolcock 1992), lung resection (Wright et al 1983) and autopsy (Niewoehner et al 1974).

MacNee: In the study I showed of David Lamb's (Gillooly & Lamb 1993), the cigarette smokers who did not show any change outwith the normal ageing change in mean distal airspace had a degree of mild centrolobular emphysema, since the mean airspace size is not affected by mild centrolobular emphysema. However, the 20% of smokers who showed a fall below 'abnormal' enlargement of distal airspaces had in addition panlobular emphysema. In a paper that we have succeeded in publishing, we produced similar results using CT scanning to assess emphysema. The hypothesis would be that all smokers develop some centrolobular emphysema, and those that are susceptible to COPD develop in addition panlobular emphysema.

Hogg: I don't want to get tied up in the terminology. The reason people gave up measuring internal surface area is because it isn't sensitive enough. Dr Dunnill suggested many years ago that we should be measuring SA:V ratio. The experiments that the Massaros published in *Nature Medicine* showed that the earliest change is in the surface:volume ratio and that the surface area remains the same (Massaro & Massaro 1997). Data reported by Coxson et al (1999) from our laboratory showed that SA:V decreases in early disease and is followed by a reduction in surface area when the disease becomes more severe. This means that the first change is an increase in lung volume. And why the volume goes up is not a simple question to answer.

Massaro: Do you have any idea why the SA:V ratio breaks down in the normals?

Hogg: I think this is an interesting question: it is much discussed in our laboratory, but we still don't know the answer.

Massaro: The reason I ask is that the symptoms and physiological measurements you get will depend on where you start. I was struck by a paper by Angus & Thurlbeck (1972) in which they counted the number of alveoli. At the same lung volume, some people had twice as many alveoli as others. This fits with the breakdown in the normal levels. With coronary artery disease, what happens early on has a large influence on what occurs later. Perhaps the same is true for COPD.

Hogg: This could well be the case. There is a fair amount of variability in how the alveoli change in number and in size with age. If we try to explain the data we have using an equation to relate SA:V to volume of air/gram tissue into the normal range, the concept of adding alveoli and decreasing the size of alveoli doesn't explain the increase in SA:V we observed (Coxson et al 1999). What this might reflect is some unfolding of the alveolar surface but we don't have any direct evidence for that in our studies.

MacNee: Jan Stolk may wish to comment on his studies of CT scanning in emphysematous patients. You have both the lungs and the CT scan. However, in order to utilize these measurements of emphysema in the clinical or research studies, we need to standardize measurements of emphysema based on the CT scans alone.

Stolk: We are starting to learn that CT measurements can not only be used for cross-sectional studies, but also for longitudinal studies. We have been doing a follow-up study in PI Z patients, with α_1-antitrypsin treatment and placebo treatment. Much to our surprise, it turned out that dependent on the outcome parameter that you choose from the histogram, the percentile method is a robust enough measurement for follow-up studies over a three year period to be able to discriminate groups in treatment versus placebo. Then the question arises as to how this relates to normal ageing. Since these patients are quite young, we are trying to build up a database of patients with normal lungs, for instance from people who go for osteosarcoma surgery and are checked by CT. With the data we have so far we see little decline in our percentile method parameter compared to the dataset we have obtained in the last three years of 57 α_1-antitrypsin patients.

MacNee: What measurement do you use?

Stolk: We calculate all sorts of parameters, including decline in weight, decline in percentile points, as well as area under the curve below a selected threshold. It turns out that a selected percentile point between 10 and 20 on the y axis corresponding with a Hounsfield unit level on the x axis is the most robust. With the introduction of multislice CT scanners, it may even be possible to look at changes occurring in the bronchial tree, to see if this can be quantified.

Calverley: We keep returning to this question of why only 20% of smokers get COPD. The issue we have to think about—and it does integrate all the information we have heard—is that more than 20% of smokers get the disease if you wait long enough. People are dying from other causes. There is clearly an interaction between host susceptibility and exposure here. If you choose an arbitrary point, you are certainly going to have some people who have got this illness to a significant degree, and others who won't. We have to try to fight shy of the idea of a binary distribution, that there are those with COPD and those who are without it. This fits in with what Jim Hogg was saying about latent adenovirus: it may take different exposures to lead to a degree of expression. The data about inflammation in the lungs and the lung volume resection are very exciting, but we must be sure that they do not reflect a selection bias as only patients with predominant emphysema underwent surgery. I find these ideas very attractive, but I want to make sure that I am standing on firm ground before going forward.

Nadel: We have just heard a paper on structure and about the pathology of the lung. In the past, people have tended to look at obvious structural features. First, the submucosal glands dominated the field of hypersecretion. This probably was due to the fact that the outlets of gland ducts are located in close proximity to cough receptors. Thus, when glandular hypersecretion occurs it stimulates cough. Therefore, when epidemiologists found patients with *cough* and sputum production, they were looking at the effects of gland hypersecretion. Secretion from surface epithelial goblet cells in peripheral airways may not cause such profound symptoms and therefore they were not likely to be studied by epidemiologists! Major effects of goblet cells in peripheral airways were therefore not studied in detail. Furthermore, degranulation of goblet cells could easily plug peripheral airways, whereas they could be cleared more easily by cough in larger bronchi. Another issue is the hypersecretion is taking place dynamically. Goblet cells can develop and mature within three days, and they can degranulate in minutes to hours. Thus, timing then becomes very important. For example, in COPD, exacerbations could rapidly cause goblet cell growth and degranulation, which may not be obvious clinically but could have profound pathophysiologic effects. Furthermore, cells such as activated neutrophils could cause goblet cell growth 24 or 48 h after the neutrophils have disappeared from the airway epithelium (by inducing gene and protein expression in cells such as goblet cells).

Hogg: I think it is extraordinarily difficult to come to grips with the dynamics. As we sit here at rest with an average cardiac output of 5 l/min, 7800 litres of blood pass through our lungs in 24 h and each litre contains 10^9 neutrophils. It is relatively easy to consider the neutrophils because these are the cells that can't divide. The monocytes and lymphocytes are more difficult because they can divide after they migrate and this makes the understanding of their kinetics even more difficult. Only about 2% of delivered neutrophils migrate into the alveoli even when powerful stimuli such as bacteria are present (Doerschuk et al 1994). We have to try to come to grips with leukocyte kinetics in the tissue in cigarette smokers in a more meaningful way to understand how they may destroy the lung tissue in emphysema.

Jeffery: In asthma, I suppose our handle on the dynamics is looking at the response to allergen challenge.

Hogg: This gives clues about the changes in the tissue but it doesn't tell you much about the dynamics. You don't know whether the cells are there because they migrated more, or whether they are dying at a slower rate — you have no idea of the traffic in the tissue other than that there are more there.

Nadel: The reason I asked the question had to do with the fact that I am very 'visual': I like pictures. Jim Hogg showed a picture with many small plugs in an airway of a subject with COPD. I was thinking about what was in that plug? What was in the wall? How they were interrelating and how we can get information out of the morphology? The problem is that people do biopsies in loci that are accessible, so we obtain information mainly in large airways. Thus, changes that are occurring dynamically, such as during exacerbations and in the periphery of the lungs, obtaining information is not easy! Another question is what happens to airways that become plugged? Do they become 'unplugged', and if so, how?

Rennard: We do ourselves a disservice by trying to make COPD or emphysema a categorical kind of classification. All the data suggest that it is continuous. There are undoubtedly different kinds of susceptibility, and some people have COPD worse than others. When we say only 15% of people get COPD, this would be like saying that only 50% of people get atherosclerosis based on when they have their strokes or heart attacks. Yet we recognize that the pathological entities exist at a much higher frequency. This does us a political disservice in terms of getting the disease recognized, getting funding and so forth. It also does us a conceptual disservice, because COPD is clearly not categorical.

With regard to the issue of susceptibility, it is not just an epidemiological issue, with some people susceptible and others not. Instead, in the tissues that you showed from some individuals there are areas of the lung that are thoroughly destroyed and areas that are strikingly well preserved. What is different about the histology in those two areas? Can you identify areas where you think there may be destruction in progress?

Hogg: We are trying to do this at the moment as an ongoing study in patients that are having lung reduction surgery, comparing the relatively normal areas with the abnormal areas.

Rennard: Do you see just as many inflammatory cells in the unaffected areas?

Hogg: I don't think it is possible to tell, unless you express this per unit surface area.

Rennard: Does the adenoviral expression tend to be higher within an individuals lungs in areas that are worse?

Hogg: We don't have good data on this. However, in groups of patients that have worse disease there is more adenovirus. And as the disease progresses there is also more adenovirus E1A protein expressed.

Nadel: One thing that needs to be kept in mind is that cells that live a long time (e.g. eosinophils) have a better chance of being counted than cells with a shorter half-life (e.g. neutrophils). Counting alone may give limited or false information.

Lomas: I was struck by Jim Hogg's photographs of the usual centrolobular versus panlobular emphysema. Does centrolobular emphysema progress to panlobular emphysema, and therefore it is all a spectrum? Or is panlobular emphysema a different disease process associated with antitrypsin deficiency? And what about inflammation, proteinases/antiproteinases and latent adenoviruses in the two types? Is this classification of any value?

Hogg: I don't know whether it is of any value or not. With panacinar emphysema there are cases that have the same pathology as is seen in α_1-antitrypsin deficiency. Joel Cooper from St Louis talks about these as the 'non-α_1 α_1': they look like they should have α_1-antitrypsin deficiency but they don't. There must be some other mechanism that is important in these situations. When the lobule is completely destroyed, I don't think you can really tell what type of emphysematous process did it.

Lomas: If you look at patients with severe centrolobular disease and compare them with patients with more panlobular disease, what is the distribution of cells in your latent adenovirus studies?

Hogg: We don't have data that are good enough to answer that yet.

Campbell: I want to raise a hypothesis that has been advanced by Dr Norbert Voelkel in the USA. His work hasn't reached the literature yet, but he has raised the hypothesis that some event or events in cigarette smokers who are developing obstructive lung disease lead to apoptosis of the pulmonary capillary endothelial cells, following the loss of expression of vascular endothelial growth factor (VEGF). He feels that the loss of alveolar structures may occur because the alveolar walls are necrosing because of loss of blood supply. It seems that this is an attractive alternative explanation for how alveolar septal injury might occur.

Hogg: I know that work and I have discussed it with the authors at a recent Aspen conference. I think it is an interesting hypothesis that has to be examined. It is interesting that if you go back far enough in the literature, it is well described in textbooks of pathology in the 1930s: this is what they thought happened when there was overinflation of the lung. We are currently working with the TUNEL stains to see whether we can reproduce the result which they reported at the 1999 Aspen conference.

Agusti: Following up on that kind of idea, and thinking about what Dr Nadel said before about the dynamics of this inflammatory response, we might want to consider the fact that the pulmonary and bronchial circulations are different. One is venous and the other is arterial. Accordingly there might not just be one type of inflammatory response. There may be at least two: one coming from the venous blood in the pulmonary circulation, and the other coming from the arterial blood in the bronchial circulation. I will show in my presentation that there is evidence of activation of inflammatory cells in the systemic (arterial) circulation.

References

Angus GE, Thurlbeck WM 1972 Number of alveoli in the human lung. J Appl Physiol 32:483–485

Auerbach O, Hammond EC, Garfinkel L, Benante C 1972 Relation of smoking and age to emphysema: whole-lung section study. New Engl J Med 286:853–857

Coxson HO, Rogers RM, Whittall KP et al 1999 A quantification of lung surface area in emphysema using computed tomography. Am J Respir Crit Care Med 159:151–156

Doerschuk CM, Markos H, Coxson HO, English D, Hogg JC 1994 Quantitation of neutrophil migration in acute bacterial pneumonia in rabbits. J Appl Physiol 87:2593–2599

Gillooly M, Lamb D 1993 Microscopic emphysema in relation to age and smoking habit. Thorax 48:491–495

Hunninghake G Crystal RG 1983 Cigarette smoking and lung destruction: accumulation of neutrophils in lungs of cigarette smokers. Am Rev Respir Dis 128:833–838

Massaro GD, Massaro D 1997 Retinoic acid treatment abrogates elastase-induced pulmonary emphysema. Nat Med 3:675–677

Niewoehner DE, Kleinerman J, Reiss DB 1974 Pathologic changes in the airways of young cigarette smokers. N Engl J Med 291:755–758

Ollenshaw GL, Woolcock AL 1992 Characteristics of inflammation in biopsies of large airways in subjects with asthma and chronic bronchitis. Am Rev Respir Dis 145:922–927

Petty TL, Ryan SF, Mitchell RS 1967 Cigarette smoking and the lungs: relation to postmortem evidence of emphysema, chronic bronchitis, and black lung pigmentation. Arch Environ Health 14:172–177

Wright JL, Lawson LM, Pare PD, Wiggs BR, Kennedy S, Hogg JC 1983 Morphology of the peripheral airways in current and ex-smokers. Am Rev Respir Dis 127:474–477

Overview of current therapies

Peter M. A. Calverley

University Clinical Departments, University Hospital Aintree, Longmoor Lane, Liverpool L9 7AL, UK

Abstract. The therapy of chronic obstructive pulmonary disease has been comprehensively reviewed in a number of international treatment guidelines. There is consensus about what elements should be included, but the purposes of therapy and the timing of its introduction remain poorly defined. Major factors limiting effective treatment beyond those associated with the biology of the condition itself are poor diagnostic methodology, failure to identify relevant co-morbidities and reluctance to devote appropriate resources to maximizing patient gain. Too many patients are identified at the end-stages of their illness when treatment is relatively limited. Most therapy is directed at reducing the impact of the disease in terms of symptoms, exercise performance and exacerbations on the individual and only smoking cessation modifies the evolution of the disease. Treatment of hypoxaemic patients with domiciliary oxygen improves mortality and slows the development of pulmonary hypertension. Effective smoking cessation is relevant at all stages of the disease. It depends on the willingness of the individual to participate, and quit rates can be improved by the use of nicotine replacement therapy and possibly bupropion. Inhaled bronchodilator drugs palliate symptoms and improve exercise performance in pharmacologically predictable ways. In patients with severe disease, reduction in operating lung volumes is more important than 'broncodilitation' and is better sustained by long acting beta agonists and anticholinergics. Inhaled corticosteroids reduce exacerbation rates and improve health status in established disease but do not modify disease evolution. Pulmonary rehabilitation improves exercise performance and health status without changing underlying pulmonary mechanics. Whether hospitalizations and exacerbations can be modified is still to be established. Nutritional therapy is in its infancy but calorie supplementation alone is insufficient to improve patient well being. Selected individuals can undergo lung volume reduction surgery with benefits extending up to two years but the risks are dependent on the skill of the operators and the appropriateness of patient selection. Lung transplantation is symptomatically helpful but does not modify the natural history of the disease. Hospitalization due to exacerbations of disease is frequent and their treatment with bronchodilators, antibiotics and corticosteroids now have a basis in randomized trial data. Mortality reflects the incidence of respiratory acidosis and non-invasive ventilation has a role in safely managing patients outside of the intensive care unit. Effective prevention of exacerbation should be possible with newer antiviral agents but data are presently lacking.

2001 Chronic obstructive pulmonary disease: pathogenesis to treatment. Wiley, Chichester (Novartis Foundation Symposium 234) p 27–44

TABLE 1   Objectives of chronic management in COPD

Prevention of disease progression
Relief of symptoms
Improvement of exercise tolerance
Improvement of health status
Prevention and treatment of exacerbations
Prevention and treatment of complications
Prevention of mortality
Minimization of side effects from treatment

WHO/NHLBI Global Initiative for Obstructive Lung Disease.

Chronic obstructive pulmonary disease (COPD) is a leading cause of death and disability worldwide and is likely to increase in importance in the next 30 years (Murray & Lopez 1997). This reflects the increasing use of tobacco-related products globally, the control of which remains a major health problem. Both the size of the task and the chronicity of the disorder can paralyse the individual clinician when faced with the seemingly intractable problems of specific COPD patients. The feeling that this is a 'self-inflicted' disorder provides a convenient excuse for our failure to easily alleviate symptoms and disability. The situation is worsened by the current lack of clarity about what individual therapies should achieve and when to employ them for optimum benefit.

The bulk of this symposium is focused on the mechanisms central to the development and progression of this disorder. Such understanding is urgently needed if effective new therapies are to be developed. This review will emphasize the extent to which such primary treatment is absent but also the need to translate improvements in the cellular and biochemical processes underlying COPD into improvements in clinically relevant end points, which are not immediately related to these mechanistic considerations.

Principles of COPD management

A series of desirable treatment goals have been suggested in a number of guideline documents and those to be incorporated in the forthcoming WHO/NHLBI Guidelines are listed in Table 1. Many of these are achievable in part with present therapies, at least in the clinical trial setting. However, translating this into routine practice requires considerably more organizational resource than is presently devoted to the task. It may appear self evident that a firm clinical diagnosis should be established before commencing treatment, but too often empirical therapy is offered without doing this. Measurement of spirometry to confirm

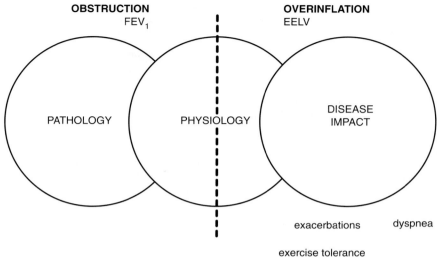

FIG. 1.   Interrelationships in COPD.

relatively fixed airflow limitation offers useful prognostic information and provides an approximate guide to disease severity. This can be helpful when considering the relative roles of COPD and other co-morbidities, e.g. heart failure, but is not a complete description of the patient's disability. Recognition of this has led to the development of a disease-specific health status questionnaire (Jones et al 1992) which assesses the impact of COPD on the patient's daily life, activity and symptoms. In general, spirometric impairment explains around 5% of the between-subject variability in disease impact. Health status is related to the use of health resources, e.g. hospitalizations (Osman et al 1997) and GP attendance, more closely than is FEV_1 (forced expiratory volume in one second). The changes in FEV_1 may modify disease progression but bear an imprecise relationship to other end points such as exacerbation rate, especially in more advanced disease (Fig. 1).

Despite these limitations, spirometric classifications of disease severity are usually used to indicate when specific treatment should begin. A typical scheme is shown in Fig. 2. In COPD, therapy is cumulative, symptomatic responses to treatment are often as reliable as spirometric end points and, in some cases, e.g. pulmonary rehabilitation, dramatic improvement in patient well-being occurs without any change in lung mechanics. Substantial bodies of evidence now exist for most of the therapy used in COPD which are briefly considered below.

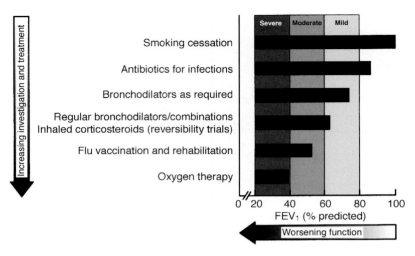

FIG. 2.   A general scheme for managing chronic obstructive pulmonary disease modified from the British Thoracic Society guidelines. Treatment is cumulative and staged. There is an advantage in stopping smoking even in patients with severe disease.

Smoking cessation

Stopping smoking is the only intervention known to modify the rate of decline of FEV_1 in COPD (Anthonisen et al 1994). Patients at all stages of the disease benefit from this but it is difficult to achieve by advice alone, especially when the patient is unwilling to accept the need to change their behaviour. Supportive therapy, both in smoking cessation groups and pharmacologically to deal with withdrawal symptoms, improves short and long term quit rates. Nicotine-replacement therapy, whether with gum, patches or inhaler shows consistent benefits (Raw et al 1998) and may be usefully augmented by bupropion (Jorenby et al 1999), which is now internationally licensed to aid smoking cessation. At present, smoking cessation is more effective among economically advantaged individuals.

Bronchodilator drugs

These drugs do not modify the rate of decline in FEV_1 (Anthonisen et al 1994) but reduce airway smooth muscle tone and promote improved lung emptying. Unlike bronchial asthma, in COPD inhaled anticholinergic agents like ipratropium bromide are as effective as inhaled β-agonists and these effects appear additive. These drugs are useful for short-term symptom relief, unlike oral β-agonists and theophyllines, which have a higher side effect profile. Long acting inhaled β-agonists like salmeterol improve health status in COPD (Jones & Bosh 1997) although similar results appear possible with frequent doses of anticholinergics (Mahler et al 1999). A new very long-acting inhaled anticholinergic, tiotropium bromide, has now been developed and offers the prospect of once daily maintenance therapy. Side effects with inhaled drugs are relatively few although palpitations and tremor can be a problem at higher doses with β-agonists.

The size of the change in FEV_1 after the bronchodilator drug is a poor guide to subsequent improvement in exercise performance (Hay et al 1992). Recent data, including those from our laboratory, suggest that bronchodilators improve the operating lung volumes without greatly changing airflow resistance, at least in severe disease. In these patients this 'broncho-deflating' action may explain the paradox of better exercise performance with little change in forced expiratory flows.

Inhaled corticosteroids

A clear picture of the role of inhaled corticosteroids has now emerged. These drugs do not modify the rate of decline in FEV_1 at any stage in COPD (Burge, 1999; Pauwels et al. 1999; Vestbo et al. 1999) contrary to earlier hopes. In patients with severe disease, especially those complaining of exacerbations, regular inhaled corticosteroids reduce the frequency of exacerbations and the rate of decline in health status. Whether they have a synergistic action with other drugs such as long-acting β-agonists is still to be determined.

Oxygen therapy

The mortality of stable COPD patients with a resting $PO_2 < 7.3$ kPa can be reduced by using supplementary oxygen for 15 or more hours per day. This is most easily delivered by an oxygen concentrator and should include overnight therapy as hypoxaemia is worsened by sleep (Anonymous 1980, 1981). Treatment of less hypoxaemic patients appears to be of no value (Górecka et al 1997). Oxygen delivered while exercising improves endurance times and slightly reduces the severity of breathlessness. Although widely prescribed, short burst oxygen treatment for acute breathlessness has little scientific basis at present.

Pulmonary rehabilitation

This involves participation in a multidisciplinary programme of education and exercise training designed to maximize the physiological and social functioning of the patient. There are good data from randomized clinical trials that this is effective (Goldstein et al. 1994) and possible even in severe disease (Berry et al 1999), although features such as social isolation and low mood can limit participation in outpatient programmes (Young et al 1999b). Benefits can last for many months and may reduce subsequent hospitalization, although appropriate prospective studies to confirm this observation have not yet been reported.

Nutritional depletion is common in COPD, is associated with increased mortality (Górecka et al 1997) and maybe reversible (Schols et al 1998). The mechanisms underlying this appear to be more than simple muscle disuse and there are now data to support a systemic component to the COPD illness, although the basis of this is still being explored.

Surgery

Surgery for large isolated bullae is known to produce substantial improvements in breathlessness and exercise tolerance. This approach has now been extended to the removal of areas of poorly perfused emphysematous lung in patients with an increased functional residual capacity (FRC) without CO_2 retention or significant co-morbidity — a procedure known as lung volume reduction surgery (LVRS). Despite a lack of randomized trial data this has rapidly become the most frequent thoracic surgical procedure in North America, leading to a moratorium on funding by some healthcare agencies. Undoubtedly in selected patients LVRS improves chest wall mechanics, increases forced expiratory flows and produces significant improvement in maximum and self paced exercise together with a reduction in breathlessness as assessed by the MRC dyspnoea scale (Young et al 1999a). The duration of the benefits is controversial and is probably in the range of 18–24 months, with the risk of surgery depending greatly on the experience of the surgeon.

LVRS does not appear to be a contraindication to lung transplantation, which is also effective in palliating symptoms. However, transplantation does not modify long-term mortality in COPD, at least using current donor sources and immunosupressive regimes (Hosenpud et al 1998).

Exacerbations of COPD

These are reviewed in detail elsewhere. Therapy remains relatively stereotyped and consists of bronchodilators (usually by nebulization) (Moayyedi et al 1995), appropriate antibiotics if sputum is purulent and symptoms severe (Anthonisen et al 1987), a short course of oral prednisolone (Davies et al 1999), and, in those hospitalized, supplementary oxygen. Mortality relates to the development of hypercapnic acidosis and in such patients nasal positive pressure ventilation provides an acceptable alternative to intubation with a shorter hospital stay and a lower risk of pneumonic complications (Brochard et al 1995). For the normocapnic patient hospital at home care may offer a cost-effective alternative (Gravil et al 1998).

References

Anonymous 1980 Continuous or nocturnal oxygen therapy in hypoxemic chronic obstructive lung disease: a clinical trial. Nocturnal Oxygen Therapy Trial Group. Ann Intern Med 93:391–398

Anonymous 1981 Long term domiciliary oxygen therapy in chronic hypoxic cor pulmonale complicating chronic bronchitis and emphysema. Report of the Medical Research Council Working Party. Lancet 1:681–686

Anthonisen NR, Manfreda J, Warren CP, Hershfield ES, Harding GK, Nelson NA 1987 Antibiotic therapy in exacerbations of chronic obstructive pulmonary disease. Ann Intern Med 106:196–204

Anthonisen NR, Connett JE, Kiley JP et al 1994 Effects of smoking intervention and the use of an inhaled anticholinergic bronchodilator on the rate of decline of FEV1. The Lung Health Study. JAMA 272:1497–1505

Berry MJ, Rejeski WJ, Adair NE, Zaccaro D 1999 Exercise rehabilitation and chronic obstructive pulmonary disease stage. Am J Respir Crit Care Med 160:1248–1253

Brochard L, Mancebo J, Wysocki M et al 1995 Noninvasive ventilation for acute exacerbations of chronic obstructive pulmonary disease. N Engl J Med 333:817–822

Burge PS 1999 EUROSCOP, ISOLDE and the Copenhagen City Lung Study. Thorax 54:287–288

Davies L, Angus RM, Calverley PMA 1999 Oral corticosteroids in patients admitted to hospital with exacerbations of chronic obstructive pulmonary disease: a prospective randomised controlled trial. Lancet 354:456–460

Goldstein RS, Gort EH, Stubbing D, Avendano MA, Guyatt GH 1994 Randomised controlled trial of respiratory rehabilitation. Lancet 344:1394–1397

Górecka D, Gorzelak K, Sliwinski P, Tobiasz M, Zielinski J 1997 Effect of long-term oxygen therapy on survival in patients with chronic obstructive pulmonary disease with moderate hypoxaemia. Thorax 52:674–679

Gravil JH, Al-Rawas OA, Cotton MM, Flanigan U, Irwin A, Stevenson RD 1998 Home treatment of exacerbations of chronic obstructive pulmonary disease by an acute respiratory assessment service. Lancet 351:1853–1855

Hay JG, Stone P, Carter J et al 1992 Bronchodilator reversibility, exercise performance and breathlessness in stable chronic obstructive pulmonary disease. Eur Respir J 5:659–664

Hosenpud JD, Bennett LE, Keck BM, Edwards EB, Novick RJ 1998 Effect of diagnosis on survival benefit of lung transplantation for end-stage lung disease. Lancet 351:24–27

Jones PW, Bosh TK 1997 Quality of life changes in COPD patients treated with salmeterol. Am J Respir Crit Care Med 155:1283–1289

Jones PW, Quirk FH, Baveystock CM, Littlejohns P 1992 A self-complete measure of health status for chronic airflow limitation. The St George's Respiratory Questionnaire. Am Rev Respir Dis 145:1321–1327

Jorenby DE, Leischow SJ, Nides MA et al 1999 A controlled trial of sustained-release bupropion, a nicotine patch, or both for smoking cessation. N Engl J Med 340:685–691

Mahler DA, Donohue JF, Barbee RA et al 1999 Efficacy of salmeterol xinafoate in the treatment of COPD. Chest 115:957–965

Moayyedi P, Congleton J, Page RL, Pearson SB, Muers MF 1995 Comparison of nebulised salbutamol and ipratropium bromide with salbutamol alone in the treatment of chronic obstructive pulmonary disease. Thorax 50:834–837

Murray CJ, Lopez AD 1997 Global mortality, disability, and the contribution of risk factors: Global Burden of Disease Study. Lancet 349:1436–1442

Osman IM, Godden DJ, Friend JA, Legge JS, Douglas JG 1997 Quality of life and hospital re-admission in patients with chronic obstructive pulmonary disease. Thorax 52:67–71

Pauwels RA, Löfdahl C-G, Laitinen LA et al 1999 Long-term treatment with inhaled budesonide in persons with mild chronic obstructive pulmonary disease who continue smoking. European Respiratory Society study on chronic obstructive pulmonary disease. N Engl J Med 340:1948–1953

Raw M, McNeill A, West R 1998 Smoking cessation guidelines for health professionals. A guide to effective smoking cessation interventions for the health care system. Thorax (suppl) 53:S1–S19

Schols AM, Slangen J, Volovics L, Wouters EF 1998 Weight loss is a reversible factor in the prognosis of chronic obstructive pulmonary disease. Am J Respir Crit Care Med 157:1791–1797

Vestbo J, Srensen T, Lange P, Brix A, Torre P, Viskum K 1999 Long-term effect of inhaled budesonide in mild and moderate chronic obstructive pulmonary disease: a randomised controlled trial. Lancet 353:1819–1823

Young J, Fry-Smith A, Hyde C 1999a Lung volume reduction surgery (LVRS) for chronic obstructive pulmonary disease (COPD) with underlying severe emphysema. Thorax 54:779–789

Young P, Dewse M, Fergusson W, Kolbe J 1999b Respiratory rehabilitation in chronic obstructive pulmonary disease: predictors of nonadherence. Eur Respir J 13:855–859

DISCUSSION

Rogers: You showed a decline in lung function with time, both for sustained quitters and people who continue to smoke. It is my experience with friends and colleagues that when they hit the age of 30 they start trying to give up smoking, and then enter a cycle of quitting, re-starting and quitting again. I'm reminded of work by Rosemary Jones and Lynne Reid, in which they exposed rats to cigarette smoke (Jones & Reid 1978). Rats were exposed for five days each week, with the weekend off. With the continual exposure for five days there was an increase in mitotic index in the respiratory epithelium, which faded by the end of the week. When the rats were re-exposed to cigarette smoke after the weekend break, there was another increase in the mitotic index. These rats may have had a greater pathology, therefore, than rats which are continuously exposed to smoke. Are there any data comparing sustained quitters with intermittent quitters?

Calverley: Not long-term, as far as I am aware. One of the problems with these data is that the Lung Health Study was a negative trial. Everybody cites this because it makes the point nicely that if you do keep off cigarettes your lung function decline is different. The purpose of the trial was actually to try to see whether if one took a group of people and gave them intensive smoking intervention, one could produce a benefit. The answer was that it didn't produce a benefit in enough people long term to really influence the rate of decline of lung function. The data I showed were from a post-hoc analysis that was taken out of that study to make the point that we all want to make: that cigarettes are bad for you and that in this context, the people who did quit did do better. There was of course a subgroup who came on and off cigarettes, and this group showed an improvement in FEV_1 on quitting, and when they re-started it dropped again. What I am not sure is whether the statistics were capable of analysing a global rate of decline for these intermittent quitters. They are a group that most people want to exclude because they complicate the analysis.

Rennard: I think there are data from the Lung Health Study showing that particularly in women who started and stopped smoking, multiple episodes of quitting and re-starting may be quite bad. This was a post-hoc analysis. The idea is that if you quit, you get a little bit of improvement, but when you re-start you rapidly decline again, presumably because you have re-started inflammatory processes. If I remember correctly, women may be particularly susceptible to the disadvantageous effects of these multiple episodes. This raises an interesting question. Smoking is obviously not a continuous practice: even heavy smokers don't smoke cigarettes all the time. Intermittent exposures to toxins at high levels in this fashion may be a different kind of phenomenon than continuous exposures. Disease may therefore be episodic over many different time scales.

MacNee: There are two questions that always seem to arise when we discuss this issue. First, these studies are in mild disease: there may be a different effect of quitting on decline in FEV_1 in patients who have more severe disease. Second, recent studies show persistence of inflammation after cessation of cigarette smoking. There may also be differences in these cases between moderate or severe disease (FEV_1 less than 50% of predicted) compared to the mild disease of patients in the Lung Health Study (FEV_1 70% of predicted).

Rennard: This is a complicated matter; I don't know the answer. Peter Calverley raised the question of the benefits of smoking cessation in the more severe COPD population. We frequently oversimplify things and say that lung function declines at a rate of so many millilitres in normals and so many millilitres in smokers. In fact, it is age dependent, and the rate of decline accelerates with age. The benefits of smoking cessation are also age dependent. In the studies that I am familiar with, once you reach a certain age the benefits of smoking cessation are not as easily detectable (Camilli et al 1987). This may be because subjects have already reached the accelerated rate of lung decline. But the question is a valid one: it may not be possible to extrapolate the benefits of smoking cessation from mild COPD patients to patients with more severe disease or older patients. Having said this, smoking cessation is still a good idea, and I don't think any of us would recommend that people continue to smoke.

MacNee: What about the persistence of inflammation upon smoking cessation?

Rennard: There are several studies that have addressed this. Perhaps the best come from Magnus Sköld in Stockholm. There are a number of parameters that are present in smokers, including inflammation. Whether they correlate with disease or not is unclear. Macrophage numbers, for example, are reliably increased in smokers. These will decrease with smoking cessation, but it can take as long as two years for macrophage numbers to return to normal levels. These macrophages have been shown to phagocytose fluorescent particles presumably derived from the tars in the cigarette smoke, and it can take several years for these fluorescent particles to disappear. The macrophage has a half-life in the alveoli of around three months, so the concept is that the macrophages continue to cycle and phagocytose the debris in the lungs, and it takes some time for this to be cleared out. There is an interesting but somewhat speculative concept that if macrophages take up too much debris, they can become incapacitated. It could be that if you are a relatively young smoker with relatively mild disease and you quit smoking, with several generations of macrophages you can clean out your alveoli over a period of years. But, if you have smoked enough and accumulated enough rubbish in your lungs, even if you quit smoking, although the macrophages will be recruited to the lungs, the disease may not be nearly so reversible. This is a plausible and partially testable concept.

MacNee: We don't have any information on the dynamics of this process.

Calverley: What is important then is that we should make some measurements. Just having a concept is fairly irrelevant. We have a clinical conundrum. The lung function of the ex-smokers in the ISOLDE study — who have established disease, and from who we have three years of detailed FEV_1 data — is declining less rapidly than the continuing smokers, but is not back to normal. This is important information. There are a small number of people (about 90) who stopped and started in that period, and we have not analysed them in detail, but we may be able to find out whether this group had a more unfavourable rate of decline. There may be a statistical fluke here but these are the only sub-group to have a change in decline of FEV_1 with inhaled steroids, but the data have yet to be properly tested.

Hogg: Dr Jody Wright has looked at the effect of stopping smoking on the severity of the inflammatory process in the airways and didn't show much of a change (Wright et al 1983). Looking at this in a slightly different way, the smokers do get a stimulation in their bone marrow and an increase in their white cell count. The decline in FEV_1 correlates with the peripheral blood leukocyte count (Weiss et al 1985), particularly the neutrophils (Young & Buncio 1984). It could be that there is a difference in the systemic response to the cigarettes with a bigger bone marrow stimulus in some cases than others.

Rogers: Do we have any data on people who have reduced their smoking habit?

Calverley: The problem with these data is that many people who reduce the number of cigarettes they smoke, smoke those that they do harder.

Lomas: Sten Eriksson published in the *European Respiratory Journal* the Swedish registry data of the α_1-antitrypsin-deficient patients (Piitulainen & Eriksson 1999). This registry has the greatest number of patients with antitrypsin deficiency ascertained through family studies. His group found that when smokers stopped, their rate of decline of lung function returned to that of the non-smoking group.

Agustí: If we accept that inflammation persists after the cessation of smoking, how should we interpret the initial bump that occurs in all these studies when people stop smoking?

Calverley: There is more than one process occurring, and more than one consequence of inflammation in the lungs.

Rennard: I don't think that it is true that inflammation persists after you quit smoking. Inflammation *can* persist after you stop smoking, but the data suggest that inflammation will tend to go away. However, it may take a few years.

Agustí: But that initial bump is within the first three to six months.

Hogg: The criteria that we used to evaluate airway inflammation did not change up to five years after stopping smoking (Wright et al 1983). It could be that what we measured was not responsible for the improvements in airway function that occurs with stopping smoking. Many years ago, Macklem et al (1970) showed that surfactant is important to the stability of the peripheral airways. Therefore, a small amount of exudate with much higher surface tension than lung surfactant could substantially change the surface tension of the airways, making them unstable. I don't think that this could be measured by histology but it might come with smoking and clear up when smoking stopped. Unfortunately, it is a difficult hypothesis to test morphologically.

Agustí: This may be an important issue to pursue. It could open up new opportunities for therapy: if we forget about inflammation and treat the surfactant deficit, we may influence lung function.

Hogg: If there are interventions that clear up the exudate, which would then alter lung function, this is a hypothesis that could be tested.

Jeffery: We are interested in the bump in the data from the recent ISOLDE and Euroscope trials of the effects of inhaled steroids in COPD; the improvements in FEV_1 decline which occurred during the first three to six month period. Although the conclusion from these studies is that there is no change in the rate of decline long term (i.e. over three years) with inhaled steroids, there is clearly an effect in the bronchoscopic study that we have recently completed in which we have examined bronchial biopsies. This, of course, samples a different site—the larger airway rather than the lung parenchyma. The data are that steroids taken over three months reduce inflammation in the larger airway: that is the CD8:CD4 ratio and, particularly, mast cell numbers. Our interpretation would be that this is affecting mucus hypersecretion, for example, which on a day-to-day basis would affect FEV_1, at least in the short term. Having improved that maximally, one then sees the underlying decline in FEV_1 re-appearing and continuing. On the other point, that there is inflammation in all smokers but not all necessarily have COPD, Dr Saetta's studies (Turato et al 1995) have shown that if you have a group that give up smoking but continue to be productive of sputum, then there is no reduction in inflammation. There seems to be a mismatch between inflammation and decline in

FEV_1, but a good match between inflammation and large airway function (i.e. mucus hypersecretion). The question is what relationship does inflammation have, if any, to long-term accelerated decline.

Calverley: This comes back to the point Jay Nadel raised earlier on: the term 'inflammation' is not actually that informative. It is a step forward, but it certainly is not a uniform entity. If we look at the consequences of individual pathological change in COPD, there is no simple relationship between tissue damage and symptoms: it is not like having a narrowing of the coronary artery where you get reproducible chest pain when you walk a given distance. The pathophysiology in that situation is relatively straightforward. The consequences we are looking at in COPD are more diffuse and downstream, and different processes will produce different degrees of change. The inhaled steroid studies show that it is possible to have a short-term effect modifying outcome of the disease, but this doesn't change the core process.

Jeffery: It is not just the nature and amount of the inflammation that is important here, but also where in the lung the inflammation is occurring.

Nadel: If there is inflammation in the airways following the cessation of smoking, this is a very interesting area to investigate: if inflammation is occurring in the absence of an obvious stimulus, the mechanism must be important.

Paré: I was struck by Peter Calverley's data from the ISOLDE study showing a discordance between symptoms and changes in FEV_1 and a relative concordance between symptoms and changes in inspiratory capacity, which is a measure of end expiratory lung volume. Is this a reflection of sites of airway narrowing and closure? Why do some people develop more hyperinflation than others despite similar levels of lung function as assessed by FEV_1?

Calverley: I share your interest. I would love to work out why these effects occur. This will be an important research topic, and it would be nice to combine this with something a little more mechanistic rather than just physiological description. There seem to be a mixture of changes in the passive characteristics of the respiratory system and others more relevant to exercise such as dynamic hyperinflation. Clearly, in some people the latter predominate. We have shown that the changes that occur in dynamic hyperinflation are predictable from the time constants of the respiratory system. If you have relatively long time constants for lung emptying, and you increase the respiratory rate, you are going to overinflate your lungs. This will lead to a change in end expiratory lung volume, which seems to be what produces the secondary changes in respiratory muscle activation and increased breathlessness. How drugs affect this needs to be readdressed. I have a nagging feeling that we are going to have to revisit some of these areas to look at a concept we have rather forgotten: that this is a very heterogeneous disease. What happens when some of these drugs are inhaled is that regional lung mechanics are altered. They may open up units which have had relatively long time constants and are effectively acting as an area of trapped gas in the lung. These units can be deflated without much change in the overall airway resistance. If drugs effectively produce a low grade 'lung volume reduction', this may explain the changes that we are seeing in symptoms. This needs to be properly explored.

Dunnill: Is there a sex difference in the response to smoking?

Calverley: Almost certainly. The Copenhagen group have clearly shown that the rates of COPD in women in Denmark are at least as high as in males, and the mortality may be greater (Prescott et al 1997). The Lung Health Study data looking at bronchial hyper-responsiveness as a prospective marker for future COPD — the so-called Dutch hypothesis — found a very high prevalence of non-specific bronchial hyper-responsiveness in women as compared to men. There are a clutch of data from around the world that are beginning to find that women have a worse mortality experience than men in terms of their smoking. They find it harder, possibly for reasons of the female psyche, to give up smoking. Smoking has become very much part of young women's lifestyle, and it is now sadly the case in the UK that more women in their early 20s smoke than young men.

Paré: I thought Ben Burrows and colleagues have shown that women smokers have a better prognosis (Sherrill et al 1993).

Calverley: The confounding factor in that study was that his atopic group did better. The problem with Ben's other studies is that now people are much better at quantifying the smoking history. Historically, women smoked much less than men.

MacNee: I'm confused about the gender issue. Is there no good evidence of a gender difference?

Silverman: In our series there was a marked predominance of females among severe early-onset COPD patients. I think that in previous series of severe COPD, there has traditionally been a male predominance, not in population-based studies but just in collections of COPD patients. We need to factor in the temporal changes in cigarette smoking patterns that have occurred. It may only be that with the relatively recent increase in cigarette smoking in women we might be able to see an increased susceptibility to develop COPD in women.

Stolk: It appears from your data that quality of life measurements can pick up signals that pulmonary function tests regularly don't. Does this mean that we have to go through this item of correlating these quality of life parameters with pathology? Do you think the regulatory authorities would like to see that?

Calverley: The issue of quality of life measurements has become another hurdle to jump over. It is being driven by the fact that existing physiology measurements such as FEV_1 don't actually mean a whole lot to individual patients. Practical doctors have developed quality of life questionnaires to measure the impact of the disease on life. The difficulty with that is that although you might establish some structural associations, these measurements assess other things as well. For instance, mood is an important determinant of health status, and can be improved by pulmonary rehabilitation. This influences quality of life, but I can't produce a structure–function correlation with mood. At the moment the quality of life is most valuable in assessing treatments of a palliative nature.

MacNee: There have been studies in COPD showing a relationship between quality of life and markers of inflammation.

Lomas: The poor correlation between changes in FEV_1 in response to bronchodilators and the effect on exercise tolerance suggests that what we do in routine clinical practice is wrong. Should we do away with the BTS/ATS guidelines? Your data suggest that measuring FEV_1 bronchodilator responses is of little clinical value and instead we should be measuring six minute walks in response to bronchodilator challenges.

Calverley: FEV_1 is not useless, but it is not useful for what we tend to use it for! It isn't useless because FEV_1 will pick up asthmatics. FEV_1 is like using a hand lens: it doesn't have good resolution. Using FEV_1 to detect small changes is like using a hand lens and trying to guess what the scanning electron micrograph would look like. The difficulty about six minute walk tests is that they are time consuming.

Paré: What about inspiratory capacity? That is an easy measurement.

Calverley: It is an easy measurement, and I'm more impressed by this, but we are only just beginning the studies to look at this. The Kingston Ontario group think that this is a robust measurement.

Lomas: The follow on from your argument is that everyone gets everything.

Calverley: That is sort of what it comes down to. The best way of picking up people without spending a fortune on fancy physiology is to ask them whether they think that this makes any difference to their symptoms. If you give them three months for the placebo effect to wear off, and apply some clinical judgement about who will unreasonably accept and want everything, it is a surprisingly robust approach.

Wedzicha: It is important to remember that the positive quality of life results have come mainly from large studies. The St George's Respiratory Questionnaire (SGRQ) has a close relationship to exacerbation frequency. There you can get the relationship with just 40 or 50 patients. However, when we went back and tried to look at sputum IL-6 and IL-8, there was no relationship with SGRQ. We have to be careful about what quality of life actually represents. It represents some function of the effect of exacerbations on patients, but the airway changes that give that exacerbation frequency are completely different. We could waste a lot of time trying to find relationships between these outcome measures.

Calverley: Health status questionnaires are a sort of 'epidemiological ESR': they are a pretty non-specific integrative marker of all the consequences of this illness.

This is quite useful if you are a health economist, and it gets round some of the problems of whether what you do matters to the patient. But to take these measurements and relate them to pathological processes is a bit dubious.

Senior: The age when an individual starts to smoke is probably important in the long term effects on lung structure and function. Starting to smoke regularly in the early teens may be more harmful than starting to smoke regularly at age 20 or later. A study using CT scanning to quantify lung growth in teenagers who are smoking and those who are not would be interesting.

Hogg: We need to get the measurements worked out a lot better before we embark on any large studies. Coxson and colleagues have been able to get very good CT scans on guinea pigs, and will try to measure lung growth in small animals using CT. If the method is sensitive enough to pick up the changes in lung growth, the study you suggest might be very useful.

Calverley: That is also relevant in one other way. When I was talking about cigarette smoking and other interventions, I said that we were very interested in rate of decline of lung function. This an easy shorthand for something that we have up until now been able to measure with difficulty in humans but that we think is related to the key pathological process. But, cigarette smoking will influence the peak lung growth that can be achieved. There are epidemiological data that show that instead of rising to a peak, lung growth is slurred and starts to decline in individuals who begin to smoke in their early teens. The corollary of this is that if you start off with lower lung function and don't achieve your full developmental potential, you don't need such a rapid rate of decline to reach the point where symptoms and exercise limitation develop. We are always going to have a mixture of people in our studies, some of who are rapid decliners, and others who didn't get to that peak growth because they were smoking in childhood or because of other illness. This is probably why COPD is such a big global problem at the moment, because lung damage due to tobacco occurs on a background of people who are not well nourished in childhood, are subject to severe atmospheric pollution and do not reach their maximum lung function.

Jeffery: It would be interesting to do the ISOLDE study again, but this time investigating the modifying effects of inhaled steroids on smoke-induced retardation of growth (i.e. increasing FEV_1 in childhood and young adult smokers).

Rogers: In terms of end points, FEV_1 is important. But what do people think about surrogate markers in terms of disease progression, such as induced sputum or exhaled gases?

Barnes: There are several non-invasive markers that can be used to monitor the inflammatory process and oxidative stress in airways. For example, looking at sputum tells you about inflammatory components in the airways. We have used this approach to demonstrate that corticosteroids do not seem to suppress the inflammatory process or the increased proteolytic activity in COPD. There are several markers in exhaled air that reflect oxidative stress in the lungs. We use 8-isoprostane concentrations in exhaled condensate, and ethane, which is a marker of lipid peroxidation. These markers may be useful for assessing some treatments before commencing large clinical trials.

Nadel: Anything you can produce by coughing (e.g. sputum) is likely to be a manifestation of disease in airways with a large cross-sectional area. Cough clearance of sputum is likely to clean secretions primarily from the large airways. These are not necessarily the airways where the major pathophysiology occurs.

MacNee: We know already that the cellular profile in sputum is different from the cellular profile in bronchial biopsies.

Jeffery: If you look at the profile of inflammation, not the extent, it seems to be surprisingly similar if you compare central and peripheral airways.

MacNee: To follow up on this, Peter Barnes' studies have shown no effect of inhaled or oral corticosteroids of inflammatory markers in induced sputum. However, other studies show changes in sputum neutrophil counts in response to corticosteroids.

Barnes: There is a study from Italy showing a small effect of inhaled steroids on neutrophils, but it was not placebo-controlled and the sputum had a high number of eosinophils, suggesting, that they were studying a mixture of COPD and asthma.

Calverley: We need some more descriptive studies that try to tell us about some of these variables. It may be that those eosinophils indicate an asthmatic element, which would be very useful to know. We know a lot more about the pathology of asthma. Is this just random coexistence, or is the presence of eosinophils in a subset of COPD patients a marker of a different balance of disease processes which have primarily been triggered by the COPD. I don't think we know.

Nadel: I wanted to ask a plumbing question. Peter Calverley showed data that the FEV_1 was markedly decreased, and the FVC was not nearly as decreased. As an old plumber, I wanted to know whether that has special meaning. Is it just an artefact of the way the ratios are put together? In the old days no one examined FEV_1 to determine whether airway obstruction existed. I always thought that FVC was a better yardstick of restriction.

Calverley: In a sense, yes. If you think about the pressure–volume relationships of the respiratory system, people are going to be high up on that pressure–volume relationship. Once you get to this severity of disease, the effect of a bronchodilator drug or a bronchoconstrictor like histamine, is to change the operating lung volumes, rather than to change airway resistance. In that sense, if you want to use the term 'restrictive' in that way, that is right. If you take $FEV_1/$ FVC ratios, they haven't proven to be particularly robust in big studies. This is largely because people don't do the FVC correctly. The ratios themselves have been used for defining mild disease, but have not proven of great value in figuring out who does what in the severe stage 3 disease. Consequently they have tended to be ignored, as you implied.

Nadel: In exacerbations or severe disease, is it possible that you are looking at a lot of peripheral airway disease (e.g. plugging)?

Calverley: Yes.

MacNee: It seems to me that there is a need to have a population study that characterizes patients in terms of the measurements we have been discussing. One powerful tool is CT scanning. However, we have still to properly validate measurements of emphysema by CT scanning to enable us to perform population studies. There are also many markers of inflammation in both biopsy and non-invasive samples (such as breath and sputum) that we haven't characterized as well as we could do. If we did both of these things, then we need to look again at the population of COPD patients and response to the treatment in subgroups of patients characterized in this way. If we apply treatments to a general population of COPD patients — some with inflammation, some without; some with eosinophils, some without; some with reversibility, some without; some with emphysema, some without — we are unlikely to get any meaningful results. Surely we need some population studies using markers of inflammation and other techniques such as CT scanning.

Calverley: Effectively, what you are saying is that we need to define a sort of pulmonary Framingham study, where you measure a lot of relatively robust markers and follow people up over time and see which ones come out best. One of the reasons our knowledge of cardiovascular disease has progressed is that people did these sorts of things. And since these cohort studies are on-going, as new ideas have developed they have been studied in the established cohorts. For example, polysomnography has just been added into six of the US cohorts to see whether sleep apnoea is a predictor of cardiovascular disease.

Rennard: I see it differently. To come back to Jan Stolk's question, should we be getting information about the pathological correlates of the quality of life data? This is an important question. We deal with COPD as if it were a single entity, and yet we know that it is extraordinarily heterogeneous. We have these interesting data from the ISOLDE study that quality of life by some parameters got better with inhaled glucocorticoids. We haven't the faintest idea what this means at a histological level. It is appealing to speculate that this is because the people have fewer exacerbations, but I think it is an extraordinarily important unanswered question to know how this is happening.

References

Camilli AE, Burrows B, Knudson RJ, Lyle SK, Lebowitz MD 1987 Longitudinal changes in forced expiratory volume in one second in adults. Am Rev Respir Dis 135:794–799

Jones R, Reid L 1978 Secretory cell hyperplasia and modification of intracellular glycoprotein in rat airways induced by short periods of exposure to tobacco smoke, and the effect of the antiinflammatory agent phenylmethyloxadiazole. Lab Invest 39:41–49

Macklem PT, Proctor DF, Hogg JC 1970 The stability of peripheral airways. Resp Physiol 8:191–203

Piitulainen E, Eriksson S 1999 Decline in FEV_1 related to smoking status in individuals with severe α_1-antitrypsin deficiency (PiZZ). Eur Respir J 13:247–251

Prescott E, Bjerg AM, Andersen PK, Lange P, Vestbo J 1997 Gender difference in smoking effects on lung function and risk of hospitalization for COPD: results from a Danish longitudinal population study. Eur Respir J 10:822–827

Sherrill DL, Lebowitz MD, Knudson R J, Burrows B 1993 Longitudinal methods for describing the relationship between pulmonary function, respiratory symptoms and smoking in elderly subjects: the Tucson Study. Eur Respir J 6:342–348

Turato G, Di Stefano A, Maestrelli P et al 1995 Effect of smoking cessation on airway inflammation in chronic bronchitis. Am J Respir Crit Care Med 152:1262–1267

Weiss ST, Segel MR, Sparrow D, Wager C 1985 Relationship of FEV_1 and peripheral blood leukocyte counts. Am J Epidemiol 142:493–498

Wright JL, Lawson LM, Pare PD, Wiggs BR, Kennedy S, Hogg JC 1983 Morphology of the peripheral airways in current and ex-smokers. Am Rev Respir Dis 127:474–477

Young MC, Buncio AD 1984 Leukocyte count, smoking and health. Am J Med 76:31–37

Genetics of chronic obstructive pulmonary disease

Edwin K. Silverman

Channing Laboratory/Pulmonary and Critical Care Division, Brigham and Women's Hospital, 181 Longwood Avenue, Boston, MA 02115, USA

Abstract. The marked variability in the development of chronic obstructive pulmonary disease (COPD) in response to cigarette smoking has been known for decades, but severe α_1-antitrypsin deficiency (PI Z) remains the only proven genetic risk factor for COPD. With cigarette smoking, PI Z subjects tend to develop more severe pulmonary impairment at an earlier age than non-smoking PI Z individuals. However, PI Z individuals exhibit wide variability in pulmonary function impairment, even among individuals with similar smoking histories. Therefore, other genes and environmental exposures are also likely involved. The role of heterozygosity for the Z allele as a risk factor for COPD remains controversial, but accumulating evidence suggests that at least some PI MZ individuals are at increased risk of developing airflow obstruction. In individuals without α_1-antitrypsin deficiency, familial aggregation of COPD has been reported in several studies. To study novel genetic determinants of COPD, our research group enrolled 44 severe, early-onset COPD probands ($FEV_1 < 40\%$, age < 53 yrs, non-PI Z) and 266 of their relatives. A marked female predominance was noted among the early-onset COPD probands. In addition, increased risk to current or ex-smoking first-degree relatives of early-onset COPD probands for reduced FEV_1, chronic bronchitis and spirometric bronchodilator responsiveness has been demonstrated. These data strongly support the genetic basis for the development of COPD and the potential for gene-by-environment interaction. A variety of studies have examined candidate gene loci with association studies, comparing the distribution of variants in genes hypothesized to be involved in the development of COPD in COPD patients and control subjects. For most genetic loci which have been tested, there have been inconsistent results. Genetic heterogeneity could contribute to difficulty in replicating associations between studies. In addition, case-control association studies are susceptible to supporting associations based purely on population stratification, which can result from incomplete matching between cases and controls — including differences in ethnicity. No association studies in COPD have been reported which used family-based controls, a study design which is immune to such population stratification effects. More importantly, no linkage studies have been published in COPD to identify regions of the genome which are likely to contain COPD susceptibility genes — regions in which association studies are likely to be more productive.

2001 Chronic obstructive pulmonary disease: pathogenesis to treatment. Wiley, Chichester (Novartis Foundation Symposium 234) p 45–64

The Human Genome Project will soon provide a complete DNA sequence of the human genome. This knowledge will offer tremendous opportunities for research into the mechanisms of complex diseases like coronary artery disease, diabetes mellitus, asthma and chronic obstructive pulmonary disease (COPD), which are likely influenced by multiple genetic and environmental factors. However, identifying the genetic determinants of such complex disorders will remain very challenging, due to genetic heterogeneity, environmental phenocopies of genetically determined traits, genotype-by-environment interaction, incomplete penetrance, and multilocus effects, as well as misclassification of the phenotype.

The frequent development of COPD in individuals with severe α_1-antitrypsin deficiency (e.g. PI Z), the one proven genetic risk factor for COPD, has provided a foundation for the protease–antiprotease hypothesis for the pathogenesis of emphysema (Janoff 1985). However, most subjects who develop COPD are not α_1-antitrypsin deficient.

It is well known that cigarette smoking is a major risk factor for the development of COPD. In 1977, Burrows and colleagues demonstrated a dose–response relationship between percentage predicted FEV_1 (forced expiratory volume in one second) and pack-years of cigarette smoking (Burrows et al 1977). Heavier smokers were more likely to develop airflow obstruction, indicated by reduced FEV_1. However, many smokers had pulmonary function within the normal range. In the Burrows study, pack-years was the smoking-related variable which correlated most closely with FEV_1, but it only accounted for 15% of the variability in FEV_1. A study of lung pathology in smokers demonstrated that microscopic emphysema, assessed by quantitative measurements of airspace wall surface area per unit volume of lung tissue, was present in only 26% of smokers (Gillooly & Lamb 1993). In part, this relates to competing risks; cigarette smokers may die from other smoking-related illnesses such as coronary artery disease and lung cancer prior to the development of COPD. However, genetic factors are also likely to influence the variable susceptibility to develop COPD. Because cigarette smoking is a proven major environmental risk factor for COPD, which can be readily assessed with questionnaires, COPD offers unique opportunities and challenges to incorporate environmental influences in the study of genetics of complex diseases.

α_1-antitrypsin deficiency

A small percentage of COPD patients (estimated at 1–2%) inherit severe α_1-antitrypsin deficiency (Lieberman et al 1986). We will discuss α_1-antitrypsin deficiency in some detail, because it is the one proven genetic risk factor for COPD, and because it can serve as a model of the manner in which genetic and environmental factors can interact to lead to COPD.

α_1-antitrypsin, specified by the PI (protease inhibitor) locus, is the major plasma protease inhibitor of leukocyte elastase — one of the enzymes which has been hypothesized to play a role in the development of emphysema (Travis & Salvesen 1983). The *PI* locus is polymorphic; in Caucasian populations the most common alleles are *M*, which includes 95% of the alleles and is associated with normal antitrypsin levels; *S*, which includes 2–3% of the alleles and is associated with mildly reduced antitrypsin levels; and *Z*, which includes 1% of the alleles and is associated with severely reduced antitrypsin levels. Isoelectric focusing of serum can accurately determine PI type, which reflects the genotype at the *PI* locus for these common alleles.

A small percentage of subjects inherit null alleles, which lead to the absence of any α_1-antitrypsin production through a heterogeneous collection of mutations (Brantly et al 1988). Individuals with two *Z* alleles or one *Z* and one null allele are referred to as PI Z. PI Z individuals have approximately 15% of normal plasma antitrypsin levels, because the Z protein polymerizes within the endoplasmic reticulum of hepatocytes (Mahadeva & Lomas 1998). Several large series of α_1-antitrypsin-deficient individuals have concluded that PI Z subjects who smoke cigarettes tend to develop more severe pulmonary impairment at an earlier age than non-smoking PI Z individuals (Larsson 1978, Tobin et al 1983, Janus et al 1985). PI Z individuals occur with prevalence of approximately 1/3000 in the United States (Silverman et al 1989a).

However, the development of COPD in PI Z subjects is not absolute. In a study performed at Washington University in St Louis, we assembled 52 PI Z subjects (Silverman et al 1989b); significant variability in pulmonary function was found (Fig. 1). In Fig. 1, index PI Z subjects, who were tested for α_1-antitrypsin deficiency because they had COPD and who were the first PI Z identified in their family, all had significantly reduced FEV_1 values. Non-index PI Z subjects, who were ascertained by a variety of other means, including family studies, population screening and liver disease, suggest a much different natural history for severe α_1-antitrypsin deficiency than index PI Z subjects. Many non-index PI Z subjects have preserved pulmonary function.

Part of the variability in pulmonary function among PI Z individuals is explained by cigarette smoking; however, some smokers maintain normal FEV_1 values at least into middle age, when some non-smokers have already developed significant airflow obstruction (Silverman et al 1989b). With subjects identified from the Danish α_1-antitrypsin Register, Seersolm and colleagues also found significantly higher FEV_1 values in non-index PI Z subjects compared to index PI Z subjects despite similar ages and smoking histories (Seersholm et al 1995). Consequently, the natural history of PI Z subjects in the general population remains uncertain. There are likely three groups of unidentified PI Z subjects in the general population: (1) PI Z subjects with diagnosed COPD who have not been tested for α_1-antitrypsin deficiency (possibly because they were not diagnosed with *early-onset* COPD); (2) PI Z subjects who have significant airflow obstruction but who

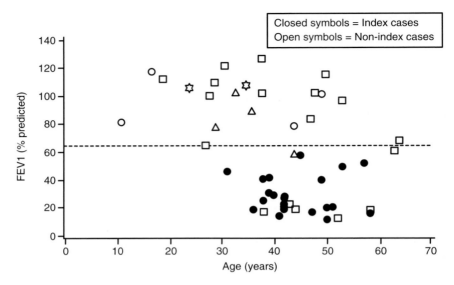

FIG. 1. Effect of ascertainment bias on FEV_1 (percentage predicted) among severely α_1-antitrypsin-deficient (PI Z) subjects in the St. Louis α_1-antitrypsin study. Closed circles correspond to index PI Z subjects (individuals diagnosed with α_1-antitrypsin deficiency because they had COPD, who were the first PI Z subject identified in their family). Open symbols correspond to non-index subjects, who were ascertained by liver disease (open circles), family studies (squares), population screening (triangles) and other pulmonary symptoms (stars). From Silverman et al (1989b).

have not been diagnosed with COPD; and (3) PI Z subjects with normal pulmonary function. However, the relative proportions of these groups are unknown.

Our primary interest in the St Louis study was to determine if genetic factors other than PI type influenced the variable development of lung disease among PI Z subjects. Therefore, we enrolled first-degree relatives of the 52 PI Z subjects. Despite comparable smoking history, parents of PI Z subjects with reduced FEV_1 tended to have lower FEV_1 values themselves compared to parents of PI Z subjects with preserved FEV_1. This difference in FEV_1 (% predicted), 95% vs. 75.3%, was of borderline statistical significance, with $P=0.05$. These results, as well as the findings with segregation analysis from the St Louis study, suggested that additional genetic factors influence the development of airflow obstruction in α_1-antitrypsin deficiency. Moreover, we demonstrated that significant genotype-by-environment interaction between PI type and pack-years of smoking was present (Silverman et al 1992).

Recently, Novoradovsky and colleagues published a study in which they assessed whether variation in the endothelial nitric oxide synthase (*NOS3*) gene could be one of the genetic factors which influences the variable development of airflow obstruction in PI Z subjects (Novoradovsky et al 1999). They identified polymorphisms in the *NOS3* gene, and they tested for genetic association of these variants with airflow obstruction in 55 PI Z subjects with FEV_1 <35% predicted, 122 PI Z subjects with FEV_1 > 35% predicted, and 93 control subjects. Two polymorphisms in the coding region, which likely do not lead to functionally important changes in the NOS3 protein, were associated with severe airflow obstruction in PI Z subjects. Further work to replicate this finding and to identify a functionally important variant will be required, but NOS3 is an intriguing possible contributor to the variable development of airflow obstruction in PI Z subjects.

The risk of lung disease in heterozygous PI MZ individuals has been a subject of considerable controversy for many years. PI MZ individuals do have reduced serum α_1-antitrypsin concentrations (approximately 60% of PI M levels). However, random population surveys (including a large study by Bruce and colleagues that matched subjects for age, race, sex and smoking history), have typically found no difference in pulmonary function between PI MZ and PI M individuals (Morse et al 1977, Bruce et al 1984). On the other hand, case-control studies comparing the prevalence of the PI MZ type in patients with COPD and control subjects have usually discovered an excess of PI MZ individuals among COPD patients (Bartmann et al 1985, Lieberman et al 1986). The basis for these inconsistencies remains unresolved.

A recent study by Sandford and colleagues provides additional evidence that an elevated prevalence of PI MZ is found among individuals with airflow obstruction (Sandford et al 1999). They assessed PI type in 266 individuals who were undergoing surgical resection of lung cancer; 193 subjects had significant airflow obstruction and 73 subjects did not have airflow obstruction. PI MZ was found in 12/193 subjects with airflow obstruction and 0/73 subjects without airflow obstruction. However, the actual risk associated with PI MZ remains unclear. In addition, it is uncertain if PI MZ represents a slight risk for all PI MZ subjects or a significant risk for a few PI MZ subjects because of gene–gene or gene–environment interactions.

Risk to relatives for COPD

A variety of studies of pulmonary function measurements performed in the general population and in twins have suggested that genetic factors influence variation in pulmonary function. Redline studied spirometry in 256 monozygotic and 158 dizygotic twins who were not selected for respiratory problems (Redline et al 1989). For FEV_1, higher correlation in monozygotic twins (0.72) than dizygotic twins (0.27) suggested that genetic factors influence variation in pulmonary function. However, such general population studies do not address whether genetic factors influence the development of significant airflow obstruction.

Studies in relatives of COPD patients have also supported a role for genetic factors. Several studies in the 1970s reported higher rates of airflow obstruction in first-degree relatives of COPD patients than in control subjects. For example, Larson compared spirometry in 156 first-degree relatives of COPD patients to 86 spouse controls with similar pack-years of smoking (Larson et al 1970). Airflow obstruction was found in 23% of first-degree relatives, but only 9% of control subjects. Although this study did show familial aggregation for airflow obstruction, it did have several weaknesses. α_1-antitrypsin deficiency was not rigorously excluded as a potential contributor to airflow obstruction. In addition, a higher percentage of the first-degree relatives than controls were smokers, so at least some of the observed differences may relate to differences in smoking behaviour. None the less, genetic predisposition to COPD was suggested by this study.

Kueppers et al (1977) studied 114 subjects with COPD and control subjects matched based on age, gender, occupation and smoking history. Siblings of COPD and control subjects were also included; the mean FEV_1 among siblings of COPD subjects (90% of predicted) was significantly lower than the siblings of control subjects (103% of predicted). This difference in pulmonary function between sibs of COPD and control subjects remained significant after adjustment for smoking history.

A large study of COPD in families was performed by Cohen et al (Cohen et al 1977, Cohen 1980). Familial aggregation for airflow obstruction was demonstrated (Cohen et al 1977, Beaty et al 1987). Segregation analysis of their data provided support for the existence of a major gene influencing FEV_1 in families with COPD (Rybicki et al 1990). However, they performed segregation analysis with class A regressive models, which subsequently have been shown to be susceptible to false positive assignments of major genes when they are applied to quantitative traits like FEV_1 (Demenais & Bonney 1989).

In an effort to identify novel genetic risk factors for COPD; we have focused our efforts on subjects with severe, early-onset COPD. In several previous successful studies of complex trait genetics, focusing on early-onset cases led to the identification of susceptibility genes for breast cancer, Alzheimer's disease, glaucoma and diabetes mellitus (maturity-onset diabetes of the young; MODY) (Hall et al 1990, Goate et al 1991, Vionnet et al 1992, Wooster et al 1994, Stone et al 1997). By enrolling severe, early-onset COPD probands and their relatives, the population studied may be enriched for genetic influences.

Probands in this Boston COPD study had $FEV_1 < 40\%$ predicted at age less than 53 years, without severe α_1-antitrypsin deficiency (Silverman et al 1998). Probands were recruited from Lung Transplant and Lung Volume Reduction Surgery

Program referrals and from pulmonary clinics. All available first-degree relatives and spouses were invited to participate. Focusing on older second-degree relatives, all available aunts, uncles and grandparents were also included.

The initial phase of the Boston COPD Study included 44 early-onset COPD probands, with 204 first-degree relatives, 54 second-degree relatives and 20 spouses. In addition, we recruited 20 control families with 83 individuals from previous population-based studies at the Channing Laboratory. Control probands were matched to early COPD probands based on age, gender and smoking status. Recent information on pack-years of smoking was not available before enrolment, so matching of control subjects to probands was not performed based on pack-years.

The 44 severe, early-onset COPD probands are compared to the 20 control probands recruited from previous population-based studies in Table 1. Because the control probands had significantly lower pack-years than the COPD probands, a second control group, which includes all control probands and spouses of early-onset COPD probands who had greater than 10 pack-years of smoking, is also presented. This second control group matches well for age and pack-years to the early-onset COPD probands. The dramatically reduced levels of FEV_1 and $FEV_1/$ FVC (forced vital capacity) among the early-onset COPD probands are evident.

We were surprised to find such a high percentage of females, 80%, among our early COPD probands. This female predominance differs from previous studies of severe COPD, which have typically found a male predominance (Damsgaard & Kok-Jensen 1974, Postma et al 1979, O'Donnell & Webb 1992, Wegner et al 1994). We have started to examine possible contributors to the female predominance in our study, which includes a group of probands who are younger, more severely affected and more recently collected than prior series of severe COPD subjects (Silverman et al 2000). Reduced survival of male subjects

TABLE 1 Comparison of early COPD probands with controls for spirometry and demographics

	Age	Pack-years	FEV_1 (% pred)	FEV_1/FVC (% pred)	Gender (% female)
Early COPD probands (n = 44)	47.2±5.6	38.8±22.5,	16.9±6.1	37.2±11.8	79.6
Control probands (n=20)	49.3±6.8	24.6±16.7*	89.3±15.9*	91.4±11.0*	75
Smoking-matched controls (n=30)	49.6±6.2	36.2±26.1	86.5±16.8*	90.1±11.0*	50*

*Indicates $P < 0.05$ compared to early-onset COPD probands.
From Silverman et al (1998).

with severe COPD could contribute to the observed female predominance; however, it is certainly possible that there is a biological basis for the increased female susceptibility, due to hormonal or other factors, that could mediate a genotype-by-gender interaction in our early-onset COPD pedigrees.

Spirometric and demographic data for the first-degree relatives of early COPD probands, stratified by smoking status, are shown in Table 2. Highly significant differences in FEV_1 and FEV_1/FVC were found when current or ex-smoking first-degree relatives of early-onset COPD probands were compared to control subjects. No significant differences in age or pack-years of smoking were noted. The distribution of FEV_1 values in current and ex-smoking first-degree relatives of early-onset COPD probands and control subjects is shown in Fig. 2; a trend toward lower FEV_1 values in the first-degree relatives is evident.

No significant differences in FEV_1 or FEV_1/FVC were found when lifelong non-smoking first-degree relatives of early COPD probands were compared to lifelong non-smoking control subjects. In fact, the mean FEV_1 values were 93.4% of predicted in both groups of non-smokers. This pattern would be consistent with genetic risk factors which interact with smoking to result in COPD. A similar pattern of smoking-related susceptibility was also seen for chronic bronchitis (Silverman et al 1998).

From the standpoint of genetic studies, the key issue is whether relatives of early-onset COPD probands have increased risk for reduced FEV_1 and chronic bronchitis beyond the effects of smoking. To account for potential familial correlations and for the effects of age and pack-years of smoking, generalized estimating equations were used to calculate odds ratios of developing chronic

TABLE 2   Age, smoking history, and spirometry in first-degree relatives of early-onset COPD probands compared to control subjects

Group	n	FEV_1/FVC (%pred)	FEV_1 (%pred)	Age	Pack-years
Smoking first-degree relatives	112	83.5±16.1*	76.1±20.9*	45.9±17.3	28.5±26.6
Smoking control subjects	48	94.3±10.3	89.2±14.4	48.6±13.9	22.1±22.1
Non-smoking first-degree relatives	92	92.7±7.6	93.4±12.9	34.4±18.9	0.00
Non-smoking control subjects	35	95.5±7.2	93.4±14.2	39.9±18.2	0.00

*Indicates $P < 0.01$ compared to control subjects.
From Silverman et al (1998).

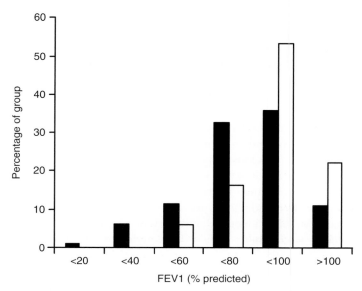

FIG. 2.   FEV$_1$ (percentage predicted) in current or ex-smoking first-degree relatives of early-onset COPD probands and current or ex-smoking control subjects. Solid bars correspond to first-degree relatives of early-onset COPD probands; open bars correspond to control subjects. Although 7% of smoking first-degree relatives had FEV$_1$ values below 40% predicted, this degree of severe impairment was not observed in the control group. From Silverman et al (1998).

bronchitis and various levels of reduction in FEV$_1$. When all first-degree relatives of early COPD probands were compared to all controls, an increased risk of FEV$_1 < 80\%$ was seen. However, stratification by smoking status revealed that this risk was exclusively found in smoking first-degree relatives; with significant odds ratios of 4.5 for FEV$_1$ below 80% and 3.6 for chronic bronchitis, and nearly significant odds ratio of 3.5 for FEV$_1$ below 60%. Lifelong non-smokers had no increased risk for reduced FEV$_1$ or chronic bronchitis.

We also measured bronchodilator responsiveness and total serum IgE levels in first-degree relatives of early-onset COPD probands, and we compared bronchodilator responsiveness and IgE in adult first-degree relatives of early-onset COPD probands with adult control subjects (Celedon et al 1999). Among current or ex-smoking adult first-degree relatives of COPD probands, higher levels of bronchodilator responsiveness were noted than in current or ex-smoking adult control subjects, assessed as the absolute volume increase in FEV$_1$ (120 ± 130 ml in first-degree relatives vs. 60 ± 110 ml in controls, $P < 0.05$), the increase in FEV$_1$ as a percentage of baseline FEV$_1$ ($5.8\pm8.1\%$ in first-degree relatives vs. $2.9\pm5.1\%$ in controls, $P < 0.01$), and the increase in FEV$_1$ as a percentage of predicted FEV$_1$ ($3.6\pm4.1\%$ in first-degree relatives vs. $2.2\pm3.9\%$

in controls, $P < 0.05$). Significant differences in bronchodilator responsiveness between adult smoking first-degree relatives and adult smoking control subjects were also found in multivariate analysis with generalized estimating equations, adjusting for age, pack-years of smoking and gender. No significant differences in bronchodilator responsiveness were found between non-smoking adult first-degree relatives and adult control subjects in univariate or multivariate analyses.

First-degree relatives of early-onset COPD probands did not have elevated total serum IgE levels compared to control subjects. Although the control probands were recruited from previous population-based respiratory studies, and all of their first-degree relatives were invited to participate, a very high rate of physician-diagnosed asthma in the control subjects (21.7%) suggests that asthmatic subjects were more likely to volunteer for this study.

In summary, we have identified a variety of phenotypes which demonstrate smoking-related susceptibility in first-degree relatives of early COPD probands including FEV_1, FEV_1/FVC, chronic bronchitis, and bronchodilator responsiveness. These phenotypes will likely be useful in further genetic studies of early-onset COPD.

Association studies in COPD

A variety of association studies have compared the distribution of variants in genes hypothesized to be involved in the development of COPD in COPD patients and control subjects. A recent study examined a dinucleotide short tandem repeat polymorphism in the 5′ flanking region of the haeme oxygenase 1 gene (Yamada et al 2000). The alleles at this polymorphic marker were grouped as S (<25 repeats), M (25–29 repeats) and L (>29 repeats). The classes of alleles were compared in 101 Japanese emphysema patients and 100 smoking controls who were also Japanese. A higher frequency of class L alleles was noted in emphysema patients (21%) than control subjects (10%). The authors performed transient-transfection assays to assess the functional significance of polymorphism repeat length; they demonstrated up-regulation with hydrogen peroxide exposure of the 16 and 20 repeat alleles, but not with the 29 or 38 repeat alleles. Although this study is interesting, it is not clear if grouping repeat length alleles in this manner is appropriate; it would have been more compelling if this classification scheme were selected from an independent population. In addition, although they examined several alleles for functional significance, it is not clear if there is a definite relationship between repeat length and function with each dinucleotide repeat allele.

A variety of candidate genetic loci have been studied with the case-control association approach. A non-exhaustive list of loci which have alleles that have been associated with COPD is presented in Table 3. A representative study supporting the association is shown for polymorphic variants located beyond the 3' end of the α_1-antitrypsin gene, the vitamin D binding protein, the cystic fibrosis transmembrane regulator (CFTR) gene, ABO blood group and α_1-antichymotrypsin (Cohen 1980, Kalsheker et al 1990, Poller et al 1992, Gervais et al 1993, Schellenberg et al 1998). In some cases, more than one study exists to support an association. However, in each case, at least one study refutes the association (Vestbo et al 1993, Artlich et al 1995, Sandford et al 1997, 1998, Kauffmann et al 1983).

Several factors could contribute to the inconsistent results of case-control genetic association studies in COPD. Genetic heterogeneity, or different genetic mechanisms, in different populations could contribute to difficulty in replicating associations between studies. In addition, false positive or false negative results could contribute as in any study design. A potentially important factor is that case-control association studies are susceptible to supporting associations based purely on population stratification. Population stratification can result from incomplete matching between cases and controls, including differences in ethnicity and geographic origin. No association studies in COPD have been reported which used family-based controls, a study design which is immune to such population stratification effects. In addition, no linkage studies have been published in COPD to identify regions of the genome which are likely to contain COPD susceptibility genes — regions in which association studies may be more fruitful. In summary, a variety of candidate genes have been examined in COPD, but no genetic loci other than α_1-antitrypsin have been proven as risk factors for COPD.

Conclusions

In conclusion, severe α_1-antitrypsin deficiency is the only proven genetic risk factor for COPD, but the development of COPD in PI Z subjects is variable — and genetic factors likely contribute to this variability. Case-control association studies have suggested a variety of candidate gene variants as potential contributors to the

TABLE 3   Case-control association studies in COPD — conflicting evidence

	Support association	*Do not support association*
A1AT 3' flanking region	Kalsheker et al (1990)	Sandford et al (1997)
Vitamin D binding	Schellenberg et al (1998)	Kauffmann et al (1983)
CFTR	Gervais et al (1993)	Artlich et al (1995)
ABO blood group	Cohen (1980)	Vestbo et al (1993)
α_1-antichymotrpsin	Poller et al (1992)	Sandford et al (1998)

development of COPD, but the results have not been consistent across studies. Linkage analysis and family-based association studies have the potential to be valuable tools in the identification of new genetic risk factors for COPD. Further research will be required to determine the optimal study designs, phenotypes, and analytical methods to identify the genetic determinants of COPD.

Acknowledgements

This work was supported by R01 HL61575 from the National Institutes of Health and by a Research Grant from the American Lung Association.

References

Artlich A, Boysen A, Bunge S, Entzian P, Schlaak M, Schwinger E 1995 Common CFTR mutations are not likely to predispose to chronic bronchitis in northern Germany. Hum Genet 95:226–228

Bartmann K, Fooke-Achterrath M, Koch G et al 1985 Heterozygosity in the Pi-system as a pathogenetic cofactor in chronic obstructive pulmonary disease (COPD). Eur J Respir Dis 66:284–296

Beaty TH, Liang KY, Seerey S, Cohen BH 1987 Robust inference for variance components models in families ascertained through probands: II. Analysis of spirometric measures. Genet Epidemiol 4:211–221

Brantly M, Nukiwa T, Crystal RG 1988 Molecular basis of α-1-antitrypsin deficiency. Am J Med 84:13–31

Bruce RM, Cohen BH, Diamond EL 1984 Collaborative study to assess risk of lung disease in Pi MZ phenotype subjects. Am Rev Respir Dis 130:386–390

Burrows B, Knudson RJ, Cline MG, Lebowitz MD 1977 Quantitative relationships between cigarette smoking and ventilatory function. Am Rev Respir Dis 115:195–205

Celedon JC, Speizer FE, Drazen JM et al 1999 Bronchodilator responsiveness and serum total IgE levels in families of probands with severe early-onset COPD. Eur Respir J 14:1009–1014

Cohen BH 1980 Chronic obstructive pulmonary disease: a challenge in genetic epidemiology. Am J Epidemiol 112:274–288

Cohen BH, Ball WC Jr, Brashears S et al 1977 Risk factors in chronic obstructive pulmonary disease (COPD). Am J Epidemiol 105:223–232

Damsgaard T, Kok-Jensen A 1974 Prognosis in severe chronic obstructive pulmonary disease. Acta Med Scand 196:103–108

Demenais FM, Bonney GE 1989 Equivalence of the mixed and regressive models for genetic analysis. I. Continuous traits. Genet Epidemiol 6:597–617 (erratum 1990 Genet Epidemiol 7:103)

Gervais R, Lafitte JJ, Dumur V et al 1993 Sweat chloride and delta F508 mutation in chronic bronchitis or bronchiectasis. Lancet 342:997

Gillooly M, Lamb D 1993 Microscopic emphysema in relation to age and smoking habit. Thorax 48:491–495

Goate A, Chartier-Harlin M-C, Mullan M et al 1991 Segregation of a missense mutation in the amyloid precursor protein gene with familial Alzheimer's disease. Nature 349:704–706

Hall JM, Lee MK, Newman B et al 1990 Linkage of early-onset familial breast cancer to chromosome 17q21. Science 250:1684–1689

Janoff A 1985 Elastases and emphysema. Current assessment of the protease–antiprotease hypothesis. Am Rev Respir Dis 132:417–433

Janus ED, Phillips NT, Carrell RW 1985 Smoking, lung function, and α1-antitrypsin deficiency. Lancet 1:152–154

Kalsheker NA, Watkins GL, Hill S, Morgan K, Stockley RA, Fick RB 1990 Independent mutations in the flanking sequence of the α-1- antitrypsin gene are associated with chronic obstructive airways disease. Dis Markers 8:151–157

Kauffmann F, Kleisbauer J-P, Cambon-de-Mouzon A et al 1983 Genetic markers in chronic airflow limitation: a genetic epidemiologic study. Am Rev Respir Dis 127:263–269

Kueppers F, Miller RD, Gordon H, Hepper NG, Offord K 1977 Familial prevalence of chronic obstructive pulmonary disease in a matched pair study. Am J Med 63:336–342

Larson RK, Barman ML, Kueppers F, Fudenberg HH 1970 Genetic and environmental determinants of chronic obstructive pulmonary disease. Ann Intern Med 72:627–632

Larsson C 1978 Natural history and life expectancy in severe α$_1$-antitrypsin deficiency, Pi Z. Acta Med Scand 204:345–351

Lieberman J, Winter B, Sastre A 1986 α1-antitrypsin Pi-types in 965 COPD patients. Chest 89:370–373

Mahadeva R, Lomas DA 1998 Genetics and respiratory disease. 2. α1-antitrypsin deficiency, cirrhosis and emphysema. Thorax 53:501–505

Morse JO, Lebowitz MD, Knudson RJ, Burrows B 1977 Relation of protease inhibitor phenotypes to obstructive lung diseases in a community. N Engl J Med 296:1190–1194

Novoradovsky A, Brantly ML, Waclawiw MA et al 1999 Endothelial nitric oxide synthase as a potential susceptibility gene in the pathogenesis of emphysema in α1-antitrypsin deficiency. Am J Respir Cell Mol Biol 20:441–447

O'Donnell DE, Webb KA 1992 Breathlessness in patients with severe chronic airflow limitation. Physiologic correlations. Chest 102:824–831

Poller W, Faber JP, Scholz S et al 1992 Mis-sense mutation of α1-antichymotrypsin gene associated with chronic lung disease. Lancet 339:1538

Postma DS, Burema J, Gimeno F et al 1979 Prognosis in severe chronic obstructive pulmonary disease. Am Rev Respir Dis 119:357–367

Redline S, Tishler PV, Rosner B et al 1989 Genotypic and phenotypic similarities in pulmonary function among family members of adult monozygotic and dizygotic twins. Am J Epidemiol 129:827–836

Rybicki BA, Beaty TH, Cohen BH 1990 Major genetic mechanisms in pulmonary function. J Clin Epidemiol 43:667–675

Sandford AJ, Spinelli JJ, Weir TD, Paré PD 1997 Mutation in the 3′ region of the α-1-antitrypsin gene and chronic obstructive pulmonary disease. J Med Genet 34:874–875

Sandford AJ, Chagani T, Weir TD, Paré PD 1998 α1-antichymotrypsin mutations in patients with chronic obstructive pulmonary disease. Dis Markers 13:257–260

Sandford AJ, Weir TD, Spinelli JJ, Paré PD 1999 Z and S mutations of the α1 -antitrypsin gene and the risk of chronic obstructive pulmonary disease. Am J Respir Cell Mol Biol 20:287–291

Schellenberg D, Paré PD, Weir TD, Spinelli JJ, Walker BA, Sandford AJ 1998 Vitamin D binding protein variants and the risk of COPD. Am J Respir Crit Care Med 157:957–961

Seersholm N, Kok-Jensen A, Dirksen A 1995 Decline in FEV1 among patients with severe hereditary α1- antitrypsin deficiency type PiZ. Am J Respir Crit Care Med 152:1922–1925

Silverman EK, Miletich JP, Pierce JA et al 1989a α-1-antitrypsin deficiency. High prevalence in the St Louis area determined by direct population screening. Am Rev Respir Dis 140:961–966

Silverman EK, Pierce JA, Province MA, Rao DC, Campbell EJ 1989b Variability of pulmonary function in α1-antitrypsin deficiency: clinical correlates. Ann Intern Med 111:982–991

Silverman EK, Province MA, Campbell EJ, Pierce JA, Rao DC 1992 Family study of α1-antitrypsin deficiency: effects of cigarette smoking, measured genotype, and their interaction on pulmonary function and biochemical traits. Genet Epidemiol 9:317–331

Silverman EK, Chapman HA, Drazen JM et al 1998 Genetic epidemiology of severe, early-onset chronic obstructive pulmonary disease. Risk to relatives for airflow obstruction and chronic bronchitis. Am J Respir Crit Care Med 157:1770–1778

Silverman EK, Weiss ST, Drazen JM et al 2000 Gender differences in severe, early onset chronic obstructive pulmonary disease. Am J Respir Crit Care Med, in press

Stone EM, Fingert JH, Alward WLM et al 1997 Identification of a gene that causes primary open angle glaucoma. Science 275:668–670

Tobin MJ, Cook PJL, Hutchison DC 1983 α1-antitrypsin deficiency: the clinical and physiological features of pulmonary emphysema in subjects homozygous for Pi type Z. A survey by the British Thoracic Association. Br J Dis Chest 77:14–27

Travis J, Salvesen GS 1983 Human plasma proteinase inhibitors. Annu Rev Biochem 52:655–709

Vestbo J, Hein HO, Suadicani P, Sørensen H, Gyntelberg F 1993 Genetic markers for chronic bronchitis and peak expiratory flow in the Copenhagen Male Study. Dan Med Bull 40:378–380

Vionnet N, Stoffel M, Takeda J et al 1992 Nonsense mutation in the glucokinase gene causes early-onset non-insulin-dependent diabetes mellitus. Nature 356:721–722

Wegner RE, Jörres RA, Kirsten DK, Magnussen H 1994 Factor analysis of exercise capacity, dyspnoea ratings and lung function in patients with severe COPD. Eur Respir J 7:725–729

Wooster R, Neuhausen S, Mangion J et al 1994 Localization of a breast cancer susceptibility gene, BRCA2, to chromosome 13q12–13. Science 265:2088–2090

Yamada N, Yamaya M, Okinaga S et al 2000 Microsatellite polymorphism in the heme oxygenase-1 gene promoter is associated with susceptibility to emphysema. Am J Hum Genet 66:187–195

DISCUSSION

Paré: It is often said that in patients who develop COPD, the mechanism(s) could either be more rapid decline in lung function, and/or a lower starting lung function at the onset of decline. If there was a genetic contribution to the latter mechanism, I would have thought that the non-smoking relatives of your severe COPD cases should have had lower lung function expressed as FEV_1 percentage predicted.

Silverman: It depends on why individuals who start from a lower maximally attained threshold got there in the first place. If they got there because they started smoking when they were 13, then the non-smoking relatives will not share that effect. If they got there because they had smaller lungs due to genetic factors, then the relatives may share that effect. The mean FEV_1 of the non-smoking, first degree relatives of severe early-onset COPD probands was 93.4% of predicted: it wasn't 100%. We tried to go out to previous population based studies and get a control group, but there was a very high percentage of asthmatics in our control group; it wasn't representative of the general population. It is possible that our control group may have been reduced in pulmonary function, so that we would not find differences in FEV_1 between non-smoking, first-degree relatives of early-onset COPD probands and non-smoking controls. I think either explanation is possible.

Jackson: You showed the distribution of female versus male lung function. Couldn't that be as a result of maximally attained lung function at adulthood? With the males starting off with a larger lung function, it will take longer before they actually present with COPD-like symptoms. They may well die before they get significant large-scale lung function reduction. If females have a similar rate of lung function decline as males, but have lower lung function at maturity and similar life expectancy to males, one might expect a higher proportion of female smokers to present with COPD.

Silverman: All of the FEV_1 values that I showed are percentage predicted, adjusted for age, height and gender. It is reasonable to ask how accurate these predicted values are at the extremes. We worry about their accuracy at extremes of age, and perhaps also at extremes of pulmonary function. This would be a more likely explanation, since these are percentage predicted values.

Lomas: For the past two years we have been doing a similar study, but we ascertained patients differently. We ascertained them as having emphysema rather than COPD. Our mean age is about 52, whereas Ed Silverman's was in the high 40s. We don't see any sex difference at all — the male:female ratio is 1:0.9. This may just be that our group is older or that we are ascertaining them through different criteria.

Paré: Ed Silverman, did you try using ΔFEV_1 in your simulations? This is another phenotype that is a lot harder to get, because it requires longitudinal data, but it may be the most robust phenotype, especially if different genes are responsible for rate of decline and for the FEV_1 at age 30.

Silverman: We would love to look at it but we don't have it from these data. All of the probands have had previous pulmonary function tests, but we are a little reluctant to rely on spirometric values obtained from a clinical laboratory as hard endpoints to look at relatively modest declines.

Calverley: I support your reluctance to use that information. It is likely to be more misleading than helpful. If you want to measure rate of decline you need to know that the tests were done to robust criteria.

Paré: Probably the best data on longitudinal decline are from the Lung Health Study. We had the opportunity to get the DNA from the 300 individuals who had the most rapid decline in lung function in that study (a staggering 150 ml per year!) and the 300 individuals who had the least rapid decline (plus 20 ml per year). Interestingly, we found that the PI MZ came out as a significant predictor of being in the rapid decline group with an odds ratio of about 2. The microsomal epoxide haplotype which is associated with a slow metabolism of epoxides also came out with an odds ratio of about 2. However, the vitamin D binding protein polymorphisms which have previously been shown to be associated with COPD had no influence on the rate of decline, but were associated with the subjects' FEV_1 when they entered the cohort. These data suggest that different genes can influence the sub-phenotypes making it even more complicated.

MacNee: To me, the interesting group seem to be those individuals with α_1-antitrypsin deficiency who smoked and who did not have a low FEV_1. What do we know about these individuals?

Silverman: We tried to compare PI Z individuals who had preserved FEV_1 with PI Z individuals who had reduced FEV_1. We looked at a variety of epidemiological characteristics. The PI Z subjects with reduced FEV_1 were more likely to have chronic bronchitis; the PI Z subjects with reduced FEV_1 were more likely to have been diagnosed with asthma and to have had pneumonia. I don't know whether a PI Z smoker who happens to avoid those things has preserved pulmonary function for those reasons, or whether they have additional protective genetic factors, or even the absence of an additional detrimental genetic factor that causes some PI Z subjects to have reduced pulmonary function. We have such a warped view of the natural history of PI Z individuals because of the way most PI Z subjects are identified. There are undoubtedly three populations of PI Z subjects out there that we don't know about. There are those who have severe early-onset COPD who haven't been tested. There are those who have later-onset COPD who haven't been tested because they don't seem any different from other late-onset patients. Finally, there are PI Z individuals who have normal pulmonary function. We don't know what the relative sizes of these three groups are. We know that smoking is certainly bad for PI Z individuals, but we don't know even among PI Z smokers what the natural history is.

Agustí: What we are discussing here is the issue of susceptibility to COPD. I would like to add an extra level of complexity to this. We are talking of susceptibility to disease development, but we forget that not all smokers have the same ability to quit smoking. It seems that some can quit relatively easily, and others who can't. This brings in the susceptibility to becoming addicted to smoking. If you go to the other end of the disease, and you are a smoker and you have COPD, not all COPDs lose weight, for instance. As Emiel Wouters has shown, weight loss is a prognostic factor in COPD. The worst scenario for a patient would then be being susceptible to smoking addiction, COPD and weight loss: this results in very severe disease.

Rennard: In Laurell & Eriksson's (1963) series, they found five individuals who were deficient in α_1-antitrypsin, and four of them had severe emphysema. From your review of a population-based study, those results are a little implausible.

Silverman: Laurell & Eriksson (1963) noted that three of the initial five severely α_1-antitrypsin deficient individuals had evidence of emphysema. However, these were blood samples from individuals which had been sent to the hospital laboratory, so it was not necessarily representative of the general population.

Rennard: It is not a random population. It wouldn't be as wide a population base as you would have had in the St Louis study. This raises the question of gene–gene interactions. There have been a number of candidate genes: Peter Paré, have you done any analysis looking at gene–gene interactions?

Paré: 300 fast and 300 slow declines sounds like a large cohort, but when you start looking at interactions of gene polymorphisms, your *n* goes down, and we weren't able to detect any gene interaction among the polymorphisms that we studied. That there are additional genetic factors in our cohort was suggested by the observation that if the MZ and microsomal epoxide polymorphic subjects had a family history of COPD then relative risk increased from about 2 to the 8–10 range. This suggests that there are other genetic risk factors. There will likely be a whole bunch of interacting genes involved in COPD pathogenesis. The worry is that there will be too many to ever make anything out of the genetics. The hope is that there will be few enough with large enough effects that we will be able to select individuals particularly at risk.

Silverman: We are talking about identifying genetic risk factors for complex diseases. This is something that is taking place not just in COPD, but also in other diseases such as asthma, diabetes and coronary artery disease. It is clearly a difficult task. There are a variety of issues. Should we use a candidate gene approach, or some sort of genome screen? As work proceeds with the Human Genome Project and all genes become identified, are those two pathways going to merge? Clearly, there are going to be cases where there are Mendelian sub forms: the α_1-antitrypsin deficiency equivalents. These are major genes that may not account for a large percentage of the variation, but could point us towards important pathways. These may be major gene effects. If the model of other diseases such as breast cancer holds true, then perhaps a small number of these genes could be illuminating.

Paré: This work might also be able to provide risk stratification for clinical studies.

Rennard: In correcting for smoking, how do you deal with this as a variable?

Silverman: It is hard, but there are a variety of ways to consider adjustment for smoking effects. If one is using a linkage or association analysis technique that allows the inclusion of environmental covariates (something that the model accounts for in the attempt to determine if evidence for linkage or association is present), then inclusion of smoking measurements as covariates may work. Another approach is to perform regression analysis in advance for smoking-related variables such as pack-years or the dichotomous classification of ever smoker/never smoker. Finally, one could limit the analysis to smokers only. In our St Louis study of PI Z individuals, pack-years was strongly predictive of reduced FEV_1, but there are a variety of other variables that one could consider looking at, such as age started smoking, maximum smoking intensity, and depth of inhalation while smoking.

Rennard: The reason I ask is that all of those measures of smoking are OK in statistical groups, but people can smoke quite variably. Pack-years is not going to be the same thing as dose of toxins inhaled. When you do the correction for smoking, there will be inherent errors. Is it worthwhile to measure some objective biochemical measure of smoking, such as cotinine levels?

Silverman: It is worth thinking about. In our early-onset COPD series, the vast majority of probands have quit smoking, so we can no longer get these data. We could do this in a small subset to see how reproducibly the number of cigarettes smoked per day relates to cotinine levels, but there are likely to be genes involved in nicotine metabolism that may influence cotinine levels as well.

Agustí: How variable is the pattern of smoking of a given individual through time? Do smokers smoke the same way all the time?

Rennard: First of all, not all smokers smoke the same way. 15–20% can smoke a few today and then abstain for a while, whereas the majority maintain a constant adult habit once they have developed a mature addiction. The number of cigarettes they smoke and how they smoke them is fairly constant from day to day, but not all of those cigarettes during the course of the day may be smoked in the same way. It is likely that some of those cigarettes are 'optional', so if you tell a person to smoke fewer cigarettes they will do so, but their nicotine intake measured by cotinine or exhaled CO will not change. Presumably they are adjusting their smoking behaviour to maintain relatively constant nicotine intake.

Barnes: Have there been any studies on non-smoking COPD? One might hypothesize that these people are particularly susceptible genetically. This may be a fruitful group to explore.

Silverman: I agree. Black & Kueppers (1978) and Seersholm & Kok-Jensen (1998) have both reported studies on non-smoking PI Z subjects, which have shown substantial variation in the degree of pulmonary function impairment in PI Z non-smokers. Among patients with giant bullous emphysema (vanishing lung syndrome), some non-smokers have been described (Stern et al 1994).

Agustí: Does non-smoking COPD exist?

Barnes: Yes, particularly in developing countries.

Silverman: In our 84 probands we had three lifelong non-smokers, so it is certainly rare in the severe COPD group, but it happens.

Calverley: It is probably the sort of issue that is not best addressed in Western communities. In all the big series of COPD, you know they are being honest with the data when about 3% haven't smoked.

Jeffery: Can you be sure about the absence of occupational air pollution?

Lomas: In the 150 probands that we have looked at, they have all smoked.

Barnes: Non-smoking COPD patients may be the most interesting individuals: having the least environmental influence may highlight a greater genetic predisposition.

Dunnill: What do you know about the genetics of addiction? It seems to me that from a preventive viewpoint, if you could identify those people and stop them smoking, this would take out a large proportion of the population with the disease.

Rennard: There have been several studies (see Rossing 1998). Many addictions may work through dopaminergic transmission in the locus coeruleus and the mesolimbic system, so people have focused on this. Several dopaminergic receptors, including the DRD2 and DRD4, have been related to smoking in association studies (Noble et al 1994). There are also several dopamine transporters implicated. The problem with these association studies is that they can be connected for other reasons, such as population stratification. It is controversial. One study of particular interest is from Toronto (Pianezza et al 1998). The major enzyme that metabolizes nicotine is a P450 called CYP2A6. Null mutations exist in this enzyme, which means that carriers cannot metabolize nicotine at a normal rate. In an association study, individuals with this null mutation were shown to be less likely to become smokers. In the field, researchers believe that there is a clear genetic basis for addiction, but this will involve multiple genes, and the estimate is that no single gene will account for more than 3–4% of the variance.

Agustí: Blocking CYP2A6 would then be a great smoking cessation therapy.

Rennard This is being investigated.

Shapiro: With regard to the non-smoking COPD, I have talked with paediatricians who take care of many asthmatics, and they swear that they have a subgroup of their population who either aren't responsive to bronchodilators or who, after maximal therapy still have significant obstruction. Some of these patients have emphysema with bronchopulmonary dysplasia, but there is another group that is different, i.e. children with COPD. Due to the complexity of human emphysema, animal models would be ideal to study genetic susceptibility. In animals one can compare genetics between susceptible vs. resistant inbred strains.

Jeffery: There are genetic aspects which influence the response to irritation by cigarette smoke. Not only the level of inflammation but also the type of inflammation (e.g. CD4:CD8 ratio) will vary on the basis of genetic background. This could be an important factor.

MacNee: I'd like to return to an issue we raised earlier, which is the assessment of emphysema by computed tomography (CT) scanning. In α_1-antitrypsin patients, have there been some studies looking at the risk of developing emphysema, rather than just of COPD?

Silverman: Chest CT scan data has been reported in severe α_1-antitrypsin deficiency by Guest & Hansell (1992), who assessed the chest CT scan appearance at baseline, and by Dirksen et al (1997), who assessed longitudinal changes in CT scans of PI Z subjects. In our study of non-α_1-antitrypsin deficiency severe COPD, it turns out that most of the probands had chest CT scans, because they were under evaluation for lung transplantation or lung volume reduction surgery. In almost every case there was some emphysema, but it varied from mild to very severe.

Stolk: Out of 100 patients that I have CT data on, only one patient had upper lobe emphysema with none on the lower lobes. Most patients are referrals from chest physicians, so they come in with COPD. I now have only two families where smoking history is the same among the probands. In one, the patient had severe emphysema while her sister had a normal CT scan and normal pulmonary function tests. These are rare identities.

Paré: To address that question, David Lomas and Ed Silverman could compare their studies. David, your probands are identified based on low diffusing capacity, whereas Ed's are identified by low FEV_1. Have you looked at your data to see whether there are any differences in familial pattern?

Lomas: We ascertain our probands as having an airflow obstruction and a gas transfer factor of less than 70% of predicted. Many have CTs for the same reasons that Ed describes. When they have, the CT results match the physiology. When we looked at the siblings, about 32% of the siblings who smoked have airflow obstruction. About 26% of these meet fairly stringent criteria of emphysema by lung function tests.

Rennard: Have either of you looked at glucocorticoid responsiveness in COPD patients, in terms of stratifying phenotypes?

Calverley: That is very difficult to do.

References

Black LF, Kueppers F 1978 Alpha 1-antitrypsin deficiency in nonsmokers. Am Rev Respir Dis 117:421–428

Dirksen A, Friis M, Olesen KP, Skovgaard LT, Sorensen K 1997 Progress of emphysema in severe alpha 1-antitrypsin deficiency as assessed by annual CT. Acta Radiologica 38:826–832

Guest PJ, Hansell DM 1992 High resolution computed tomography (HRCT) in emphysema associated with α1-antitrypsin deficiency. Clin Radiol 45:260–266

Laurell C-B, Eriksson S 1963 The electrophoretic α1-globulin pattern of serum in α1-antitrypsin deficiency. Scand J Clin Lab Invest 15:132–140

Noble EP, St Jeor ST, Ritchie T et al 1994 D2 dopamine receptor gene and cigarette smoking: a reward gene? Med Hypotheses 42:257–260

Pianezza ML, Sellers EM, Tyndale RF 1998 Nicotine metabolism defect reduces smoking. Nature 393:750

Rossing MA 1998 Genetic influences on smoking: candidate genes. Environ Health Perspect 106:231–238

Seersholm N, Kok-Jensen A 1998 Clinical features and prognosis of life time non-smokers with severe α1-antitrypsin deficiency. Thorax 53:265–268

Stern EJ, Webb WR, Weinacker A, Muller NL 1994 Idiopathic giant bullous emphysema (vanishing lung syndrome): imaging findings in nine patients. Am J Roentgenol 162:279–282

Mucus hypersecretion in chronic obstructive pulmonary disease

Duncan F. Rogers

Department of Thoracic Medicine, National Heart & Lung Institute (Imperial College), Dovehouse Street, London SW3 6LY, UK

Abstract. Most patients with chronic obstructive pulmonary disease (COPD) exhibit characteristics of airway mucus hypersecretion, namely sputum production, increased luminal mucus, submucosal gland hypertrophy and goblet cell hyperplasia. The clinical consequences of hypersecretion are impaired gas exchange and compromised mucociliary clearance, which encourages bacterial colonization and associated exacerbations. However, the extent of the contribution of mucus to pathophysiology of COPD is controversial. Early epidemiological studies found little evidence for the involvement of mucus in the age-related decline in lung function and mortality associated with COPD and concluded that chronic airflow obstruction and mucus hypersecretion were independent processes. Later studies found positive associations between phlegm production and decline in lung function, hospitalization and death. Thus, although not diagnostic for the condition, mucus hypersecretion contributes to morbidity and mortality in certain groups of patients with COPD. This suggests that it is important to develop drugs that inhibit mucus hypersecretion in these patients. Unfortunately, ambiguity in clinical studies of mucoactive drugs means that mucolytics are not recommended in clinical management. Future research should determine whether there is an intrinsic abnormality in mucus in COPD, which will determine development of appropriate inhibitors, which in turn can be used in 'proof of concept' and in treatment.

2001 Chronic obstructive pulmonary disease: pathogenesis to treatment. Wiley, Chichester (Novartis Foundation Symposium 234) p 65–83

Excessive secretion of airways mucus is a traditional concept characterizing patients with chronic bronchitis. Together with emphysema, and possibly some cases of chronic asthma, chronic bronchitis is one component of chronic obstructive pulmonary disease (COPD). The relative contribution of each component to the pathophysiology of COPD in any one patient is difficult to determine. Chronic bronchitis originated as a clinical term based upon long-standing sputum production, invariably with recurrent or persistent cough (Ciba Guest Symposium 1959, American Thoracic Society 1962, Medical Research Council 1965, Fletcher & Pride 1984). In each of the latter reports, productive cough was considered to indicate 'excessive mucous secretion in the bronchial tree.' Although not an unreasonable supposition, there was little or no direct evidence for a correlation between sputum production and mucus hyper-secretion. Consequently, 'mucus hypersecretion' is omitted from definitions in current international guidelines on COPD (American Thoracic Society 1995, British Thoracic Society 1997, Canadian Thoracic Society Workshop Group 1992, Jenkins et al 1995, Siafakas et al 1995). The present chapter reviews the evidence for a role for mucus hypersecretion in the pathophysiology and clinical progression of COPD, and considers the value of mucoactive drugs in treatment. For completeness, a brief overview of airway mucus and mucus secreting cells will be given first.

Airway 'mucus'

The airway lumen is lined with a thin film of viscoelastic liquid which under normal conditions protects the epithelial surface (Widdicombe & Widdicombe 1994; Fig. 1). The liquid is often referred to as 'mucus' and is a 1–2% aqueous solution of glycoconjugates (predominantly mucous glycoproteins, termed mucins), proteoglycans, electrolytes, enzymes, anti-enzymes, antioxidants, antibacterial agents, other plasma-derived proteins, lipids and various cellular mediators (Table 1). The liquid is in intimate contact with the surface epithelial cells, the most abundant of which are the ciliated cells. This arrangement allows mucus to function as an escalator for removal of inhaled particles from the lung. The efficiency of mucociliary clearance is directly related to the elasticity and inversely proportional to the viscosity of the gel layer, and is dependent upon the depth of the aqueous layer. Consequently, airway hypersecretory pathophysiology is a result of dysfunctional ciliary motility and/or of abnormal mucus, either its quantity or viscoelastic properties.

Mucins

The viscoelastic properties of airway mucus are attributed largely to high molecular weight glycoproteins known as mucins which are secreted by specialized cells in the epithelium and submucosa. Mucins consist of a peptide backbone, termed apomucin, to which multiple oligosaccharide side chains are bound (Thornton et al 1997). Apomucins are 100–400 kDa peptides which account for 10–20% of the total glycoprotein weight. They are encoded by several genes (Rose & Gendler 1997) and are expressed in goblet cells and submucosal glands. Currently, nine main human mucin genes are recognized, namely *MUC1–4*, *MUC5AC*, *MUC5B* and *MUC6–8*, of which the MUC5AC product is a predominant mucin in secretions from normal healthy children and in sputum samples from patients with chronic bronchitis (Hovenberg et al 1996).

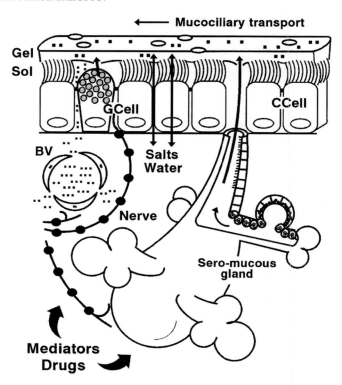

FIG. 1. Production of airway 'mucus'. Airway mucus forms an upper gel phase, in which inhaled particles are trapped and removed by mucociliary clearance, and a lower sol phase in which the cilia beat (CCell, ciliated cell). Mucins, secreted by surface epithelial goblet cells (GCell) and mucous cells in the submucosal glands, play a vital role in conferring the correct viscoelasticity on the mucus for optimal mucociliary transport. Plasma proteins, exuded from the bronchial microvasculature (BV, blood vessel) also contribute to production of the surface mucus. The hydration of the mucus is controlled by active transepithelial transport of electrolytes and water. The processes which contribute to the production of airway mucus are under neural and humoral control, and may also be influenced by pharmaceutical compounds.

MUC5AC in the latter secretions was large, polydisperse in size, subunit-based and gel-forming. In contrast, the biophysical properties of other *MUC* gene products is either poorly defined or has yet to be determined.

Sources of airway mucus

The principal mucus-secreting cells of the airways are the surface epithelial goblet cells and the mucous cells of the submucosal glands (Jeffery & Li 1997; Fig. 1).

TABLE 1   Composition of airway 'mucus'

Water

Electrolytes

High molecular weight mucous glycoproteins (mucins)

Lipids

Anti-microbial enzymes (lactoferrin, lactoferricin, peroxidases, transferrin)

Immunoglobulins (secretory IgA, IgG, IgM)

Antiproteases (α_1-antitrypsin, α_2-macroglobulin, elafin, SLPI, TIMP)

Enzymatic antioxidants (catalase, glutathione peroxidase, superoxide dismutase)

Non-enzymatic antioxidants (ascorbic acid, reduced glutathione, uric acid, α-tocopherol)

Albumin

Proline-rich proteins

DNA

Cell-derived proteinases (e.g. neutrophil elastase)

Cell-derived inflammatory mediators

Plasma/interstitial fluid-derived inflammatory mediators (e.g. bradykinin)

Endogenous anti-inflammatory molecules (e.g. lipocortin)

Ig, immunoglobulin; SLPI, secretory leukoprotease inhibitor; TIMP, tissue inhibitor of metalloproteinase.

Other secretory cells in the airways are the surface epithelial serous cells, Clara cells and possibly ciliated cells, and the serous cells of the submucosal glands. In diseases associated with airway mucus hypersecretion, including COPD, goblet cell hyperplasia and submucosal gland hypertrophy are associated with reduced numbers of serous and ciliated cells, and possibly Clara cells.

Goblet cells are the principal mucus-secreting cells of the surface epithelium and as such contribute to 'first-line' defence of the airways (Rogers 1994). Submucosal glands comprise secretory tubules lined with mucous, and more distally, serous cells, a collecting duct and a ciliated neck. They are predominant in large cartilaginous airways where the mucous cells are responsible for most of the mucus secretion. The mucous cells line the 'mucous tubules' in the submucosal gland and contain an abundance of acid-staining secretory granules. Morphologically, they are similar to goblet cells. In normal human airways, the ratio of submucosal gland to goblet cells has been calculated as 40:1, indicating that goblet cells do not play a significant role in hypersecretion. However, goblet cells undergo hyperplasia and metaplasia in disease conditions (see below), and predominate in distal airways where excess 'abnormal' mucus may be harder to remove. It is in the pathophysiology of chronic airway hypersecretory disease that the importance of goblet cells becomes apparent.

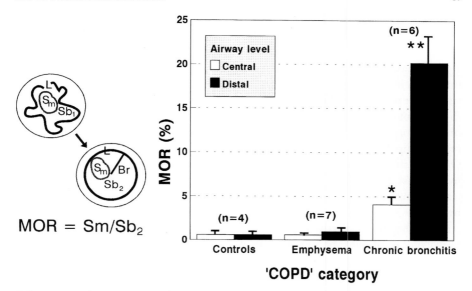

$$MOR = Sm/Sb_2$$

FIG. 2.   Morphometric quantification of airway mucus hypersecretion in chronic obstructive pulmonary disease (COPD). The amount of Elastica Goldner-stained mucus in the airway lumen of Japanese patients at autopsy was expressed as a mucus occupying ratio (MOR), where S_m is the area of mucus, and the internal area of the bronchus (Sb_2) is derived from the length (L) of the basement membrane (when digitized to a circle) and its predicted radius (Br). $*P < 0.05$, $**P < 0.01$ compared with either control or emphysema groups. Redrawn after Aikawa et al (1989).

Evidence for mucus hypersecretion in COPD

Patients with COPD usually have chronic sputum production, which can vary from just a few millilitres after waking, to more than 100 ml produced during the day (Snider et al 1994). The mucus content of the sputum is not reported. However, the sputum may represent excess mucus secretion. Reid (1954) noted an excess of mucus in the air passages in both early and advanced cases of chronic bronchitis. Morphometric quantification in autopsied lungs confirms significantly more mucus in the airway lumen, particularly in distal airways, of patients dying of COPD compared with patients with emphysema or controls without respiratory disease (Aikawa et al 1989; Fig. 2). ELISA analysis (using a monoclonal 'mucin' antibody) of bronchial lavage fluid demonstrated a fourfold increase in mucin-like material in comparatively healthy smokers compared with non-smokers (Steiger et al 1994). There was also a trend to an increase in *MUC2* mRNA in the smokers. The significance of the latter observation is unclear because MUC2 protein does not appear to be an important component of airway secretions from normal subjects or patients with chronic bronchitis (Hovenberg et al 1996).

Increased luminal mucus is associated with increases in amount of mucus secreting tissue. Submucosal gland hypertrophy was noted by Reid (1954), and was later confirmed by quantitative measures (Reid 1960, Restrepo & Heard 1963, Aikawa et al 1989). A consistent observation was an increase in number of mucous cells relative to serous cells. In the study of Aikawa et al (1989), there was a significant positive correlation between the amount of submucosal gland and both the amount of mucus in the airway lumen and the daily sputum volume. The latter observation, although not absolute for causality, certainly suggests a strong relationship between gland hypertrophy and mucus hypersecretion in COPD. One of the cardinal pathophysiological features of COPD is goblet cell hyperplasia, particularly in the bronchioles in which they are normally scarce, a feature emphasised by Reid (1954) for both early and advanced cases of chronic bronchitis.

From the above, it would appear that there is mucus hypersecretion in COPD. However, it should be noted that not all patients exhibit all clinical and pathological characteristics of COPD/chronic bronchitis. Not all patients expectorate (Widdicombe 1990), and there is an overlap in gland size, not only with healthy non-smokers but also between sputum producers and non-producers (Hayes 1969, Thurlbeck & Angus 1964, Thurlbeck et al 1963, Restrepo & Heard 1963). In a small group of Japanese patients with COPD, goblet cell hyperplasia could not be demonstrated when compared with controls (Aikawa et al 1989). From the above discussion, it may be seen that although not diagnostic for the condition, mucus hypersecretion delineates different populations of COPD patients.

Theoretical clinical consequences of mucus hypersecretion

The principal site of airflow obstruction in COPD is distal airways of less than 3 mm diameter (Hogg et al 1968). The obstruction may be accompanied by the presence of excess mucus, fibrosis and obliteration of the airways. Theoretically, the clinical consequences of hypersecretion and mucus plugging are ventilation–perfusion mismatch, with impairment of gas exchange, and compromised mucociliary clearance, which encourages bacterial colonization and leads to a vicious cycle of exacerbations. Also, coughing up sputum embarrasses and exhausts patients. However, the importance of mucus hypersecretion in COPD is debated (see following section).

Importance of mucus hypersecretion in COPD

The extent of the contribution of mucus to pathophysiology and clinical symptoms in COPD is controversial. For example, a number of influential epidemiological studies in the late 1970s and 1980s found little or no evidence for the involvement of mucus in either the age-related decline in lung function (forced expiratory volume in one second; FEV_1) or mortality associated with COPD. In a study of 792 London working men, Fletcher & Peto (1977) concluded that chronic 'phlegm' production did not cause airflow obstruction to progress more rapidly. Phlegm was also excluded as a causal factor in early stage airflow obstruction in 575 Paris working men, although its role was not excluded at later stages (Kauffmann et al 1979). In two much larger studies, of 2718 British men (Peto et al 1983) and of 17 717 London male civil servants (Ebi-Kryston 1988), phlegm production was not significantly associated with death from COPD when other factors (for example age or smoking habits) were controlled for. In a study of 2955 men and women in the Tecumseh community (USA), although incidence rates of developing COPD were higher in men and women who reported phlegm production, the differences were not significant (Higgins et al 1982). The consensus of the above studies was that chronic airflow obstruction and mucus hypersecretion were largely independent disease processes. In contrast, a number of studies in the late 1980s and 1990s found positive associations between phlegm production and decline in lung function, hospitalization and death. A 22-year follow-up of the above Paris working men found a weak but significant relationship between chronic phlegm production and death (Annesi & Kauffmann 1986). The authors did not consider the relationship trivial because of the high prevalence of chronic phlegm production in the population. Analysis of 9000–14 000 men and women from the general population of Copenhagen (Denmark) found that chronic phlegm production was significantly and consistently associated with an accelerated decline in FEV_1 and subsequent hospitalization due to COPD (Vestbo et al 1996) and, in patients with impaired lung function, with an increased risk of death from COPD (Lange et al 1990). Similarly, analysis of 4000–8000 men and women in six cities in eastern USA found that men with phlegm production showed a greater decline in FEV_1 (Speizer et al 1989, Sherman et al 1992) with a greater risk of death from COPD (Speizer et al 1989). It should be noted that in a number of the latter studies, although statistically significant, the associations for mucus were less impressive than for other measures, for example between mortality and the FEV_1/VC (vital capacity) ratio (Annesi & Kauffmann 1986). The reason(s) for the contradictory observations of the two groups of publications are unclear. Use of occupational cohorts, exposed to high levels of industrial pollution, in some of the earlier studies may not equate with studies in the general population. Use of FEV_1 and mortality, at either end of the disease 'spectrum', as primary outcome measures may have less relevance than intermediate measures such as hospitalization. For all of the above studies, the relationship between sputum production and mucus hypersecretion, particularly in the small airways, the main site of obstruction

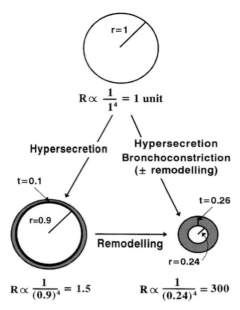

FIG. 3.   Contribution of luminal mucus to airflow limitation in chronic obstructive pulmonary disease (COPD): prediction using Poiseuille's law. Resistance to air flow (R) is proportional to the reciprocal of the radius raised to the fourth power (upper segment). With mucus hypersecretion alone, the reduction in luminal radius may have little impact on R (lower left). However, with airway remodelling, or with a combination of bronchoconstriction and remodelling, a similar degree of hypersecretion markedly increases R (lower right). Differences in the contribution of mucus hypersecretion to pathophysiology in different groups of patients may be due in part to the relative contribution of each component of airway narrowing.

(Hogg et al 1968), is not reported. From the above discussion, it may be seen that although not strongly associated with disease progression in all cases, mucus hypersecretion clearly contributes to morbidity and mortality in certain groups of patients with COPD (Fig. 3). This suggests that it is important to develop drugs that inhibit mucus hypersecretion in these patients.

Mucoactive drugs in treatment of COPD

The clinical symptoms of cough and sputum production, coupled with a perception of the importance of mucus in pathophysiology of a number of severe lung conditions, including COPD, has driven the development and use of drugs affecting mucus. At present, over 50 compounds have potentially beneficial actions on mucus or its secretion, of which over 15 are listed worldwide in publications for prescribing physicians (Nightingale & Rogers 2000; Table 2). Of the latter, seven

TABLE 2   'Mucoactive' drugs

N-Acetylcysteine (mucolytic)

Ambroxol (expectorant)

Bromhexine (expectorant)

Carbocysteine (mucoregulator)

Dornase-alpha (degrades DNA)

Eprazinon hydrochloride (mucoregulator)

Erdosteine (mucoregulator)

L-Ethylcysteine (mucolytic)

Guaifenesin (expectorant)

Letosteine (mucoregulator)

Sodium 2-mercaptoethane sulfonate (MESNA) (mucolytic)

Methylcysteine hydrochloride (mucolytic)

Sobrerol (expectorant)

Stepronine ([a]mucolytic)

Thiopronine (mucolytic)

Fifteen compounds categorized worldwide as 'mucolytics and expectorants' in prescribing booklets for clinicians (after Nightingale & Rogers 2000). Mucolytic, compound containing a 'free' sulfhydryl group which will break disulfide bonds in mucins to reduce viscosity; mucoregulator, compound without a 'free' sulfhydryl group which, therefore, does not break disulfide bonds to reduce mucus viscosity — mechanism of action unspecified; expectorant, compound which increases cough and expectoration — may induce mucus secretion.

[a]Non-mucolytic pro-drug which is metabolized endogenously to form a mucolytic moiety.

are listed consistently, namely N-acetylcysteine, ambroxol, bromhexine, carbocysteine, dornase-alpha (recombinant human DNase), guaifenesin and MESNA (sodium 2-mercaptoethane sulfate). In contrast to the availability of mucoactive drugs, 'mucolytic' therapy is not generally recommended in current guidelines on clinical management of COPD. For example, neither the British Thoracic Society (1997) nor the European Respiratory Society (Siafakas et al 1995) recommend mucolytic drugs in treatment. The Canadian Thoracic Society (Canadian Thoracic Society Workshop Group 1992) considers mucolytics as a treatment 'under investigation' and does not specifically recommend them in management. The Thoracic Societies of America (1995) and of Australia and New Zealand (1995) suggest 'trying' mucolytics in exacerbations associated with production of viscous mucus. The discrepancy between the abundance of mucoactive drugs and the caution in recommending them in treatment is related to inconsistencies in clinical studies on their effectiveness. A recent meta-analysis of clinical trials of mucolytics in COPD excluded 57 out of 72 papers because they were either not controlled or not correctly controlled, or did not provide information on primary outcome measures, or did not give error estimates for

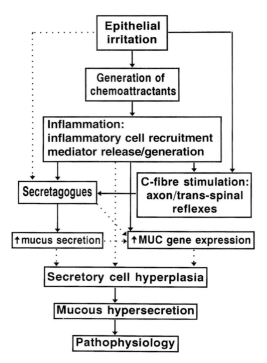

FIG. 4. Schema of contribution of mucus hypersecretion to pathophysiology of chronic obstructive pulmonary disease (COPD). Epithelial irritation (for example by cigarette smoke) produces secretagogues, either directly or via activation of nerves and humoral inflammatory processes, which in turn increase mucus output. Inflammatory processes also increase mucin (*MUC*) gene expression. Secretory cell hyperplasia leads to mucous hypersecretion which, in many patients, is linked to morbidity and mortality. However, the relationship between acute secretion/*MUC* gene expression and hyperplasia and pathophysiology is poorly defined (dashed lines).

outcomes (Poole & Black 1999). Effective clinical trials should be double-blind, placebo controlled and randomized, with well-defined primary endpoints, and include the effects of drugs over short and long periods. Rational assessment of the place of mucolytics in management of COPD will require such studies. The high cost of these studies will have to be weighed against any perceived long-term clinical benefit.

Conclusions and future directions

From the above it may be seen that there is controversy concerning both the pathophysiological and clinical significance of mucus hypersecretion in COPD, and the therapeutic value of drugs affecting mucus properties. Although not diagnostic for the condition, mucus hypersecretion clearly contributes to morbidity and mortality in certain groups of patients with COPD. This suggests that it is important to develop drugs that inhibit mucus hypersecretion in these patients, although without affecting normal mucus secretion and mucociliary clearance. Before these questions can be addressed, considerably more needs to be known about the biochemical and biophysical nature of airway mucins in normal healthy subjects, as well as which mucin gene products predominate. Following on from this, investigations need to be directed at determining whether or not there is an intrinsic abnormality in mucus in COPD, and whether any abnormality is specific for COPD compared with other hypersecretory conditions of the airways such as asthma. The factors which regulate *MUC* gene expression in health and disease, and the relationship between this regulation and the development of the hypersecretory state, will also need to be determined. The above data can then be used in deciding therapeutic targets leading to rational design of anti-hypersecretory drugs for COPD.

References

Aikawa T, Shimura S, Sasaki H, Takashima T, Yaegashi H, Takahashi T 1989 Morphometric analysis of intraluminal mucus in airways in chronic obstructive pulmonary disease. Am Rev Respir Dis 140:477–482

American Thoracic Society 1962 Definitions and classification of chronic bronchitis, asthma, and pulmonary emphysema. Am Rev Respir Dis 85:762–768

American Thoracic Society 1995 Standards for the diagnosis and care of patients with chronic obstructive pulmonary disease. Am J Respir Crit Care Med (suppl) 152:S77–S121

Annesi I, Kauffmann F 1986 Is respiratory mucus hypersecretion really an innocent disorder? A 22-year study mortality study of 1061 working men. Am Rev Respir Dis 134:688–693

British Thoracic Society 1997 BTS guidelines for the management of chronic obstructive pulmonary disease. The COPD Guidelines Group of the Standards of care of the BTS. Thorax (suppl 5) 52:S1–S28

Canadian Thoracic Society Workshop Group 1992 Guidelines for the assessment and management of chronic obstructive pulmonary disease. Can Med Assoc J 147:420–428

Ciba Guest Symposium 1959 Terminology, definitions, and classification of chronic pulmonary emphysema and related conditions. Thorax 14:286–299

Ebi-Kryston KL 1988 Respiratory symptoms and pulmonary function a predictors of 10-year mortality from respiratory disease, cardiovascular disease, and all causes in the Whitehall study. J Clin Epidemiol 41:251–260

Fletcher C, Peto R 1977 The natural history of chronic airflow obstruction. Br Med J 1:1645–1648

Fletcher CM, Pride NB 1984 Definitions of emphysema, chronic bronchitis, asthma, and airflow obstruction: 25 years on from the Ciba symposium. Thorax 39:81–85

Hayes JA 1969 Distribution of bronchial gland measurements in a Jamaican population. Thorax 24:619–622

Higgins MW, Keller JB, Becker M et al 1982 An index of risk for obstructive airways disease. Am Rev Respir Dis 125:144–151

Hogg JC, Macklem PT, Thurlbeck WM 1968 Site and nature of airway obstruction in chronic obstructive lung disease. N Engl J Med 278:1355–1360

Hovenberg HW, Davies JR, Herrman A, Lindén C-J, Carlstedt I 1996 MUC5AC, but not MUC2, is a prominent mucin in respiratory secretions. Glycoconj J 13:839–847

Jeffery PK, Li D 1997 Airway mucosa: secretory cells, mucus and mucin genes. Eur Respir J 10:1655–1662

Jenkins C, Mitchell C, Irving L, Frith P, Young I, for Thoracic Society of Australia and New Zealand 1995 Guidelines for the management of chronic obstructive pulmonary disease. Mod Med Australia 38:132–146

Kauffmann F, Drouet D, Lellouch J, Brille D 1979 Twelve years spirometric changes among Paris area workers. Int J Epidemiol 8:201–212

Lange P, Nyboe J, Appleyard M, Jensen G, Schnohr P 1990 Relation of ventilatory impairment and of chronic mucus hypersecretion to mortality from obstructive lung disease and from all causes. Thorax 45:579–585

Medical Research Council 1965 Definition and classification of chronic bronchitis for clinical and epidemiological purposes. Lancet i:775–779

Nightingale JA, Rogers DF 2000 Should drugs affecting mucus properties be used in COPD? Clinical evidence. In: Similowski T, Whitelaw W, Derenne J-P (eds) Clinical management of stable COPD. Marcel Dekker, New York, in press

Peto R, Speizer FE, Cochrane AL et al 1983 The relevance in adults of air-flow obstruction, but not of mucus hypersecretion, to mortality from chronic lung disease. Results from 20 years of prospective observation. Am Rev Respir Dis 128:491–500

Poole PJ, Black PN 1999 Mucolytic agents for chronic bronchitis (Cochrane Review). In: The Cochrane Library, Issue 2. Oxford: Update Software

Reid L 1954 Pathology of chronic bronchitis. Lancet i:275–278

Reid L 1960 Measurement of the bronchial mucous gland layer: a diagnostic yardstick in chronic bronchitis. Thorax 15:132–141

Restrepo G, Heard BE 1963 The size of the bronchial glands in chronic bronchitis. J Pathol Bacteriol 85:305–310

Rogers DF 1994 Airway goblet cells: responsive and adaptable front-line defenders. Eur Respir J 7:1690–1706

Rose MG, Gendler SJ 1997 Airway mucin genes and gene products. In: Rogers DF, Lethem MI (eds) Airway mucus: basic mechanisms and clinical perspectives. Birkhäuser Verlag, Basel, p 41–66

Siafakas NM, Vermiere P, Pride NB et al 1995 Optimal assessment and management of chronic obstructive pulmonary disease (COPD). The European Respiratory Society Taskforce. Eur Respir J 8:1398–1420

Sherman CB, Xu X, Speizer FE, Ferris BG Jr, Weiss ST, Dockery DW 1992 Longitudinal lung function decline in subjects with respiratory symptoms. Am Rev Respir Dis 146:855–859

Snider GL, Faling LJ, Rennard SI 1994 Chronic bronchitis and emphysema. In: Murray JF, Nadel JA (eds) Textbook of respiratory medicine. WB Saunders, Philadelphia, PA, p 1331–1397

Speizer FE, Fay ME, Dockery DW, Ferris BG Jr 1989 Chronic obstructive pulmonary disease mortality in six US cities. Am Rev Respir Dis 140:S49–S55

Steiger D, Fahy J, Boushey H, Finkbeiner WE, Basbaum C 1994 Use of mucin antibodies and cDNA probes to quantify hypersecretion *in vivo* in human airways. Am J Respir Cell Mol Biol 10:538–545

Thornton DJ, Davies JR, Carlstedt I, Sheehan JK 1997 Structure and biochemistry of human respiratory mucins In: Rogers DF, Lethem MI (eds) Airway mucus: basic mechanisms and clinical perspectives. Birkhäuser Verlag, Basel, p 19–39

Thurlbeck WM, Angus GE 1964 A distribution curve for chronic bronchitis. Thorax 19:436–442

Thurlbeck WM, Angus GE, Paré JAP 1963 Mucous gland hypertrophy in chronic bronchitis, and its occurrence in smokers. Br J Dis Chest 57:73–78

Vestbo J, Prescott E, Lange P 1996 Association of chronic mucus hypersecretion with FEV1 decline and chronic obstructive pulmonary disease morbidity. Copenhagen City Heart Study Group. Am J Respir Crit Care Med 153:1530–1535

Widdicombe JG 1990 A critical look at mucus markers. In: Persson CGA, Brattsand R, Laitinen LA (eds) Inflammatory indices of chronic bronchitis. Birkhäuser Verlag, Basel, p 269–279

Widdicombe JH, Widdicombe JG 1994 Regulation of human airway surface liquid. Respir Physiol 99:3–12

DISCUSSION

MacNee: In the first part of your talk you dealt with the question of whether mucus hypersecretion is important, and looked at the evidence for two opposing views on this issue. It seems to me that the studies give contradictory results because they study different populations. Peter Lange's work (Lange et al 1990) shows that as airflow limitation develops, the presence of mucus hypersecretion enhances the risk for hospitalization or death from COPD. The previous studies by Fletcher & Peto (1977) were in subjects with no or mild airflow obstruction. There is now good evidence of an association of mucus hypersecretion and increased morbidity and mortality as airways obstruction develops.

Rogers: In Francine Kauffmann's studies, she used almost the same population as in her earlier paper (Kauffmann et al 1979) and then re-evaluated these people in a later paper (Annesi & Kauffmann 1979). As the patients got older, the mucus hypersecretion seemed to be coming more into play. In the early papers, mucus hypersecretion did not seem to be markedly involved in disease progression. In the subsequent paper in which she looked at almost the same group of patients, mucus hypersecretion became significantly related to mortality.

Calverley: It is very important that we re-evaluate these things, but I just want to stress that sputum production and mucus hypersecretion are not necessarily the same things. One of the issues in COPD is the problem of moving from quantitative data to qualitative data. Many of the things that we talk about are associative. We say that there is a stimulus that produces a cytokine that has the potential to do something, and this is helpful in terms of building models. When you look at the epidemiology side of things you want to make it slightly more precise. Sputum production is relatively speaking a binary variable: the only thing that you have got is some confidence that people regularly do cough up something or not. The amount of the stuff they cough up, and even the colour of it, is not actually that helpful in COPD. It doesn't have quantitative aspects unless it becomes very severe. The evidence is increasing that this signifies something. It is my clinical impression that cough and sputum production becomes less important as the disease gets more advanced. The person who might have some data about this is Jim Hogg, because he has resection specimens from people who underwent lung cancer surgery who had relatively modest COPD, and now he has the more severe COPD patients from the lung volume reduction. Jim, do you see glandular hypertrophy or luminal plugging with mucus? This might be a good way of strengthening or weakening that association.

Hogg: The measurements reported from Montreal were made using a grading system (Cosio et al 1978). The most reproducible measurement in that panel was the percentage of small airway lumen that was occluded with mucus. However, mucus plugging of the small airways did not correlate with either abnormality in the small airways tests or the decline in FEV_1. Studies of chronic bronchitis reported by Mullen and colleagues from Vancouver showed that cough and sputum production correlated with inflammation of the gland and there was no relationship between cough and sputum and small airways disease.

Rogers: The sputum issue is very important. We know that the sputum people cough up contains mucus. What we need to know is how much mucus hypersecretion there is down in the small airways. Presumably the mucus in the sputum is coming from large airways and may be predominantly glandular in nature. It is vital to know whether there is the same relationship between sputum and the mucus content in the small airways.

Nadel: I want to comment on the 'wiring' of this system. Anatomically, the gland ducts are principally at bifurcations of large conducting airways. The highest concentration is in the larynx, the second highest concentration is at the bifurcation of the trachea, and then at the bifurcation of the subsequent major airways. Cough receptor endings are also located at airway bifurcations. Thus, hypersecretion of glands, by stimulating cough, clears the secretions, which are presumably produced by an instant stimulus. Cough and sputum production (chronic bronchitis) can affect the quality of life, but it is not surprising that chronic bronchitis is not a good predictor of deterioration or death from COPD: cough and sputum production are not well related to mortality, because when you cough you can eject sputum. On the other hand, cough is not likely to remove mucus in peripheral airways, because of the mechanical properties of the small airways. The peripheral airways have small diameters, so goblet cell metaplasia can have major effects when secretory cells degranulate, the mucus is hydrated and expands, perhaps 1000-fold. Thus, we need to distinguish goblet cell metaplasia and degranulation in peripheral airways from gland secretion in major airways. In peripheral airways, we have to design ways of detecting mucus plugging. There are no cough receptors in these airways, and the symptoms may not be evident until late in the process.

MacNee: This is an important issue in terms of a target for therapy. Does anyone disagree with the fact that there is now evidence from the Copenhagen City Lung

Study (Vestbo et al 1996) that mucus hypersecretion as defined in that study is associated with increased risk of hospitalisation, mortality and decline in FEV_1?

Calverley: I thought that the relationship with FEV_1 decline was the weakest bit.

Campbell: I realize that we are trying for a balance, and that there are two camps on this issue, but what is the consensus of this group?

Rogers: The consensus now is that the mucus hypersecretion is related to the decline in lung function and possible risk of hospitalization, morbidity and mortality, but only in the later stages of the disease.

MacNee: The best relationship is with mortality.

Calverley: And I think it is mortality through pneumonia rather than respiratory failure.

Campbell: Then we shouldn't leave this meeting thinking that the expectoration of sputum is a benign process.

Stockley: I would like to describe a recent study we have been doing. One of the problems with COPD is the heterogeneity of the condition that we are looking at. Probably the most homogeneous group of patients that we would call COPD are those with α_1-antitrypsin deficiency. These patients have a defined genetic defect, they have emphysema, and they do or they don't have chronic bronchitis. We have studied two groups of these patients, and we have taken out any patients who have bronchiectasis, as defined by high-resolution computed tomography (CT) scanning. We have divided the population into patients with chronic bronchitis by MRC criteria, producing sputum on a regular daily basis, and those that do not have that condition. These patients are matched for age, smoking history, sex and treatment. Concerning health status, if we take a St George's respiratory questionnaire, there is a major difference between the two groups: those with bronchitis have a worse quality of life, even though they are otherwise the same. With regard to airflow obstruction, the chronic bronchitic patients have a lower FEV_1. If we ask these patients how they are using an SF36 health status questionnaire for the question, 'How do you feel this year compared to last year?', the data polarize, with the majority of patients without chronic bronchitis saying that they feel essentially the same, but those with chronic bronchitis mainly feel somewhat worse. These data suggest that chronic bronchitis is not so benign. We are just left with one worry: have they got chronic bronchitis because they are worse, or are they worse because they have got chronic bronchitis? This is a totally different question which I am unable to answer at present

Wedzicha: It is like the exacerbations. Bronchitic symptoms are also an important predictor of exacerbation frequency.

Rennard: Can we go from this association of cough and sputum production/ mucus hypersecretion to a cause–effect relationship? Is the cough and mucus production contributing directly to this, or is it just a manifestation of something else that is going on that is related to these clinical endpoints?

Stockley: From the point of view of the data I mentioned, it is still early days. It is not just the fact that a patient coughs up sputum, but also what they cough up, which is important in the inflammatory process. Some of these sputum samples are relatively benign to the airway, and some are pretty toxic. It may be that it is a reflection of inflammation throughout the airway, and therefore what we are seeing is a surrogate of damage that has occurred in small airways or even at the alveolar level.

Rennard: I think it is important to distinguish this. You could make a case that there should be absolutely opposite therapeutic strategies: if it is the cough and sputum production that contributes to the worsening quality of life and mucus occluding the airways that contributes to the increased mortality at end-stage disease, then you would want to inhibit the mucus release. But if you thought the mucus itself was getting rid of bad things and its secretion was a manifestation of something else, this would be exactly the wrong therapeutic approach to take.

MacNee: What studies should we do to answer this?

Stockley: The first thing I'd like to do with these data is to follow what happens to the patients over a couple of years, to see whether there is a greater decline in those who have cough and sputum versus those without.

Rogers: The other thing is to investigate CT scanning and other methods to determine how much mucus there is in the small airways, and then to correlate this with sputum production.

MacNee: We also need some more epidemiological studies of the decline in FEV_1.

Calverley: If we are going to do some more epidemiology we ought to try to see whether there is anything at all to be gained by refining the symptom questions. The virtue of using the MRC approach is that it is comparable between studies, but the choice of questions about severity was very arbitrary — it was cooked up at about the same time as the original Ciba Guest Symposium on defining emphysema.

Campbell: I'm surprised when I hear people suggest that sputum production is binary: that it is either on or off. It seems to me that airway secretions are potentially complex and variable from patient to patient and time to time.

Calverley: Just to clarify what I meant by binary: it is not that you can't have different grades of secretion, it is the reproducibility of questions about this in a population study. I'm not sure whether we can grade this any better than whether we have got sputum production or not.

Paré: I would agree with Steve Rennard's earlier point. We need to decide whether this mucus hypersecretion is a marker of something bad, or whether it is intrinsically bad in itself. In Rob Stockley's study, he has matched them as well as he could, and there is a difference. This difference has to be genetic or environmental. If we are going to do studies, let's look at potential candidate genes relating to mucus hypersecretion. One source of variation that is well known is the enormous interindividual variability in the length of tandem repeats in the *MUC* genes, the domains that contain the glycosylation sites. The question is, if you have much longer MUC proteins, and much more glycosylation, could this change the physical properties of the mucus? This is a testable hypothesis.

Stockley: If someone has the appropriate tests, we have the DNA.

Rogers: Dave Thornton has been looking at this (Thornton et al 1996). He has a nice pool of normal secretions from about 20 people who were having minor dental surgery. He showed that MUC5AC seems to be the predominant mucin in those secretions. Interestingly, he has got mucus plugs from an airway of a patient who died of asthma, and 5AC seems to be predominant in those secretions also. He also has some sputum from one chronic bronchitis patient. It looks as if in these secretions 5AC is markedly reduced compared with normal and asthma. This is a little indication that there is a difference in mucus in chronic bronchitis. He is now looking at sputum from additional chronic bronchitic patients. It would be interesting to see if this relationship still holds. Then if it does, what you say about the size of the molecules and the glycosylation becomes increasingly relevant.

Agustí: If we accept that mucus plugs in peripheral airways are important pathophysiologically, and we also accept that they cannot be taken out through coughing and sputum, how do these patients get rid of them?

Jackson: We know very little. Some of the mucin can be moved by mucociliary transport, even down in the small airways, so there are clearance mechanisms there. Airflow can remove some of the mucin or help to clear it. I think it known that macrophages have enzymes capable of degrading some of the mucins, although very little work has been done on this.

Agustí: If macrophage clearing mechanisms are impaired in COPD because of smoking, for instance, this would be a mechanism whereby mucus will accumulate in the peripheral airways, contributing therefore to the pathophysiology of the disease.

Jackson: Possibly. Another mechanism might be damage to the ciliated epithelium in the small airways.

Nadel: If you have a disease, you have to be able to figure out ways of treating it. I want to mention one strategy for treatment which is presently feasible. We have published a paper in that is relevant here (Takeyama et al 1999). Healthy humans and pathogen-free animals have few goblet cells in their airways. Induction of epidermal growth factor receptors (EGFRs) and their activation gives rise to goblet cell metaplasia and hyperplasia. The process can be complete within 72 h. Furthermore, selective inhibition of EGFR activation prevents goblet cell growth. These inhibitors show promise for therapy of hypersecretion in COPD. When goblet cells are formed, they can degranulate; total degranulation occurs when activated neutrophils contact goblet cells. Complete degranulation occurs within 1–4 h. In a 15 mm airway (trachea) this might not be important, but it may totally occlude a 2 mm airway. Such a process can proceed in an exacerbation in a short period of time. In post-mortem studies in patients who died within 4 h of coming into hospital with acute asthma, we find that the peripheral airways are totally occluded by mucus. There are two approaches to this: you can look at degranulation and block it, or block the formation or metaplasia of goblet cells. I want to mention the EGFR pathway, through which allergen sensitization proceeds in experimental studies. Interleukin (IL)-4- and IL-13-induced goblet cell metaplasia and cigarette smoke-induced goblet cell metaplasia cause goblet cells to grow via the EGFR cascade. Mechanical stimuli such as alginate plugs also produce goblet cells through this pathway. The EGFR is not normally expressed in healthy airways. EGFRs are quite easily induced with tumour necrosis factor (TNF)α in experimental animals. Then you can activate EGFR with a series of EGFR ligands. When the EGFR is bound to a ligand, phosphorylation of intracellular tyrosine kinase occurs, which leads to transcription of mucin genes. This system transdifferentiates other cells in the airways. The second pathway is intracellular signalling of the EGFR. We have shown that free radicals in cigarette smoke and activated neutrophils bypass the ligand, phosphorylate the EGFR and go through the same p44 cascade leading to mucin production (Takeyama et al 2000). These are two pathways by which the EGFR can be stimulated in an autocrine fashion to cause mucin production. The most interesting part of this is the fact that there are EGFR-selective drugs which are presently used to treat cancer. One which we use completely blocks goblet cell production by every stimulus that we have been able to use.

References

Annesi I, Kauffmann F 1986 Is respiratory mucus hypersecretion really an innocent disorder? Am Rev Respir Dis 134:688–693

Cosio M, Ghezzo H, Hogg JC et al 1978 The relation between structural changes in small airways and pulmonary function tests. N Engl J Med 298:1277–1281

Fletcher C, Peto R 1997 The natural history of chronic airflow obstruction. Br Med J 1:1645–1648

Kauffmann F, Drouet D, Lelloch J, Brille D 1979 Twelve years spirometric changes among Paris area workers. Int J Epidemiol 8:201–212

Lange P, Nyboe J, Appleyard M, Jensen G, Schnohr P 1990 Relation of ventilatory impairment and of chronic mucus hypersecretion to mortality from obstructive lung disease and from all causes. Thorax 45:579–585

Takeyama K, Dabbagh K, Lee H-M et al 1999 Epidermal growth factor system regulates mucin production in airways. Proc Natl Acad Sci USA 96:3081–3086

Takeyama K, Dabbagh K, Jeong Shim J, Dao-Pick T, Ueki IF, Nadel JA 2000 Oxidative stress causes mucin synthesis via transactivation of epidermal growth factor receptor: role of neutrophils. J Immunol 164:1546–1552

Thornton DJ, Carlstedt I, Howard M, Devine PL, Price MR, Sheehan JK 1996 Respiratory mucins: identification of core proteins and glycoforms. Biochem J 316:967–975

Vestbo J, Prescott E, Lange P 1996 Association of chronic mucus hypersecretion with FEV_1 decline and chronic obstructive pulmonary disease morbidity. Copenhagen City Heart Study Group. Am J Respir Crit Care Med 153:1530–1535

Mechanisms of exacerbations

Jadwiga A. Wedzicha

Academic Respiratory Medicine, St Bartholomew's and Royal London School of Medicine and Dentistry, St Bartholomew's Hospital, Dominion House, West Smithfield, London EC1A 7BE, UK

Abstract. Exacerbations of chronic obstructive pulmonary disease (COPD) are a major cause of morbidity and mortality and hospital admission. Some patients are particularly susceptible to develop frequent exacerbations; exacerbation frequency being an important determinant of health related quality of life. Patients with frequent exacerbations (three or more exacerbations per year) have increased induced sputum cytokine interleukin (IL)-6 and IL-8 levels when stable, suggesting that frequent exacerbation is associated with increased airway inflammatory changes. Respiratory viral infections are a major cause of COPD exacerbations, with upper respiratory tract infections (colds) being associated with two-thirds of COPD exacerbations. Rhinovirus has been detected in induced sputum by PCR in 25% of exacerbations, suggesting that rhinovirus may directly infect the lower airway triggering exacerbation. The presence of an upper respiratory tract infection leads to a longer symptom recovery time at exacerbation. At exacerbation induced sputum IL-6 levels were increased compared to stable, though there were no significant increases in IL-8 or sputum cell counts. Sputum IL-6 levels were found to be higher in those patients with symptoms of a common cold. Increased airway eosinophilia has been also found at exacerbation. Other factors including bacterial colonization of the airways, temperature and interactions with environmental pollutants may also play a role in COPD exacerbation.

2001 Chronic obstructive pulmonary disease: pathogenesis to treatment. Wiley, Chichester (Novartis Foundation Symposium 234) p 84–103

Exacerbations of chronic obstructive pulmonary disease (COPD) are an important cause of the considerable morbidity and mortality found in patients with COPD (Fletcher & Peto 1977). Some patients are prone to frequent exacerbations that are an important cause of hospital admission and readmission, which may have considerable impact on activities of daily living and well-being (Osman et al 1997). There has been little information available on factors predicting development, severity or time course of COPD exacerbations and it is not known why some patients are particularly susceptible to exacerbation. COPD exacerbations are also associated with considerable physiological deterioration, yet airway inflammatory changes associated with exacerbation have received little attention, but are potentially important in planning novel therapies.

This paper will discuss epidemiological aspects of exacerbations. the nature of the associated airway inflammatory changes and aetiological factors for exacerbation.

Epidemiology of COPD exacerbation

Earlier descriptions of COPD exacerbations concentrated mainly on studies of hospital admission, though most COPD exacerbations are treated in the community and not associated with hospital admission. We have followed a cohort of moderate to severe COPD patients in East London (East London COPD Study) with daily diary cards and peak flow readings, who were asked to report exacerbations as soon as possible after symptomatic onset. The diagnosis of COPD exacerbation was based on criteria modified from those described by Anthonisen et al (1987), that require two symptoms for diagnosis, one of which must be a major symptom of increased dyspnoea, sputum volume or sputum purulence. Minor exacerbation symptoms included cough, wheeze, sore throat, nasal discharge or fever. We found that about 50% of exacerbations were unreported to the research team, despite considerable encouragement provided, and were only diagnosed from diary cards, though there were no differences in major symptoms or physiological parameters between reported and unreported exacerbations (Seemungal et al 1998). Patients with COPD are accustomed to frequent symptom changes and thus may tend to under-report exacerbations to physicians. These patients have high levels of anxiety and depression and may accept their situation (Okubadejo et al 1996). The tendency of patients to under-report exacerbations may explain the higher total rate of exacerbation at 2.7 per patient per year, which is higher than previously reported by Anthonisen and co-workers at 1.1 per patient per year (Anthonisen et al 1987). However in the latter study, exacerbations were unreported and diagnosed from patients' recall of symptoms.

Using the median number of exacerbations as a cut-off point, COPD patients in the East London Study were classified as frequent and infrequent exacerbators. Quality of life scores measured using a validated disease specific scale — the St George's Respiratory Questionnaire (SGRQ), was significantly worse in all its components (symptoms, activities and impacts) in the frequent, compared to the infrequent exacerbators (Table 1). This suggests that exacerbation frequency is an important determinant of health status in COPD and is thus an important outcome measure in COPD. Factors predictive of frequent exacerbations included daily cough and sputum and frequent exacerbations in the previous year. A previous study of acute infective exacerbations of chronic bronchitis found that one of the factors predicting exacerbation was the number in the previous year (Ball et al

TABLE 1   Relationship between mean **SGRQ (St George's Respiratory Question-naire) and exacerbation frequency**

Exacerbation frequency	Total	Symptoms	Activities	Impacts
0–2	48.9	53.2	67.7	36.3
3–8	64.1	77.0	80.9	50.4
CI	−22.3 to −7.8	−29.7 to −14.0	−21.2 to −5.3	−22.9 to −5.6
P value	<0.0005	<0.0005	0.001	0.002

SGRQ is expressed as total score, which is the sum of its components — symptoms, activities and impacts. 32 patients had an exacerbation frequency of 0–2 and 38 patients had exacerbation frequency at 3–8. CI = 95% confidence interval (Seemungal et al 1998).

1995), though this study was limited to exacerbations presenting with purulent sputum and no physiological data were available.

In a further prospective analysis of 504 exacerbations, where daily monitoring was performed, there was some deterioration in symptoms, though no peak flow changes (Seemungal et al 2000a). Falls in peak flow and FEV_1 (forced expiratory volume in one second) at exacerbation were generally small and not useful in predicting exacerbations, but larger falls in peak flow were associated with symptoms of dyspnoea, presence of colds or related to longer recovery time from exacerbations. Symptoms of dyspnoea, common colds, sore throat and cough increased significantly during the prodromal phase and this suggests that respiratory viruses have early effects at exacerbations. The median time to recovery of peak flow was 6 days and 7 days for symptoms, but at 35 days peak flow had returned to normal in only 75% of exacerbations, while at 91 days, 7.1% of exacerbations had not returned to baseline lung function. Recovery was longer in the presence of increased dyspnoea or symptoms of a common cold at exacerbation. The changes observed in lung function at exacerbation were smaller than those observed at asthmatic exacerbations, though the average duration of an asthmatic exacerbation was longer at 9.6 days (Reddel et al 1999, Tattersfield et al 1999).

The reasons for the incomplete recovery of symptoms and lung function are not clear, but may involve inadequate treatment or persistence of the causative agent. The association of the symptoms of increased dyspnoea and of the common cold at exacerbation with a prolonged recovery suggests that viral infections may lead to more prolonged exacerbations. As colds are associated with longer exacerbations, COPD patients who develop a cold may be prone to more severe exacerbations and should be considered for therapy early at onset of symptoms.

Airway inflammation at exacerbation

Although it has been assumed that exacerbations are associated with increased airway inflammation there has been little information available on the nature of inflammatory markers especially when studied close to an exacerbation, as performing bronchial biopsies at exacerbation is difficult in moderate to severe COPD. The relation of any airway inflammatory changes to symptoms and physiological changes at exacerbations is also an important factor to consider.

In one study, where biopsies were performed at exacerbation in patients with chronic bronchitis, increased airway eosinophilia was found, though the patients studied had only mild COPD (Saetta et al 1994). With exacerbation, there were more modest increases observed in neutrophils, T lymphocytes (CD_3) and tumour necrosis factor (TNF)α-positive cells, while there were no changes in CD4 or CD8 T cells, macrophages or mast cells. However the technique of sputum induction allows study of these patients at exacerbation and we have shown that it is a safe and well tolerated technique in COPD patients (Bhowmik et al 1998). Levels of inflammatory cytokines have been shown to be elevated in induced sputum in COPD patients when stable, though changes at exacerbation had not been previously studied (Keatings et al 1996).

We prospectively followed a cohort of patients from the East London COPD Study and related inflammatory markers in induced sputum to symptoms and physiological parameters both at baseline and at exacerbation (Bhowmik et al 2000). There was a relation between exacerbation frequency and sputum cytokines, in that there was increased sputum interleukin (IL)-6 and IL-8 found in patients at baseline when stable with frequent exacerbations compared to those with infrequent exacerbations (Fig. 1), although there was no relation between cytokines and baseline lung function. Sputum cell counts were not increased at baseline in patients with more frequent exacerbations suggesting that the increased cytokine production comes from the bronchial epithelium in COPD. As discussed below, exacerbations are triggered by viral infections, especially by rhinovirus that is the cause of the common cold. Rhinovirus has been shown to increase cytokine production in an epithelial cell line (Subauste et al 1995) and thus repeated viral infection may lead to up-regulation of cytokine airway expression.

At exacerbation we found increases in induced sputum IL-6 levels and the levels of IL-6 were higher when exacerbations were associated with symptoms of the common cold (Fig. 2). Experimental rhinovirus infection has been shown to increase sputum IL-6 in normal subjects and asthmatics (Fraenkel et al 1995, Grünberg et al 1997, Fleming et al 1999). However, rises in cell counts and IL-8 were more variable with exacerbation and did not reach statistical significance, suggesting marked heterogeneity in the degree of the inflammatory response at

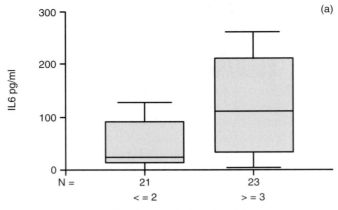

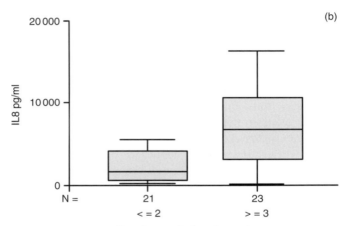

FIG. 1. (a) Induced sputum levels of IL-6 in patients who are categorized as frequent exacerbators (≥3 exacerbations in the previous year) and those who are infrequent exacerbators (≤2 exacerbations in previous year). Data are expressed as medians (interquartile range; IQR). (b) Induced sputum levels of IL-8 in patients with frequent exacerbations and infrequent exacerbations. Data are expressed as medians (IQR). (Bhowmik et al 2000).

exacerbation. The exacerbation IL-8 levels were related to sputum neutrophil and total cell counts, indicating that neutrophil recruitment is the major source of airway IL-8 at exacerbation. Lower airway IL-8 has been shown to increase with experimental rhinovirus infection in normal and asthmatic patients in some studies (Grünberg et al 1997), but not in others (Fleming et al 1999). However COPD

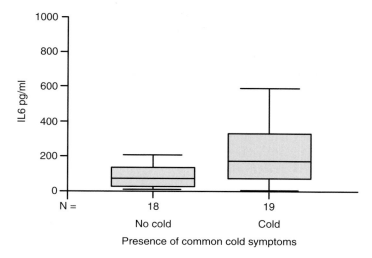

FIG. 2.   Induced sputum IL-6 levels in the absence and presence of a natural cold. Data are
expressed as medians (IQR). (Bhowmik et al 2000).

patients already have up-regulated airway IL-8 levels when stable due to their high
sputum neutrophil load (Keatings et al 1996) and further increases in IL-8 would be
unlikely. COPD exacerbations are associated with a less pronounced airway
inflammatory response than asthmatic exacerbations (Pizzichini et al 1997), and
this may explain the relatively reduced response to steroids seen at exacerbation
in COPD patients, relative to asthma (Albert et al 1980, Thompson et al 1996,
Niewoehner et al 1999, Davies et al 1999).

We did not detect an increase in eosinophil count at exacerbation, even though
our patients were sampled early at exacerbation with onset of symptoms.
Compared to the study by Saetta and colleagues where patients had mild COPD
(Saetta et al 1994), our patients had more severe and irreversible airflow
obstruction with an FEV_1 at 39% predicted. Thus it is possible that the
inflammatory response at exacerbation is different in nature in patients with
moderate to severe COPD than in patients with milder COPD.

As patients were followed with daily diary cards, we could also relate the
inflammatory markers to exacerbation recovery. There was no relation between
the degree of inflammatory cell response with exacerbation and duration of
symptoms and lung function changes. Induced sputum markers taken three to
six weeks after exacerbation showed no relation to exacerbation changes. Thus
levels of induced sputum markers at exacerbation do not predict the subsequent
course of the exacerbation and will not be useful in the prediction of exacerbation
severity.

Aetiology of COPD exacerbation

COPD exacerbations have been associated with a number of aetiological factors, including infection and pollution episodes. COPD exacerbations are frequently triggered by upper respiratory tract infections and these are commoner in the winter months, when there are more respiratory viral infections in the community. Patients may also be more prone to exacerbations in the winter months as lung function in COPD patients shows small but significant falls with reduction in outdoor temperature during the winter months (Donaldson et al 1999). COPD patients have been found to have increased hospital admissions, suggesting increased exacerbation when increasing environmental pollution occurs. During the December 1991 pollution episode in the UK, COPD mortality was increased together with an increase in hospital admission in elderly COPD patients (Anderson et al 1995). However common pollutants especially oxides of nitrogen and particulates may interact with viral infection to precipitate exacerbation rather than acting alone.

Viral infections

Viral infections are an important trigger for COPD exacerbations. Studies in childhood asthma have shown that viruses, especially rhinovirus (the cause of the common cold) can be detected by PCR from a large number of these exacerbations (Johnston et al 1995). Rhinovirus has not hitherto been considered to be of much significance during exacerbations of COPD. In a study of 44 chronic bronchitics over two years, Stott et al (1968) found rhinovirus in 13 (14.9%) of 87 exacerbations of chronic bronchitis. In a more detailed study of 25 chronic bronchitics with 116 exacerbations over 4 years, Gump et al (1976) found that only 3.4% of exacerbations could be attributed to rhinoviruses. In a more recent study of 35 episodes of COPD exacerbation using serological methods and nasal samples for viral culture, little evidence was found for a rhinovirus aetiology of COPD exacerbation (Philit et al 1992).

We have recently shown that up to about one third of COPD exacerbations were associated with viral infections, and 75% of these were due to rhinovirus, when samples were taken from nasopharyngeal aspirates (Harper-Owen et al 1999). Viral exacerbations were associated with symptomatic colds and prolonged recovery. However, we found that rhinovirus was recovered from induced sputum more frequently than from nasal aspirates at exacerbation, suggesting that wild-type rhinovirus can infect the lower airway and contribute to exacerbation inflammatory changes (Seemungal et al 2000b). We also found that exacerbations associated with the presence of rhinovirus in induced sputum had larger increases in airway IL-6 levels. Other viruses may trigger COPD exacerbation, though coronavirus was associated with only a small proportion of asthmatic exacerbations and is unlikely to play a major role in COPD (Nicholson et al 1993, Johnston et al 1995).

Bacterial colonization

Airway bacterial colonization has been found in approximately 30% of COPD patients, and has been shown to be related to the degree of airflow obstruction and current cigarette smoking (Zalacain et al 1999). Although bacteria such as *Haemophilus influenzae* and *Streptococcus pneumoniae* have been associated with COPD exacerbation, some studies have shown increasing bacterial counts during exacerbation, while others have not confirmed these findings (Monsó et al 1999, Wilson 1999). There is no evidence that patients with frequent exacerbations have increased sputum bacterial colonization to explain the higher cytokine levels observed. Cultures of bronchial epithelial cells showed increased cyoktine IL-6 production after stimulation with endotoxin (Khair et al 1994). However it is also possible that there may be interactions between viral and bacterial infection at COPD exacerbation. Other organisms such as *Chlamydia pneumoniae*, that have been associated with asthmatic exacerbation, may also play a role in COPD exacerbation. We found little evidence of *C. pneumoniae* in nasopharyngeal samples, though there was greater isolation of *C. pneumoniae* in induced sputum (Harper-Owen et al 2000).

Conclusions

This paper has described some important characteristics of COPD exacerbations. Some patients with COPD seem prone to frequent exacerbations that are an important determinant of health status. These patients have higher airway cytokine levels suggesting increased airway inflammation that could increase susceptibility to exacerbation. The inflammatory response at COPD exacerbation is variable, but rises in IL-6 at exacerbation are related to the presence of a common cold. Rhinoviral infection is the most important aetiological factor in COPD exacerbations and is an important target for preventive therapy. Reduction of COPD exacerbation will have an important impact on the considerable morbidity and mortality associated with COPD.

Acknowledgements

This work has been funded by the Joint Research Board of the St Bartholomew's Hospital Trustees and the British Lung Foundation.

References

Albert RK, Martin TR, Lewis SW 1980 Controlled clinical trial of methylprednisolone in patients with chronic bronchitis and acute respiratory insufficiency. Ann Intern Med 92:753–758

Anderson HR, Limb ES, Bland JM, Ponce de Leon A, Strachan DP, Bower JS 1995 Health effects of an air pollution episode in London, December 1991. Thorax 50:1188–1193

Anthonisen NR, Manfreda J, Warren CPW, Hershfield ES, Harding GKM, Nelson NA 1987 Antibiotic therapy in exacerbations of chronic obstructive pulmonary disease. Ann Intern Med 106:196–204

Ball P, Harris JM, Lowson D, Tillotson G, Wilson R 1995 Acute infective exacerbations of chronic bronchitis. Q JM 88:61–68

Bhowmik A, Seemungal TAR, Sapsford RJ, Devalia JL, Wedzicha JA 1998 Comparison of spontaneous and induced sputum for investigation of airway inflammation in chronic obstructive pulmonary disease. Thorax 53:953–956

Bhowmik A, Seemungal TAR, Sapsford RJ, Wedzicha JA 2000 Relation of sputum inflammatory markers to symptoms and physiological changes at COPD exacerbations. Thorax 55:114–120

Davies L, Angus RM, Calverley PMA 1999 Oral corticosteroids in patients admitted to hospital with exacerbations of chronic obstructive pulmonary disease: a prospective randomised controlled trial. Lancet 354:456–460

Donaldson GC, Seemungal T, Jeffries DJ, Wedzicha JA 1999 Effect of environmental temperature on lung function and symptoms in chronic obstructive pulmonary disease patients. Eur Respir J 13:844–849

Fleming HE, Little EF, Schnurr D et al 1999 Rhinovirus-16 colds in healthy and asthmatic subjects. Similar changes in upper and lower airways. Am J Respir Crit Care Med 160:100–108

Fletcher C, Peto R 1977 The natural history of chronic airflow obstruction. Br Med J 1:1645–1648

Fraenkel DJ, Bardin PG, Sanderson G, Lampe F, Johnston SL, Holgate ST 1995 Lower airways inflammation during rhinovirus colds in normal and in asthmatic subjects. Am J Respir Crit Care Med 151:879–886

Grünberg K, Smits HH, Timmers MC et al 1997 Experimental rhinovirus 16 infection. Effects on cell differentials and soluble markers in sputum of asthmatic subjects. Am J Respir Crit Care Med 156:609–616

Gump DW, Phillips CA, Forsyth BR, McIntosh K, Lamborn KR, Stouch WH 1976 Role of infection in chronic bronchitis. Am Rev Respir Dis 113:465–474

Harper-Owen R, Seemungal TAR, Bhowmik A, Johnston SL, Jeffries DJ, Wedzicha JA 1999 Virus and Chlamydia isolation in COPD exacerbations. Eur Respir J (suppl 30) 14:47s

Harper-Owen R, Seemungal TAR, Johnston SL, Jeffries DJ, Wedzicha JA 2000 Role of Chlamydia pneumoniae in COPD exacrebations. Am J Respir Crit Care Med, in press (abstract)

Johnston SL, Pattemore PK, Sanderson G et al 1995 Community study of the role of viral infections in exacerbations of asthma in 9–11 year old children. BMJ 310:1225–1229

Keatings VM, Collins PD, Scott DM, Barnes PJ 1996 Differences in interleukin-8 and tumour necrosis factor α in induced sputum from patients with chronic obstructive pulmonary disease and asthma. Am J Respir Crit Care Med 153:530–534

Khair OA, Devalia JL, Abdelaziz MM, Sapsford RJ, Tarraf H, Davies RJ 1994 Effect of Haemophilus influenzae endotoxin on the synthesis of IL-6, IL-8, TNF-α and expression of ICAM-1 in cultured human bronchial epithelial cells. Eur Respir J 7:2109–2116

Monsó E, Rosell A, Bonet G et al 1999 Risk factors for lower airway bacterial colonization in chronic bronchitis. Eur Respir J 13:338–342

Nicholson KG, Kent J, Ireland DC 1993 Respiratory viruses and exacerbations of asthma in adults. BMJ 307:982–986

Niewoehner DE, Erbland ML, Deupree RH et al 1999 Effect of systemic glucocorticoids on exacerbations of chronic obstuctive pulmonary disease. Department of Veterans Affairs Cooperative Study Group. N Engl J Med 340:1941–1947

Okubadejo AA, Jones PW, Wedzicha JA 1996 Quality of life in patients with chronic obstructive pulmonary disease and severe hypoxaemia. Thorax 51:44–47

Osman IM, Godden DJ, Friend JAR, Legge JS, Douglas JG 1997 Quality of life and hospital re-admission in patients with chronic obstructive pulmonary disease. Thorax 52:67–71

Philit F, Etienne J, Calvet A et al 1992 Infectious agents associated with exacerbations of chronic obstructive bronchopneumopathies and asthma attacks. Rev Mal Respir 9:191–196

Pizzichini MMM, Pizzichini E, Clelland L et al 1997 Sputum in severe exacerbations of asthma: kinetics of inflammatory indices after prednisone treatment. Am J Respir Crit Care Med 155:1501–1508

Reddel HS, Ware S, Marks G, Salome C, Jenkins C, Woolcock A 1999 Differences between asthma exacerbations and poor asthma control. Lancet 353:364–369

Saetta M, Di Stefano A, Maestrelli P et al 1994 Airway eosinophilia in chronic bronchitis during exacerbations. Am J Respir Crit Care Med 150:1646–1652

Seemungal TAR, Donaldson GC, Paul EA, Bestall JC, Jeffries DJ, Wedzicha JA 1998 Effect of exacerbation on quality of life in patients with chronic obstructive pulmonary disease. Am J Respir Crit Care Med 157:1418–1422

Seemungal TAR, Donaldson GC, Bhowmik A, Jeffries DJ, Wedzicha JA 2000a Time course and recovery of exacerbations in patients with chronic obstructive pulmonary disease. Am J Respir Crir Care Med 161:1608–1613

Seemungal TAR, Harper-Owen R, Bhowmik A, Jeffries DJ, Wedzicha JA 2000b Detection of rhinovirus in induced sputum at exacerbation of chronic obstructive pulmonary disease. Eur Resp J, in press

Stott EJ, Grist NR, Eadie MB 1968 Rhinovirus infections in chronic bronchitis: isolation of eight possible new rhinovirus serotypes. J Med Microbiol 109:117

Subauste MC, Jacoby DB, Richards SM, Proud D 1995 Infection of a human respiratory epithelial cell line with rhinovirus. Induction of cytokine release and modulation of susceptibility to infection by cytokine exposure. J Clin Invest 96:549–557

Tattersfield AE, Postma DS, Barnes PJ et al 1999 Exacerbations of asthma: a descriptive study of 425 severe exacerbations. The FACET International Study Group. Am J Respir Crit Care Med 160:594–599

Thompson WH, Nielson CP, Carvalho P, Charan NB, Crowley JJ 1996 Controlled trial of oral prednisolone in outpatients with acute COPD exacerbation. Am J Respir Crit Care Med 154:407–412

Wilson R 1999 Bacterial infection and chronic obstructive pulmonary disease. Eur Respir J 13:233–235

Zalacain R, Sobradillo V, Amilibia J et al 1999 Predisposing factors to bacterial colonization in chronic obstructive pulmonary disease. Eur Respir J 13:343–348

DISCUSSION

Jeffery: Have you any pearls of wisdom from the beginnings of your analysis on the relationship between the length of the exacerbation and the rate in decline of FEV_1 over time?

Wedzicha: We have had a look at it already, in the first three years. It does look as though the exacerbators are beginning to decline slightly faster. However, we have been told not to look at it yet, because it has not been going long enough. What I described were the first two and a half years of data. We are now getting the data for the next two years together, to have a look at some more quality of life issues, to see whether we can replicate the same decline in quality of life. We also want to look at FEV_1 and inhaled steroids. 80% of our patients are on inhaled steroids, but we have never detected any difference in exacerbation frequency between the two groups. It may be that there is some difference in the exacerbation rate over five years in the two groups. We are also interested in whether the exacerbations are different in any way in the patients on inhaled steroids. We should be able to tell that.

Jeffery: Are there any data coming out of the ISOLDE trial in relation to exacerbation frequency and rate of decline?

Calverley: Three years and 750 people is probably not enough to make those statements with great confidence. The rate of decline data are just about OK. I agree with Wisia Wedzicha that in a smaller study, the way to increase its power is to increase the duration of follow-up. If you can get five years of follow-up, that will compensate significantly for your loss in numbers. As far as the relationship between the rate of decline and FEV_1, my memory is that in the placebo-treated people, the FEV_1 decline was weakly related to exacerbation rate, but this was not true for the inhaled steroid-treated people where the exacerbation rate is modified. The worry is that even though there are many people and a lot of data, it may not be a big enough study to conclusively answer this.

MacNee: You have a large amount of data on induced sputum in exacerbations of COPD. Studies have shown differences in IL-8 levels between normals and patients with COPD. Do you find the same differences? You have also found in exacerbations that there is no clear difference in the levels of cytokines except in those that have frequent exacerbations. The problem with frequent exacerbations is that these patients may never really be clinically stable, and hence will have higher markers of inflammation.

Wedzicha: When we first started we did some normals, but we had made a decision in east London COPD not to follow a cohort of normals. This would have doubled the cost of the study, so every patient acted as their own control. Our IL-8 baseline levels were very high, but we did have a very severe group of patients. Interestingly, the IL-8 didn't do very much more: overall the levels were very similar. IL-6 was the only cytokine that seemed to show a response.

Nadel: Were neutrophil markers changed?

Wedzicha: We looked at neutrophil counts, but we didn't look at any other markers of neutrophils in these patients.

Nadel: The reason I ask is that it is reasonably well accepted that viral infections in animals and in humans causes neutrophil recruitment. I would measure both IL-8 and LTB4.

Wedzicha: The neutrophils went up from 1.8 to 2.9. One or two of our cohort have shown neutrophil rises. It is probably also a matter of numbers. If we had 150 induced sputum samples, the neutrophils may have gone up. There is enormous heterogeneity in what goes on in COPD in the airways. Again, I think one of the reasons that the ISOLDE data show what they do is because large numbers were studied.

MacNee: I have done bronchoscopies on patients with exacerbations of COPD. You don't have to be a pathologist to tell that there is inflammation in their airways. You are sampling sputum which comes from large airways and yet you do not find an increase in sputum neutrophil counts in exacerbations of COPD. What is the explanation?

Wedzicha: You are right that we are not seeing large numbers: they are not going up six or seven times as you would expect them to.

Calverley: I want to reemphasize that this is a very heterogeneous phenomenon. If you look at the clinical outcomes, such as changes in peak flow and symptoms, it was pretty clear that these measurements were following a reproducible time course. So something is happening biologically. Then, what should be the easy bit, measuring the neutrophil influx, actually turns out to be jolly hard. This is challenging in terms of what we believe induced sputum is telling us, or what we think the process is. Something is clearly happening: these people are getting more symptoms. They are even getting a measurable change in a rather crude measure of lung function. Something is happening here, but why are we seeing the changes that our models would expect us to believe?

Hogg: Before the advent of PCR, Smith et al (1980) followed a cohort of patients with COPD and looked at viral cultures and other markers of viral infection. They found that it was important to look not only at the time of exacerbation, but also between exacerbation to make sure that the virus was related to the illness and not just there for the ride. Hegele and colleagues from our laboratory have examined resected lung tissue where great pains were taken to isolate the bronchus first during the resection procedure, so that it is not going to be contaminated from the upper airway (Macek et al 1999). They found that several types of virus were present in the lower respiratory tract in patients who were free of infection at the time of surgery. Have you looked between exacerbations, and are you confident that the viruses you have found are related to the disease?

Wedzicha: In the first year of the study, all patients had monthly sampling over the winter months, from October and April, and then we did a sample in July. We had eight samples for each, together with matched blood IL-6 and fibrinogen samples. We then repeated this sampling in 1997. We found only two rhinoviruses in the baseline samples, and both of those patients had exacerbations about two days later. However, we did find some asymptomatic influenza and parainfluenza, and a few mycoplasmas, but we did not find any rhinovirus that we could confidently say outside the time were significant. We did not have repeated induced sputum at baseline on the patients. If you look at experimental rhinovirus infections in normals and asthmatics, they do produce significant inflammatory responses, while in COPD they give some response but they do not produce anywhere near the same type of response. I would have thought that considering that our patients were sampled at a very early stage of the disease, we should have seen the good airway inflammatory changes that you get with experimental rhinovirus infection. Generally, even in asthmatics, there haven't been a lot of studies where viruses have been detected at the same time as sputum has been measured.

Rennard: Jay Nadel, if you are unimpressed with the number of neutrophils and think that the sputum is being sampled in the wrong place, where would you sample from?

Nadel: I would look where the disease is. Let us suppose that 10% of the disease is in the alveoli, 90% of the disease is somewhere else, and then you have an exacerbation. Perhaps during an exacerbation, pathophysiologic changes occur acutely, and these changes could be reversible. Where are the major pathophysiologic changes? My intuition is that it is where Jim Hogg said it was some 30 years ago, in the peripheral airways.

Paré: What impressed me was how few of the exacerbations could be explained by viral infection. What other things could affect small airway calibre? It has to be small airway narrowing that is causing the transient hyperinflation that is causing the symptoms. What other processes can produce reversible worsening of the narrowing of those airways?

Wedzicha: Taking all our data together, the final figure will probably be that about half of all exacerbations will be linked to viruses. There are other candidates, such as bacterial colonization. This is a complicated subject that requires monthly sampling. The interaction with pollutant is interesting. There are data suggesting that NO_2 and viruses interact. A study from Southampton shows that exacerbations are longer in asthmatic children if there is a pollutant present.

Paré: What is the seasonality of your exacerbations?

Wedzicha: They mainly occur in the winter, between September and April. Interestingly, many of them happen in December and January. There is a steady trickle in the summer, but we have never had fewer viruses from the summer ones. The virus isolation seems to be fairly even.

MacNee: We have data that support Wisia Wedzicha's data on IL-8. We also see no change in neutrophil numbers during exacerbations. Perhaps we should discuss whether induced sputum is telling us anything about inflammation in exacerbations of COPD.

Hogg: Hegele's group has also shown that if you produce an acute respiratory syncitial virus (RSV) infection in guinea pigs, the virus survives by low level replication in alveolar macrophage long after the virus infection has apparently cleared (Dakhama et al 1997). Interestingly, it is not possible to show this with the macrophages that are present in the bronchoalveolar lavage (BAL). But when they separate macrophages from lung tissue, they can show it. He has also told me that a group in Quebec City has found the RSV in biopsies but not in BAL. I think there is something important in what Jay Nadel is saying because induced sputum and BAL may not provide the relevant cells.

Jeffery: The other point that I was going to make is that unlike bacterial infections, viruses are invading the epithelial cells, and that is the site to which the neutrophils need to get. Studies in my lab have shown *in vitro* that ICAM-1 is up-regulated in response to viral infection. Such up-regulation of ICAM-1 would retain neutrophils in the epithelium. They may not actually be trafficking through into the lumen, where you could then recover them, but they would be in the tissues, if you examined the tissue compartment. The only other comment I would make is that you are looking in the sputum for neutrophil numbers. It may be that neutrophils by that time have disintegrated, and perhaps you should also be measuring myeloperoxidase or neutrophil elastase: you may be looking at an insensitive endpoint.

Agustí: With regard to bacterial colonization, you told us that about 25% of stable COPD patients are colonized. These numbers are taken from bronchial aspirates or BAL. Might the small airways be colonized?

Wedzicha: The data on this are variable. It varies from about 15% to 50%. Most of the data have been on bronchial brush specimens. We have recently looked at our cohort with respect to colonization comparing spontaneous and induced. Our colonization rate is about 50–55% if you take out upper airway organisms that could possibly contaminate. I think this could be very important. We have also looked to see whether the frequent exacerbators have more bacterial colonization: I have not quite done enough samples yet, but it looks as this may not be the case.

Agustí: My question was more related to the location of that colonization. In my understanding, every time we talk about colonization we are talking about central colonization. Are we implying that there is also peripheral colonization?

Wedzicha: I'm not an expert on this, but I thought that the small airways were actually colonized in these patients.

Agustí: If so, shouldn't they develop pneumonia?

Wedzicha: Not necessarily.

Jackson: There was a study on lung transplant tissue suggesting that the airways can be colonized by *Haemophilus influenzae* right from the main bronchus to the lung periphery (Möller et al 1998).

Mantovani: Going back to leukocyte numbers, I got the impression that the absolute numbers were small. But lymphocytes increased quite dramatically.

Williams: Did you not find an increase in eosinophils?

Wedzicha: We have looked hard in every single study, and we have not found a rise in eosinophils. It may be that you get different types of inflammation depending on the severity of COPD. In the UK we cannot biopsy patients at exacerbation. I would maintain that if you are going to take biopsies, you have to do it early on in the natural history of the condition, before you give steroids. Because of ethical considerations it is almost impossible to get these samples without treating the patient.

Calverley: It is extremely difficult, and this is potentially going to cloud the issue. I do find Peter Paré's idea of cells being held up in the airway wall rather attractive. If one says that these are episodes where you are getting changes involving dynamic hyperinflation over four or five days returning to normal, and the disease is in the small airways, then you wouldn't need too much inflammation to produce a significant narrowing of the airway lumen. Peter and Jim have modelled the sorts of changes necessary, and have emphasized the importance of airway wall thickness as a determinant of these things. This would produce a nice story that would tie in the physiology with the mechanism. How we can sample that is clearly another issue. We would have major problems doing biopsy studies.

Rennard: There are other ways for the neutrophils to hide. They don't necessarily have to come into the lumen where they can be recovered in the induced sputum. They undergo apoptosis and phagocytosis. You would think that this actually takes place in COPD more than cystic fibrosis, since you don't see cell lysis with release of DNA nearly as much in COPD exacerbations.

MacNee: Related to this point, have you compared spontaneously produced sputum with induced sputum?

Wedzicha: Yes. We found no difference in cell counts and IL-8, except that induced sputum had much better viability. There is another aspect of COPD exacerbation which I left out of my paper: when we looked at IL-6 and fibrinogen, there is a fairly vicious acute-phase response in COPD exacerbations. Fibrinogen went up from a high baseline of 3.7 mg/l up to 4.2 mg/l, and the IL-6 had a less dramatic but significant rise. There is a significant amount of inflammation around in an exacerbation. It may be that the systemic changes are almost more important to the patient than the airway ones.

Nadel: In reticular activation, I remember the old studies of down regulation. In your studies where there are many neutrophils present containing IL-8 and LTB4, those neutrophils in the lumen could down-regulate the further recruitment of neutrophils. It is not surprising that at a certain time you may not find a lot of neutrophil recruitment molecules. Going back to Dr Mantovani's comment, if there were Th2 lymphocytes in COPD as some people report, and IL-4 is increased, IL-4 is a potent recruiter of goblet cells. If you have Th2 lymphocytes in the airways, you can get tremendous goblet cell metaplasia very rapidly. This may play a major role in exacerbations.

Jeffery: With regard to the goblet cell story, I am reminded of a study by Sanjar and colleagues in an experimental model in which they set up a system to induce first goblet cell hyperplasia and metaplasia using ovalbumin-sensitized animals (Blyth et al 1998). If you add virus, there is a triggered, marked secretagogue for mucus. If this were to occur in a small airway alone, this would have profound effects on small airway function. This might itself explain many of the effects you are seeing.

Paré: When you think of the features of mucosal inflammation that can cause small airway obstruction, there are (1) oedema, secondary to increases in microvascular permeability, (2) exudate onto the surface that will affect the surface tension in the small airways, and (3) mucus secretion by goblet cells. But the neutrophil is not directly mediating these effects. The neutrophil in BAL or induced sputum is being used as a marker of these other processes of inflammation in mucus membranes.

Nadel: I disagree: if you deliver a neutrophil chemoattractant into the airways that contain goblet cells, the neutrophil recruitment and activation occur, which mobilizes elastase to the surface of the neutrophil. When activated neutrophils move through the airway epithelium, there is adherence between ICAM-1 (on the surface of goblet cells) and Mac-1 (on the surface of neutrophils). In that closed space, elastase changes its conformation and goes to the outside surface, and there is released in a closed space and activates the goblet cell to degranulate. In addition, the neutrophil releases free radicals as it goes through, and in the presence of the epidermal growth factor (EGF) receptor cascade these free radicals are a potent *trans*-activator of tyrosine kinase phosphorylation. This increases the number of goblet cells. From both points of view, an activated neutrophil is a potent secretagougue and a potent activator of goblet cell growth.

MacNee: Does this happen in the airway lumen?

Hogg: If the goblet cells discharge, then the mucus is in the lumen.

Jeffery: The mucus is, but not the neutrophil.

Calverley: The one thing that is done in preparing induced sputum is getting rid of all the mucus. So we shouldn't get a signal on that with the measurements that we make at the moment.

Nadel: When a goblet cell degranulates, it increases the volume of that mucus about 1000-fold by hydration. If you have a 15 mm airway, this will not cause marked obstruction and cough clearance is effective. If the airway is 2 mm in diameter, total mucus plugging can occur in minutes and clearance by cough is not effective.

Jeffery: I agree, coughing will not remove mucus in a small airway, and surface tension effects will be huge.

Wedzicha: There are a number of people here who would like to say that we should have much more airway inflammation during an acute exacerbation but, after all, although steroids work, they don't really work that well. It is not like in asthma, where there are a lot of cells in the airways. This is actually a different condition.

Calverley: In asthma, there is pretty good evidence that you are turning off some of the principal mechanisms that cause the physiological effects that constitute an exacerbation. In COPD, the sorts of changes that we and others have reported in FEV_1 with moderate doses of steroids are actually fairly small but helpful. Going back to Peter Paré's analysis of why we get airways obstruction, I am inclined to say that the only part which steroids are likely to be influencing greatly is the oedema phase. I suspect that this is the component causing the small rise in FEV_1 during an acute exacerbation treated with oral steroids and in ISOLDE with inhaled steroids.

Jeffery: Relating to the oedema, there is a marked reductive effect of steroids on mast cell numbers. Perhaps mast cells are the driving force for the oedematous change, and this is the component being controlled by steroids.

MacNee: I think it would be useful to discuss what studies we should do in the future.

Calverley: We need some more information about whether exacerbations are indeed the same at different stages of the natural history of the disease. We have the suggestion from biopsy data that eosinophils might be important early on. Eosinophils may well be more evident in milder disease. Then we have Wisia Wedzicha's data, and things look rather different. Is this representative of all severe disease, or are there people coming to hospital with even worse disease doing the same things over again? Knowing what these processes are is vital.

Jeffery: I would like to know more about the CD4:CD8 ratios in exacerbations as compared with baseline. I would like to understand the molecules they are producing and how genetic background can influence the inflammatory response. I believe that genes will dictate the way the inflammation responds to a given stimulus. This might vary from individual to individual.

MacNee: How would you study that?

Jeffery: First of all, by immunostaining and gene expression in biopsies. But it is difficult to do this in a patient who is undergoing an exacerbation. The ethics of this are problematic.

Shapiro: You could step back to animal models. This may lead to more targeted studies in humans.

Calverley: We do need that kind of intermediate step. There are so many possible studies to do, and we have got to try to back winners. Peter Barnes was talking earlier about using his techniques of non-invasive assessment as a screening tool. You can only do these large studies fairly selectively. If we can have a more robust method of looking at exacerbations in a reliable animal model, that is going to be the way forward.

MacNee: Are you suggesting a model of exacerbations of COPD?

Shapiro: The combination of chronic cigarette smoke exposure plus infections or other acute inflammatory stimuli could model COPD exacerbations. There are caveats: we are limited by differences between mice and humans. But if we understand the differences, the models may prove helpful.

Nadel: I want to return to hypersecretion, this time in longer conducting airways where the submucosal glands reside. When people fall asleep, mucociliary clearance stops and the cough reflex is inhibited. In patients with chronic bronchitis, mucus glands produce excess mucus. When the patient is awake, clearance (e.g. by cough) may remove secretions. However, during sleep, they could aspirate secretions into the periphery, perhaps together with bacteria.

Calverley: A point of information: people do cough during the night. They wake up, but not necessarily fully. The cough clearance mechanism is very important in these people, because mucociliary clearance is disabled. I don't know whether the postulate you are putting up is likely to happen with macroscopic plugs, but I could see that changes in the small airways could occur. Years ago, when we started doing sleep studies in COPD patients, we looked at lung function morning and evening. Lung function in most COPD patients in the morning is at its lowest: it does drop over night. You could therefore argue that there could be some structural factor there, although this would need better studies.

Nadel: Reticular activation is needed for someone to cough. When silver nitrate drops were placed in the noses of children at night, in the morning these were in the alveoli. The idea of aspiration is not new.

Calverley: Aspiration certainly occurs overnight. The amount of gastric contents that are aspirated shows this. We have shown that people will aspirate gastric acid directly into the trachea overnight and produce an asthmatic-like reaction.

Nadel: I was just trying to think of how one would study this. It is quite possible that this is a way that bacteria get into the lower airways.

MacNee: Related to this, there have been some studies (Tsang et al 1999) showing a high prevalence of *Helicobacter pylori* in bronchiectatic patients. *H. pylori* could have a number of adverse effects in the airways as it has in the stomach. The hypothesis is that bronchiectatics are colonized by *H. pylori* as a result of aspiration of organisms from the gastrointestinal tract.

Hogg: The interaction between colonization and infection was well studied before the antibiotic era and showed that there has to be a combination of occlusion of the airways and colonization in order to get a good inflammatory reaction going. If that were the case in this situation — the lower airways were colonized and then aspiration took place — you might pick up a change in the computed tomography (CT) scan that would reflect exudation into the parenchymal airspace. It might be helpful to use CT to get information about the prevalence of parenchymal inflammation in acute exacerbations.

Wedzicha: That is an excellent idea. One approach that we have decided to take is to look at the frequent exacerbators. We are currently culturing bronchial airway epithelial cells from people with many exacerbations and without exacerbations. This is as close as we can get to the actual exacerbations. We believe that the frequent exacerbators may be our unstable patients. The idea of CT scanning them as well is an excellent idea. We just cannot study them as early in exacerbation as I would like. COPD exacerbations only last about a week, so we don't have a long time in which to do our studies.

MacNee: I was interested to learn that one of the risk factors for frequent exacerbations is hypersecretion of mucus. Getting back to induced sputum versus spontaneous sputum, in your studies if a patient spontaneously produced sputum, did you then induce sputum in these individuals?

Wedzicha: We had actually got some data showing that there was not very much difference between spontaneous and induced sputum. In our studies, we induced all the exacerbations and all the baselines.

MacNee: Rob Stockley finds a high incidence of bronchiectasis in his population of chronic of bronchitic patients.

Calverley: I think this fits in with the sort of thing that Jay Nadel was talking about. If you have localized plugging, this is the sort of pathological change you would expect. In looking for suitable people for lung volume reduction surgery, one problem is that our surgeons are concerned if even minor bronchiectasis is present. As you scan more people you see patients with small amounts of bronchiectasis, which clearly isn't the dominant pathology. It is tempting to think that that might be an important signal that we haven't used which may be relevant in terms of exacerbations. CT scanning in this condition would be helpful.

Nadel: There is an attractive idea here about bronchiectasis. The area for physiologic diffusion of oxygen in tissue at one atmosphere is about 500 mm. If you have a capillary within 500 mm, you can oxygenate tissue unless the tissue has a high metabolic rate. If a plug containing *Pseudomonas* bacteria resides in a 2 mm airway (e.g. by retrograde aspiration), what would be predicted to happen is that neutrophils would be rapidly recruited; they are aerobic, so they would die, the *Pseudomonas* within the plug is a facultative anaerobe so it would remain viable. It is easy to see how this kind of a lesion can develop into bronchiectasis.

MacNee: It seems that people who have frequent exacerbations are a group worth investigating genetically.

Silverman: It is certainly interesting. As in most complex diseases, there are likely to be modifier genes that are not related to the primary pathophysiology that influence organ specificity, and other characteristics of the disease expression. It might be harder to come up with family-based designs for a phenotype like COPD exacerbations, but case-control methods could be useful.

References

Blyth DI, Pedrick MS, Savage TJ, Bright H, Beesley JE, Sanjar S 1998 Induction, duration, and resolution of airway goblet cell hyperplasia in a murine model of atopic asthma: effect of concurrent infection with respiratory syncytial virus and response to dexamethasone. Am J Respir Cell Mol Biol 19:38–54

Dakhama A, Vitalis TZ, Hegele RG 1997 Persistence of respiratory syncytial virus (RSV) infection in a guinea pig model of acute bronchiolitis. Eur Respir J 10:20–26

Macek V, Dakhama A, Hogg JC, Green FHY, Rubin RK, Hegele R 1999 PCR detection of viral nucleic acid in fatal asthma: is the lower respiratory tract a reservoir for common viruses? Can Respir J 6:37–43

Möller LVM, Timens W, Van Der Bij W et al 1998 *Haemophilus influenzae* in lung explants of patients with end-stage pulmonary disease. Am J Respir Crit Care Med 157:950–956

Smith CB, Golden CA, Kanner RE, Renzetti AD 1980 Association of viral and mycoplasmal pneumonia infections with acute respiratory illness in patients with COPD. Am Rev Respir Dis 121:255–232

Tsang KW, Lam WK, Kwok E et al 1999 *Helicobacter pylori* and upper gastrointestinal symptoms in bronchiectasis. Eur Respir J 14:1345–1350

Epithelial cells and fibroblasts

Stephen I. Rennard

University of Nebraska Medical Center, Pulmonary and Critical Care Medicine Section, Omaha, NE 68198-5300, USA

Abstract. Chronic obstructive pulmonary disease (COPD) is characterized by acute and chronic alterations in the cellular composition structure of the airways and alveoli. Much attention has been focused on the increase in inflammatory cells present both within the airway lumen and within the airway wall. The parenchymal cells of the airway are intimately involved in the recruitment and activation of these inflammatory cells. Conversely, the behaviour of parenchymal cells can be modulated by inflammatory cells. The parenchymal cells can also alter the structural elements present within the airway leading to architecture changes which can impair lung function. Finally, epithelial cells and fibroblasts can directly modify each other's behaviour. The activity of these cells, therefore, undoubtedly can play a variety of roles in the pathophysiologic processes which underlie COPD.

2001 Chronic obstructive pulmonary disease: pathogenesis to treatment. Wiley, Chichester (Novartis Foundation Symposium 234) p 104–119

Airway epithelial cells are capable of releasing chemotactic activity for a variety of inflammatory cells including neutrophils (Shoji et al 1995), lymphocytes (Robbins et al 1989a), monocytes (Koyama et al 1989), eosinophils (Abdelaziz et al 1995) and, based on studies with a model cell line, mast cells (Shoji et al 1993). The best characterized of these activities is the ability of epithelial cells to release neutrophil chemotactic activity. In this context, several chemotactic factors including interleukin (IL)-8 (Bedard et al 1993, Choi & Jacoby 1992), HETE (hydroxyeicosatetraenoic acid) (Holtzman 1992) and leukotriene B4 (LTB4) (Holtzman et al 1983) have been reported as products of cultured epithelial cells. A variety of stimuli including cigarette smoke (Mio et al 1997), bacterial endotoxin (Koyama et al 1991), neutrophil elastase (Nakamura et al 1992a, Van Wetering et al 1997), tumour necrosis factor (TNF)α (Mio et al 1997) and a variety of other cytokines (Adachi et al 1997, Koyama et al 1995, Striz et al 1999a, Wickremasinghe et al 1999) have been reported to induce epithelial cell IL-8 production. It is likely that several of these stimuli may interact in a synergistic manner. In this context, C5a can greatly potentiate airway epithelial cell release of IL-8 in response to cigarette smoke extract (Floreani et al 1998). Transforming growth factor (TGF)β, in contrast, may be able to attenuate epithelial cell IL-8 release (Adachi et al 1997).

IL-8 is present within the intraluminal space of the airways in smokers and in patients with chronic obstructive pulmonary disease (COPD) and other inflammatory airways diseases as assessed by bronchoalveolar lavage (Mio et al 1997) and induced sputum (Keatings et al 1996, Richman-Eisenstat et al 1993). The increases in IL-8 concentration have been correlated with increases in neutrophils consistent with a role for epithelial cell production of IL-8 in the recruitment of neutrophils into the airway lumen (Mio et al 1997). It is of interest, therefore, that erythromycin can inhibit epithelial cell release of IL-8 (Takizawa et al 1997). This antibiotic has been widely used for a variety of inflammatory airways diseases (Ichikawa et al 1992). It may be that its action as an anti-inflammatory is more relevant in this regard than its potential antibiotic effects.

Airway epithelial cells are also able to express receptors which allow the adhesion of neutrophils. ICAM-1 expression appears to play a particularly important role (De Rose et al 1994, Jagels et al 1999). Normal resting airway epithelial cells can be induced to express ICAM-1 by several inflammatory stimuli (De Rose et al 1994, Striz et al 1999b). Neutrophils can adhere to ICAM-1 expressing epithelial cells through leukocyte integrin mediated mechanisms (De Rose et al 1994, Jagels et al 1999). This adhesion likely mediates several functions. Importantly, adhesion of neutrophils through this mechanism can mediate epithelial injury (De Rose et al 1994).

While fewer data are available, the ability of airway epithelial cells to recruit other inflammatory cells suggests a role in modulating the chronic inflammation within the airway as well. In this regard, epithelial cells can release mediators which are selective for lymphocyte subsets (Robbins et al 1989a). Epithelial cells can also express cell surface receptors capable of interacting with these cells. For example, in addition to ICAM-1 expression, airway epithelial cells can be induced to express HLA-DR antigens (Rossi et al 1990, Spurzem et al 1990). Through the expression of such molecules, epithelial cells may participate in the regulation of lymphocyte function. Consistent with this, epithelial cells have been reported to support antigen presentation (Kalb et al 1991, Rossi et al 1990). It is possible, therefore, that epithelial cells may play a crucial role in regulating the inflammatory milieu of both the airway wall and the airway lumen.

The airway epithelium is in a particularly strategic position to regulate inflammatory responses as it is likely the first line of defence which encounters environmental toxins and infectious agents. Unfortunately, this strategic position also makes the airway epithelium susceptible to injury either directly as a result of environmental agents or indirectly as a consequence of an inflammatory response. It is now recognized that the airway epithelium has considerable capacity to repair following injury. In this context, following mechanical injury of the airway, a provisional matrix comprised of fibronectin and plasma derived components is formed within minutes (Erjefalt et al 1995, Erjefalt & Persson 1997). Neighbouring epithelial cells de-differentiate and migrate covering the defect and re-establishing an epithelial barrier within hours (Erjefalt et al 1995). In experimental models, within 24 hours a large fraction of the cells present within the wounded area are replicating (Lane & Gordon 1974, McDowell et al 1987). This contrasts markedly with the normal replication rate in the airway epithelium where less than 1% of the cells will be in cycle (Erjefalt et al 1995).

Replication of the epithelial cells is followed by their acquisition of a columnar phenotype and, over a period of a week or two, by re-differentiation (Shimizu et al 1994). If this process proceeds successfully, a normal pseudo-stratified columnar epithelium is re-established. The cells responsible for this repair process are not completely defined. However, cells present at early stages within a wounded area express cytokeratin 14, a marker for basal cells (Shimizu et al 1994). The columnar cells present early after injury express cytokeratin 14, although this is gradually lost as cells reassume their normal cytokeratin expression. Eventually, cytokeratin 14 is expressed only in basal cells. This progression suggests that basal cells present at the borders of the wound may be responsible for mediation of repair. Alternatively, columnar cells which do not normally express cytokeratin 14 may be induced to 'de-differentiate' and express this marker.

It is likely that epithelial cells participating in a repair response are induced to express a number of genes not expressed by the normal epithelium. In this regard, cells migrating in response to a wound up-regulate the expression of a number of integrins including $\alpha 5$, αV, $\beta 5$ and $\beta 6$ (Pilewski et al 1997). Of interest in this regard is the ability of the integrin $\alpha V \beta 6$ to activate latent $TGF\beta$ to its active form (Munger et al 1999). This raises the possibility that up-regulation of $\alpha V \beta 6$ by migrating epithelial cells could lead to activation of $TGF\beta$ with further initiation of repair responses.

The migration of epithelial cells *in vitro* has been reported to depend on fibronectin and the fibronectin-binding integrins $\alpha 3$, $\alpha 5$ and $\beta 1$ (Herard et al 1996). Interestingly, $TGF\beta$ can up-regulate the expression of these integrins (Spurzem et al 1992). The ability of epithelial cells to participate in a repair response can be modulated by the local cytokine milieu. In this context, $TNF\alpha$ (Ito et al 1996) has been reported to increase the rate at which epithelial cells migrate. In contrast, $TGF\beta$ can induce epithelial cells to assume a more flattened phenotype and can increase their adhesion to the subjacent connective tissue matrix while decreasing their ability to migrate (Spurzem et al 1993). Taken together, these results suggest that epithelial wound healing is a multi-step process which is regulated by several cytokines likely requiring a series of sequential steps.

Modulation of repair may involve not only the participation of exogenous mediators, but also mediators produced by epithelial cells may function in an autocrine or paracrine manner. In this context, epithelial cells maintained in culture are capable of producing TGFβ and can release a portion of this TGFβ in its active form (Sacco et al 1991). This TGFβ can, in turn, modulate epithelial cell production of fibronectin (Romberger et al 1992, Wang et al 1991). Since this fibronectin can function as a chemoattractant for airway epithelial cells (Herard et al 1996, Shoji et al 1990), it has been suggested, therefore, that production of active TGFβ within a wound leads to the production of fibronectin which in turn mediates the recruitment of epithelial cells into an area of epithelial defect. Interestingly, the ability of epithelial cells to migrate in response to fibronectin depends on the subjacent connective tissue matrix (Rickard et al 1993). It seems likely, therefore, that repair responses will depend on the presence of mediators produced by exogenous sources, by epithelial cells present within a wound, and on the extent of injury to the structural elements present within the airway.

It is likely that epithelial mediators which play a role in normal epithelial repair can also lead to the accumulation of fibroblasts within the airway wall. In this regard, epithelial cells are capable of producing mediators which can recruit fibroblasts (Shoji et al 1989), stimulate their proliferation (Nakamura et al 1992b), induce extracellular matrix macromolecule production (Kawamoto et al 1995) and induce fibroblasts to more vigorously contract their extracellular matrix milieu (Mio et al 1998). Fibronectin, for example, is both a chemoattractant for epithelial cells (Shoji et al 1990) and for fibroblasts (Shoji et al 1989). Fibronectin can help support fibroblast proliferation. In addition, a number of other mediators derived from epithelial cells may be active in this regard (Cambrey et al 1995). Consistent with this, mechanical injury of the airway is associated not only with accumulation and proliferation of epithelial cells, but also with accumulation of proliferating fibroblasts in the area below the wound (Erjefalt et al 1995).

TGFβ can function in the repair response not only to induce epithelial cell production of fibronectin, but can also induce fibroblast production of matrix macromolecules (Fine & Goldstein 1987, Kawamoto et al 1995). TGFβ can, moreover, induce fibroblasts to assume a contractile phenotype (Montesano & Orci 1988, Pena et al 1994). In this context, fibroblasts cultured in a three-dimensional collagenous matrix can interact with and contract their surrounding connective tissue. Such a process is believed to play an important role in normal wound healing and may be responsible for the tissue contraction which characterizes many fibrotic disorders. When this develops circumferentially around an airway, for example in the peribronchiolar fibrosis which is characteristically present in individuals with COPD and moderately severe airflow limitation (Niewoehner 1998), this process could account for narrowing of the airways and, therefore, directly lead to airflow limitation. The ability of airway epithelial cells to produce TGFβ could, therefore, lead not only to the accumulation of fibroblasts and extracellular matrix within the airways, but could also contribute to airway narrowing and decreased airflow.

Other cells present within the inflammatory milieu can undoubtedly modulate the activity of fibroblasts. In this regard, both lymphocytes (Kitamura et al 1993, Postlethwaite & Seyer 1991, Sempowski et al 1996) and monocytes (Shaw & Kelley 1995) can produce a variety of factors which can modulate fibroblast proliferation and matrix production. Interestingly, co-culture of mononuclear phagocytes with fibroblasts in three-dimensional collagen gels results in inhibition of contraction (Skold et al 1998). Specifically, the monocytes in the co-culture are induced to produce TNFα and IL-1β. These monocyte derived mediators, in turn, inhibit fibroblast mediated contraction of collagen gels. This raises the possibility that monocytes present at a site undergoing tissue injury and repair could help mitigate against excessive scarring.

Mediators derived from inflammatory cells, however, could also lead to increased alteration of tissue architecture. In this context, neutrophil elastase can induce excessive fibroblast mediated contraction of collagen gels (Skold et al 1999). Consistent with this, neutrophils co-cultured with fibroblasts also induce augmented contraction. Neutrophils incubated alone in collagen gels have no effect. Interestingly, when neutrophils are incubated together with fibroblasts in collagen gels, not only is contraction augmented, but degradation of the collagen gels is stimulated. This degradation appears to be an interaction between fibroblasts and neutrophils as neither cell alone will induce degradation of the collagen gel. The final architectural result depends on collagen production, collagen degradation and collagen remodelling. Thus, it seems likely that fibroblast response to injury will be modulated not only by the activation of epithelial cells, but will also depend on the inflammatory milieu present during the repair process. This milieu could lead either to excessive or deficient repair.

Fibroblasts can not only modulate the production of extracellular matrix, but these cells can also release cytokines which can both recruit and activate inflammatory cells (Denburg et al 1991, Robbins et al 1994). Thus, like epithelial cells, it is likely that fibroblasts are active participants in regulating inflammatory responses. Finally, fibroblasts can also directly affect epithelial cells. In this regard, fibroblasts can produce growth factors which can stimulate epithelial cell proliferation (Robbins et al 1989b). When maintained in co-culture, epithelial cells and fibroblasts are capable of producing sufficient growth factors that both cells will replicate in the absence of exogenous growth factors (Nakamura et al 1995). Interestingly, in *in vitro* systems, epithelial cells and fibroblasts organize into gland-like structures where islands of epithelial cells are surrounded by a 'matrix' containing whorls of fibroblasts. It is likely, therefore, that the interactions between epithelial cells and mesenchymal cells which are crucial for the normal development of the lung are (Masters 1976), at least in part, recapitulated during wound healing.

In summary, epithelial cells in fibroblasts in the airway are capable of mutual interactions which can support proliferation and regulation of differentiated function. In addition, both epithelial cells and fibroblasts can modulate the recruitment and activation of a spectrum of inflammatory cells. These inflammatory cells, in turn, can modulate the behaviour of the epithelial cells and fibroblasts. Taken together, it is clear that the parenchymal cells present within the airways are fully integrated into the pathogenetic network responsible for tissue alterations in COPD. Finally, while this review has highlighted processes within the airways, it is likely that analogous interactions between epithelial cells, mesenchymal cells and inflammatory cells exist within the alveolar structures and contribute both to the formation of interstitial fibrosis and to the deficient repair which characterizes pulmonary emphysema. Delineation of these relationships will serve to identify novel therapeutic targets which could help ameliorate disease activity in COPD.

References

Abdelaziz MM, Devalia JL, Khair OA, Calderon M, Sapsford RJ, Davies RJ 1995 The effect of conditioned medium from cultured human bronchial epithelial cells on eosinophil and neutrophil chemotaxis and adherence *in vitro*. Am J Respir Cell Mol Biol 13:728–737

Adachi Y, Mio T, Takigawa K et al 1997 Mutual inhibition by TGF-β and IL-4 in cultured human bronchial epithelial cells. Am J Physiol 17:L701–L708

Bedard M, McClure CD, Schiller NL, Francoeur C, Cantin A, Denis M et al 1993 Release of interleukin-8, interleukin-6 and colony-stimulating factors by upper airway epithelial cells: implications for cystic fibrosis. Am J Respir Cell Mol Biol 9:455–462

Cambrey AD, Kwon OJ, Gray AJ et al 1995 Insulin-like growth factor I is a major fibroblast mitogen produced by primary cultures of human airway epithelial cells. Clin Sci (Colch) 89:611–617

Choi AM, Jacoby DB 1992 Influenza virus A infection induces interleukin-8 gene expression in human airway epithelial cells. FEBS Lett 309:327–329

De Rose V, Robbins RA, Snider RM. et al 1994 Substance P increases neutrophil adhesion to bronchial epithelial cells. J Immunol 152:1339–1346

Denburg JA, Gauldie J, Dolovich J, Ohtoshi T, Cox G, Jordana M et al 1991 Structural cell-derived cytokines in allergic inflammation. Int Arch Allergy Appl Immunol 94:127–132

Erjefalt JS, Persson CG 1997 Airway epithelial repair: breathtakingly quick and multipotentially pathogenic. Thorax 52:1010–1012

Erjefalt JS, Erjefalt I, Sundler F, Persson CG 1995 *In vivo* restitution of airway epithelium. Cell Tissue Res 281:305–316

Fine A, Goldstein RH 1987 The effect of transforming growth factor-β on cell proliferation and collagen formation by lung fibroblasts. J Biol Chem 262:3897–3902

Floreani AA, Heires AJ, Welniak LA et al 1998 Expression of receptors for C5a anaphylatoxin (CD88) on human bronchial epithelial cells: enhancement of C5a mediated release of IL-8 upon exposure to cigarette smoke. J Immunol 160:5073–5081

Herard AL, Pierrot D, Hinnrasky J et al 1996 Fibronectin and its α5 β1-integrin receptor are involved in the wound-repair process of airway epithelium. Am J Physiol 271:L726–L733

Holtzman MJ 1992 Arachidonic acid metabolism in airway epithelial cells. Ann Rev Physiol 54:303–329

Holtzman MJ, Aizawa H, Nadel JA, Goetzl EJ et al 1983 Selective generation of leukotriene B4 by tracheal epithelial cells from dogs. Biochem Biophys Res Commun 114:1071–1076

Ichikawa Y, Ninomiya H, Koga H et al 1992 Erythromycin reduces neutrophils and neutrophil-derived elastolytic-like activity in the lower respiratory tract of bronchiolitis patients. Am Rev Respir Dis 146:196–203

Ito H, Rennard SI, Spurzem JR 1996 Mononuclear cell conditioned medium enhances bronchial epithelial cell migration but inhibits attachment to fibronectin. J Lab Clin Med 127:494–503

Jagels MA, Daffern PJ, Zuraw BL, Hugli TE et al 1999 Mechanisms and regulation of polymorphonuclear leukocyte and eosinophil adherence to human airway epithelial cells. Am J Respir Cell Mol Biol 21:418–427

Kalb TH, Chuang MT, Marom Z, Mayer L et al 1991 Evidence for accessory cell function by class II MHC antigen-expressing airway epithelial cells. Am J Respir Cell Mol Biol 4:320–329

Kawamoto M, Romberger DJ, Nakamura Y et al 1995 Modulation of fibroblast type I collagen and fibronectin production by bovine bronchial epithelial cells. Am J Respir Cell Mol Biol 12:425–433

Keatings VM, Collins PD, Scott DM, Barnes PJ 1996 Differences in interleukin-8 and tumor necrosis factor-α in induced sputum from patients with chronic obstructive pulmonary disease or asthma. Am J Respir Crit Care Med 153:530–534

Kitamura A, Kitamura M, Nagasawa R et al 1993 Renal fibroblasts are sensitive to growth-repressing and matrix-reducing factors from activated lymphocytes. Clin Exp Immunol 91:516–520

Koyama S, Rennard SI, Shoji S et al 1989 Bronchial epithelial cells release chemoattractant activity for monocytes. Am J Physiol 257:L130–L136

Koyama S, Rennard SI, Leikauf GD et al 1991 Endotoxin stimulates bronchial epithelial cells to release chemotactic factors for neutrophils. A potential mechanism for neutrophil recruitment, cytotoxicity, and inhibition of proliferation in bronchial inflammation. J Immunol 147:4293–4301

Koyama S, Rennard SI, Robbins RA 1995 Bradykinin stimulates bronchial epithelial cells to release neutrophil and monocyte chemotatic activity. Am J Physiol 269:L38–L44

Lane BP, Gordon R 1974 Regeneration of rat tracheal epithelium after mechanical injury. I. The relationship between mitotic activity and cellular differentiation. Proc Soc Exp Biol Med 145:1139–1144

Masters JRW 1976 Epithelial–mesenchymal interaction during lung development: the effect of mesenchymal mass. Dev Biol 51:98–108

McDowell EM, Ben T, Newkink C, Chang B, De Luca LM 1987 Differentiation of tracheal mucociliary epithelium in primary cell culture recapitulates normal fetal development and regeneration following injury in hamsters. Am J Pathol 129:511–522

Mio T, Romberger DJ, Thompson AB, Robbins RA, Heires A, Rennard SI 1997 Cigarette smoke induces interleukin-8 release from human bronchial epithelial cells. Am J Respir Crit Care Med 155:1770–1776

Mio T, Liu X, Adachi Y et al 1998 Human bronchial epithelial cells modulate collagen gel contraction by fibroblasts. Am J Phys 274:L119–L126

Montesano R, Orci L 1988 Transforming growth factor-β stimulates collagen-matrix contraction by fibroblasts: implication for wound healing. Proc Natl Acad Sci USA 85:4894–4897

Munger JS, Huang X, Kawakatsu H et al 1999 The integrin αv β6 binds and activates latent TGF-β1: a mechanism for regulating pulmonary inflammation and fibrosis. Cell 96:319–28

Nakamura H, Yoshimura K, McElvaney NG, Crystal RG 1992a Neutrophil elastase in respiratory epithelial lining fluid of individuals with cystic fibrosis induces interleukin-8 gene expression in a human bronchial epithelial cell line. J Clin Invest 89:1478–1484

Nakamura Y, Ertl RF, Kawamoto M et al 1992b Bronchial epithelial cells modulate fibroblast proliferation: role of prostaglandin E2. Am Rev Respir Dis 145:A827

Nakamura Y, Tate L, Ertl RF et al 1995 Bronchial epithelial cells regulate fibroblast proliferation. Am J Physiol 269:L377–L387

Niewoehner DE 1998 Anatomic and pathophysiological correlations in COPD. In: Baum YL et al (eds) Textbook of pulmonary diseases. Lippincott-Raven, Philadelaphia, p 823–842

Pena RA, Jerdan JA, Glaser BM 1994 Effects of TGF-β and TGF-β neutralizing antibodies on fibroblast-induced collagen gel contraction: implications for proliferative vitreoretinopathy. Investig Ophthalmol Vis Sci 35:2804–2808

Pilewski JM, Latoche JD, Arcasoy SM, Albelda SM 1997 Expression of integrin cell adhesion receptors during human airway epithelial repair *in vivo*. Am J Physiol 273:L256–L263

Postlethwaite AE, Seyer JM 1991 Fibroblast chemotaxis induction by human recombinant interleukin-4. Identification by synthetic peptide analysis of two chemotactic domains residing in amino acid sequences 70–88 and 89–122. J Clin Invest 87:2147–2152

Richman-Eisenstat JBY, Jorens PG, Hebert CA, Ueki I, Nadel JA 1993 Interleukin-8: an important chemoattractant in sputum of patients with chronic inflammatory airway diseases. Am J Physiol 264:L413–L418

Rickard KA, Taylor J, Rennard SI, Spurzem JR 1993 Migration of bovine bronchial epithelial cells to extracellular matrix components. Am J Respir Cell Mol Biol 8:63–68

Robbins RA, Shoji S, Linder J et al 1989a Bronchial epithelial cells release chemotactic activity for lymphocytes. Am J Physiol 257:L109–L115

Robbins RA, Linder J, Stahl MG et al 1989b Diffuse alveolar hemorrhage in autologous bone marrow transplant patients. Am J Med 87:511–518

Robbins RA, Barnes PJ, Springall DR et al 1994 Expression of inducible nitric oxide in human lung epithelial cells. Biochem Biophys Res Commun 203:209–218

Romberger DJ, Beckmann JD, Claassen L, Ertl RF, Rennard SI 1992 Modulation of fibronectin production of bovine bronchial epithelial cells by transforming growth factor-β. Am J Respir Cell Mol Biol 7:149–155

Rossi GA, Sacco O, Balbi B et al 1990 Human ciliated bronchial epithelial cells: expression of the HLA-DR antigens and of the HLA-DR α gene, modulation of the HLA-DR antigens by γ-interferon and antigen-presenting function in the mixed leukocyte reaction. Am J Respir Cell Mol Biol 3:431–439

Sacco O, Rennard SI, Spurzem JR et al 1991 *In vitro* production of active transforming growth factor β by bronchial epithelial cells. Am Rev Respir Dis 143:A202

Sempowski GD, Derdak S, Phipps RP 1996 Interleukin-4 and interferon-γ discordantly regulate collagen biosynthesis by functionally distinct lung fibroblast subsets. J Cell Physiol 167:290–296

Shaw RJ, Kelley J 1995 Macrophages/monocytes. In: Phan SH, Thall RS (eds) Pulmonary fibrosis. Marcel Dekker, New York, p 405–444

Shimizu T, Nishihara M, Kawaguchi S, Sakakura Y 1994 Expression of phenotypic markers during regeneration of rat tracheal epithelium following mechanical injury. Am J Respir Cell Mol Biol 11:85–94

Shoji S, Rickard KA, Ertl RF, Robbins RA, Linder J, Rennard SI 1989 Bronchial epithelial cells produce lung fibroblast chemotactic factor: fibronectin. Am J Respir Cell Mol Biol 1:13–20

Shoji S, Ertl RF, Linder J, Romberger DJ, Rennard SI 1990 Bronchial epithelial cells produce chemotactic activity for bronchial epithelial cells: possible role for fibronectin in airway repair. Am Rev Respir Dis 141:218–225

Shoji S, Kitani S, Takizawa H, Baba M, Morita Y, Sto K 1993 Bronchial epithelial cells release a chemotactic activity for rat basophilic leukemia (RBL-2H3) cells. Am Rev Respir Dis 147: A45

Shoji S, Ertl RF, Koyama S et al 1995 Cigarette smoke stimulates release of neutrophil chemotactic activity from cultured bovine bronchial epithelial cells. Clin Sci 88:337–344

Skold CM, Liu X, Umino T, Ente R, Romberger D, Rennard SI 1998 Blood monocytes attenuate lung fibroblast contraction of three dimensional collagen gels co-culture. Am J Respir Crit Care Med 157:A190

Skold CM, Liu X, Umino T et al 1999 Human neutrophil elastase augments fibroblast-mediated contraction of released collagen gels. Am J Respir Crit Care Med 159:1138–1146

Spurzem JR, Sacco O, Rossi GA, Rennard SI 1990 MHC class II expression by bronchial epithelial cells is modulated by lymphokines and corticosteroids. Am Rev Respir Dis 141:A681

Spurzem JR, Sacco O, Veys T, Rickard KA, Rennard SI 1992 TGF-β increases expression of extracellular matrix receptors on cultured bovine bronchial epithelial cells. Am Rev Respir Dis 145:A668

Spurzem JR, Sacco O, Rickard KA, Rennard SI 1993 Transforming growth factor-β increases adhesion but not migration of bovine bronchial epithelial cells to matrix proteins. J Lab Clin Med 122:92–102

Striz I, Mio T, Adachi Y, Robbins RA, Romberger DJ, Rennard SI 1999a IL-4 and IL-13 stimulate human bronchial epithelial cells to release IL-8. Inflammation 23:545–555

Striz I, Mio T, Adachi Y et al 1999b IL-4 induces ICAM-1 expression in human bronchial epithelial cells and potentiates TNF-α. Am J Physiol 277:L58–L64

Takizawa H, Desaki M, Ohtoshi T et al 1997 Erythromycin modulates IL-8 expression in normal and inflamed human bronchial epithelial cells. Am J Respir Crit Care Med 156:266–271

Van Wetering S, Mannesse-Lazeroms SP, Dijkman JH, Hiemstra PS 1997 Effect of neutrophil serine proteinases and defensins on lung epithelial cells: modulation of cytotoxicity and IL-8 production. J Leukoc Biol 62:217–226

Wang A, Cohen DS, Palmer E, Sheppard D et al 1991 Polarized regulation of fibronectin secretion and alternative splicing by transforming growth factor. J Biol Chem 266:15598–15601

Wickremasinghe MI, Thomas LH, Friedland JS 1999 Pulmonary epithelial cells are a source of IL-8 in the response to *Mycobacterium tuberculosis*: essential role of IL-1 from infected monocytes in a NF-κB-dependent network. J Immunol 163:3936–3947

DISCUSSION

Jeffery: One of the features of cigarette smoking is that there is often extensive squamous metaplasia. Have you investigated in any way whether the effects or interrelationships when the epithelia are squamous or metaplastic are any different than when they are pseudostratified, ciliated and columnar?

Rennard: We haven't done any experiments like that. This is probably approachable, even in *in vitro* systems, because you can induce a kind of squamous metaplasia. Paul Nettesheim's group has done many studies over the years in terms of inducing this with a whole variety of toxins, including things contained in cigarette smoke (Walker et al 1987, Terzaghi et al 1978, Pai et al

1983). I don't think anyone has examined in a detailed way whether those cells are different in terms of their response in the system. Having said this, the cells we use in our co-culture experiments are not normal pseudostratified columnar epithelial cells. Our cells are clearly dedifferentiated, although they are still epithelial cells. It is possible to get these cells to differentiate in culture so that they look like normal epithelial cells, but data on these are rarer.

Dunnill: These experiments on epithelial regeneration were performed by Wilhelm (1953) and repeated by Otto & Wagner (1956), so there is nothing original about that. They both showed that repeated injury gave rise to squamous metaplasia.

The respiratory epithelium has a complex number of cells in it: it is not just one cell type. Have you tried to analyse the different types of cell in the epithelium, such as basal cells, columnar cells and goblet cells?

Rennard: None of the experiments that we have done would address that question. The regeneration following injury can restore what appears to be a fairly normal-looking cell distribution. What it is that regulates this process is an important unanswered question. There is currently a lot of work going on in terms of *in vitro* differentiation, where epithelial cells are cultured on a membrane at an air–liquid interface (Adler et al 1990). Under these conditions they will differentiate and these cells can be induced to express specific kinds of mucins. They are sort of like goblet cells. It is becoming clear that different things can drive the epithelial cells one way or another to express differentiated phenotypes. Repair is probably not going to involve the initiation of a repair response which then follows a single track; it may be that there are multiple pathways, some of which could lead to restoration of normal tissue architecture and function, and others which may lead to quite a dysfunctional structure.

Dunnill: The ciliary regeneration is important here, for the clearance of the mucus.

Jeffery: If the antibodies to TGFβ, for example, are robust enough to stain the tissues that you already have, might it be possible to see whether squamous metaplastic epithelium as opposed to pseudostratified ciliated epithelium are expressing TGFβ in similar ways? Is this feasible?

Rennard: Yes. There are actually data looking at TGFβ in interstitial disease, showing TGFβ production by several cell types (Khalil et al 1996, Broekelmann et al 1991). It is a little more complicated, however, than just TGFβ production. There are multiple levels of control for TGFβ: it is regulated by its own production, but it is also regulated by its activation in the extracellular milieu. There are several pathways that can activate TFGβ, so it is possible for there to be a lot of it bound to the extracellular matrix that is not active.

Stolk: Our *in situ* hydridization data show that TGFβ mRNA is present, and we can also stain the protein in tissues from patients with and without COPD.

We purposely looked into small airways because in previous studies, we found a higher number of macrophages in those with COPD compared to those without.

Stockley: What is the real direct evidence that there is epithelial injury in COPD, what is the mechanism involved, and how should we be monitoring it?

Dunnill: Look at the sputum. You see epithelial cells there. Presumably these have been shed. If you look at sections of a lung from a COPD patient, you will find that there are areas where the epithelium has gone and regeneration is occurring.

Jeffery: More subtle evidence is the obvious goblet cell hyperplasia and metaplasia.

Stockley: Do you think we can use the term 'injury' in these respects?

Nadel: There is a difference between epithelial sloughing (Dr Dunnill's comment) and goblet cell metaplasia (Dr Jeffrey). Goblet cell metaplasia can occur secondary to 'injury' (i.e. mechanical damage such as occurs with tracheal intubation), but it can also occur in allergic airways secondary to Th2 cell recruitment. The word 'injury' is too imprecise to describe these varied conditions. Those cascades are different cells shedding and dying.

Jeffery: But it is a response to an injurious event. Secretion of mucus is a component of inflammation, as Lord Florey proposed.

Calverley: One other piece of evidence is that one of the features of small airways disease is the presence of fibrosis. The attraction of the approach that Steve Rennard adopted is that it suggests a mechanism by which at least some of that fibrosis might arise. If you say that this is not actually what is happening and that it is coming from some other process, you need to outline what this might be. There are many potential candidates. The attraction of looking at epithelium changes is that these are a logical and biologically normal way to respond to damage. Certainly, in most cases of COPD we think we know what the principal complex chemical stimulus is, and that is first of all going to come into contact with the epithelium. At least some of those mechanisms are likely to be activated.

Jeffery: Instead of goblet cell hyperplasia, one could suggest squamous metaplasia. Would you be content with that as a response to injury? Is that indicative of an injurious process or not?

Stockley: The reason I raised this point is that I can't think of an alternative term for what we actually see in the airways of patients with COPD, where we do see mucus production, mucus gland hyperplasia, squamous metaplasia and protein leakage. You can call this inflammation if you like, but to get that sort of process happening, epithelial cells have to either be lost or separated. If this is the case, I would call that an injury.

MacNee: Increased epithelial permeability occurs in cigarette smokers, but it doesn't occur in chronic bronchitis.

Stockley: Can you clarify what you mean when you say it doesn't occur in chronic bronchitis, because that is a very important statement.

MacNee: In cigarette smokers, increased epithelial permeability develops within 24 h of starting smoking and starts to resolve within about 48 h after smoking cessation, but is not different from normals or chronic bronchitics who don't smoke.

Stockley: Is that increased permeability backwards, from the epithelium out?

MacNee: Yes.

Stockley: But in someone who has an inflamed lung, where there is protein leakage in, getting antigen or markers to go out is enormously difficult. In bronchitis fluid and proteins come into the airway.

MacNee: However, increased epithelial permeability is not present in chronic bronchitics.

Nadel: If a word helps you to get a mechanism, then I am in favour of the word. If the word clouds a mechanism, then I am opposed to the word. Recently, I have objected to words that were very important in the 1890s such as 'inflammation', but which are too general and do not tell us about sophisticated cell biological cascades. There is nothing wrong with using the word 'injury', as long as you recognize that it can be any kind of process that changes the metabolism and signalling of cells. Where I think it is dangerous is if it clouds our minds.

Senior: I want to reinforce an important point that Steve Rennard has presented, namely that contact between structural cells may be important in cell responses. An interesting example is assembly of basement membranes. In *in vitro* models using keratinocytes or alveolar epithelial cells, contact with fibroblasts appears to facilitate production of a normal appearing basement membrane (Fleischmayr et al 1998, Furuyama et al 1997).

Hogg: Aubert et al (1994a) examined TGFβ protein in resected lung specimens using immunohistochemistry and TGFβ mRNA expression by Northern analysis. They didn't find much difference between those with and without airways obstruction in terms of TGFβ (Aubert et al 1994a), but we did show differences with PDGF (Aubert et al 1994b).

Rennard: I used TGFβ and fibronectin for two reasons. First, it is helpful to have a simple model to try to create these things from a heuristic perspective. Second, these are the mediators that we study. But without doubt there are other mediators: platelet-derived growth factor (PDGF) can be released by airway epithelial cells, and one can postulate an analogous role for PDGF to that of TGFβ. Endothelin 1 can have very similar activities, as can IL-4 in terms of driving fibroblast responses. It would be naïve to think that TGFβ by itself is going to regulate repair responses. The networks of cytokines will involve multiple mediators. With respect to TGFβ activity, we have used antibodies that recognize the active site on the TGFβ molecule. With regard to the recent evidence suggesting that TGFβ can be activated through mechanisms such as interaction with the $\alpha V \beta 6$ integrin, I am not clear that this would be recognizable histologically in a tissue. Under those circumstances the TGFβ doesn't dissociate from the latency associated peptide. It somehow changes its conformation so that it can now interact with TGFβ receptor. I am not sure that the absence of staining with antibodies against active TGFβ means that there is no TGFβ activity in that tissue. As we learn more about the activation of these cytokines, I think this is an important point.

Paré: I was fascinated by the data presented by Steve Rennard showing the interaction of the neutrophil with the fibroblast can cause a decrease in collagen. We have seen many examples from Jim Hogg of thickening of the small airway wall in COPD with increased connective tissue deposition, but there are also airways with very thin walls. It has always been a paradox to me why the neutrophil — which we think is important in the airways and parenchyma in COPD — primarily causes loss of connective tissue in the parenchyma and increased connective tissue in the smaller airways.

Hogg: The data we have comes from comparing airways of patients who have lung volume reduction surgery to airways of asthmatics who died during an acute attack using airways from lung tissue resected for tumour where patients have normal lung function as controls. When we compare asthmatic airways to those from patients who have normal lung function, the asthmatic airways are thicker and tend to maintain their calibre. This contrasts sharply with COPD where the airway lumen is very much narrower and there is no thickening of the airway wall. We are currently examining the percentage of collagen in the walls. If you look at the pictures that we published with the grading system years ago, the ones with severe fibrosis tend to have thin walls (Wright et al 1985). I wonder whether the narrowing of the lumen in COPD is due to the contraction that occurs with scar formation.

MacNee: When one looks in bronchoalveolar lavage or in the small airways, there is a mixture of cells. The macrophage is the predominant cell in the earliest lesions in the small airways. Co-culture experiments are difficult to interpret. Clearly, the point is that these are nice models *in vitro*, but *in vivo* the situation is much more complex.

Rennard: In vitro systems are wonderful for asking certain kinds of questions, but they get to be difficult and cumbersome for the kinds of questions that you are asking. Every cell present is interacting with every other cell there through a whole variety of different cytokines. The final result in the tissue is going to depend on the balance of all of those cells. The point of my talk was to show that the mesenchymal and epithelial cells are equally active players in this inflammation and repair of injury as the professional inflammatory cells. Ultimately, if we are going to take this constructionist approach of adding things back in *in vitro* models, we will have to add back multiple mediators. We have tried to do those experiments, although as you can imagine they are technically rather challenging.

Calverley: You said that alveolar cells will probably behave in the same way. Is this a reasonably robust statement? We are talking about this constructionist approach of adding things together. It is very clear that there are major differences in what goes on in the small airways and the alveoli. One of the debates that has gone on about nomenclature in this disease was to try to distinguish emphysema from what we would now call small airways disease, and then we put them back together under this term COPD. Why don't we see more fibrosis in emphysema, bearing in mind the fact that the epithelial structures are basically similar?

Rennard: There are not as many data with the fibroblasts, and so there are few comparisons of alveolar fibroblasts with airway fibroblasts. In our studies we used fetal lung fibroblasts, which probably come from alveolar structures as they were made from a mince of fetal tissue (Breul et al 1980). We have done similar work with fibroblasts from airways, and we get similar kinds of responses. But there are clear differences between alveolar fibroblasts and other fibroblasts in terms of what they can do. There are more data with the epithelial cells, although it is harder to grow alveolar epithelial cells. The repertoire of alveolar epithelial cell responses is every bit as rich as is represented for the airway responses. They can release many of the same cytokines and mediators that do all of the same functions with respect to recruiting and activating inflammatory cells and mesenchymal cells. It is therefore quite plausible to believe that in the alveolae, with regard to the network of responses, the epithelial cells will play a very analogous role. In fact, this is probably true in tissues throughout the body. And, as Bob Senior mentioned, it is certainly the case in the basement membrane assembly in the skin.

Dunnill: Which 'epithelial' cells in the alveolus are you referring to?

Rennard: That is a great question. The data which I am familiar with come from either cultures in vitro of type II cells which can be isolated, or from cells such as the A549 alveolar carcinoma cell. The data are limited by the ability of people to culture cells. Type I cells are notoriously difficult to culture, but it is likely that these will also play a role.

Lomas: These are complex systems with cross-talk between different cells. One way to try to dissect out components is to use transgenic models to look at knockouts. Have people looked at knockouts of cytokines in terms of epithelial injury and repair?

Shapiro: There is one example that could be relevant here, but it does not involve a cytokine. Colleagues of mine knocked out a metalloproteinase, MMP7, which is expressed in the airway epithelium. In these knockouts, if the epithelium is injured there is decreased repair. It is believed that cells need to attach and detach from the extracellular matrix in order for cells to migrate and heal the wound.

Lomas: There are TGF knockouts, aren't there?

Shapiro: Yes, however they have marked inflammation and die too soon for these types of studies. This is a problem shared by many of the cytokine knockouts.

Rennard: The TGFα knockouts attenuate a pro-fibrotic stimulus (Madtes et al 1999). In terms of epithelial repair, I don't know of any other data besides those Steve Shapiro mentioned. I think this will be a very important way of dissecting out the roles of different molecules.

Nadel: I am trying to understand the co-culture experiments with neutrophils. How are these cells interacting? Are these 'resting' neutrophils? Is there fibronectin in the culture?

Rennard: The way the experiment is done is to take neutrophils isolated from peripheral blood without any further activation, and to mix them in with the collagen gel before it was cast. So the neutrophils are cast at the same time as the fibroblasts into the 3D gel. The fibroblasts are certainly capable of producing fibronectin, and they do produce it in 3D gels. In fact, under these growth conditions they produce fibronectin at a greater rate than they will in routine dish culture. But we don't know whether a fibroblast-derived factor is activating the neutrophils. We are interested in following Ed Campbell's concept: we know that neutrophil elastase will have similar effects on the fibroblasts as the neutrophils in the co-culture. In fact, this is the experiment we did first. We are very interested to know whether elastase is playing a role here, and whether the neutrophils are releasing elastase in a restricted kind of environment such as that you outlined yesterday.

References

Adler KB, Cheng PW, Kim KC 1990 Characterization of guinea pig tracheal epithelial cells maintained in biphasic organotypic culture: cellular composition and biochemical analysis of released glycoconjugates. Am J Respir Cell Mol Biol 2:145–154

Aubert J-D, Dalal BL, Bai TR, Roberts CR, Hayashi S, Hogg JC 1994a Transforming growth factor β_1 gene expression in human airways. Thorax 49:225–232

Aubert J-D, Hayashi S, Hards J, Bai TR, Pare PD, Hogg JC 1994b Platelet derived growth factor and its receptor in lungs from patients with asthma and COPD. Am J Physiol 266:L655–L663

Breul SD, Bradley KH, Hance AJ, Schafer MP, Berg RA, Crystal RG 1980 Control of collagen production by human diploid lung fibroblasts. J Biol Chem 255:5250–5260

Broekelmann TJ, Limper AH, Colby TV, McDonald JA 1991 Transforming growth factor beta 1 is present at sites of extracellular matrix gene expression in human pulmonary fibrosis. Proc Natl Acad Sci USA 88:6642–6646

Fleischmajer R, Perlish JS, MacDonald ED et al 1998 There is binding of collagen IV to β1 integrin during early skin basement membrane assembly. Ann NY Acad Sci 857:212–227

Furuyama A, Kimata K, Mochitate K 1997 Assembly of basement membrane *in vitro* by cooperation between alveolar epithelail cells and pulmonary fibroblasts. Cell Structure Function 22:603–614

Khalil N, O'Connor RN, Flanders KC, Unruh H 1996 TGF-β1,but not TGF-β2 or TGF-β3, is defferentially present in epithelial cells of advanced pulmonary fibrosis: an immuno-histochemical study. Am J Respir Cell Mol Biol 14:131–138

Madtes DK, Elston AL, Hackman RC, Dunn AR, Clark JG 1999 Transforming growth factor-alpha deficiency reduces pulmonary fibrosis in transgenic mice. Am J Respir Cell Mol Biol 20:924–934

Otto H, Wagner H 1956 Beitrag zur Frage der Regeneration de bronchial Epithelials. Beitr Pathol (Stuttgart) 116:5436–460

Pai B, Steele VE, Nettesheim P 1983 Neoplastic transformation of primary tracheal epithelial cell cultures. Carcinogenesis 4:369–374

Terzaghi M, Nettesheim P, Williams ML 1978 Repopulation of denuded tracheal grafts with normal, preneoplastic, and neoplastic epithelial cell populations. Cancer Res 38:4546–4553

Walker C, Nettesheim P, Barrett JC, Gilmer TC 1987 Expression of a *fms*-related oncogene in carcinogen-induced neoplastic epithelial cells. Proc Natl Acad Sci USA 84:1804–1808

Wilhelm DL 1953 Regeneration of tracheal epithelium. J Pathol 65:543–550

Wright JL, Cosio M, Wiggs B, Hogg JC 1985 A morphologic grading scheme for membranous and respiratory bronchioles. Arch Pathol Lab Med 109:163–165

Macrophage control of inflammation: negative pathways of regulation of inflammatory cytokines

Alberto Mantovani*†, Marta Muzio*, Cecilia Garlanda*, Silvano Sozzani* and Paola Allavena*

*Istituto di Ricerche Farmacologiche Mario Negri, via Eritrea 62, 20157 Milan, and †Department of Biotechnology, Section of General Pathology, University of Brescia, Italy

Abstract. The recruitment of leukocytes from the blood compartment constitutes a multistep process which involves primary and secondary inflammatory cytokines, as well as adhesion molecules expressed on leukocytes and endothelial cells. The properties of the interleukin (IL)-1 system and of chemokines, as well as their interplay, are analysed. These mediators offer new paradigms to understand diverse pathologies, and provide tools and targets for the development of novel therapeutic strategies.

2001 Chronic obstructive pulmonary disease: pathogenesis to treatment. Wiley, Chichester (Novartis Foundation Symposium 234) p 120–135

Inflammatory cytokines can be distinguished as either primary or secondary mediators (Dinarello 1996, Colotta et al 1994, Muzio 1998, Mantovani et al 2000). Primary proinflammatory cytokines are a trio: interleukin (IL)-1, tumour necrosis factor (TNF) and IL-6. Primary inflammatory cytokines are extremely pleiotropic, their spectrum of action encompassing different cells and tissues. In spite of the fact that they interact with structurally different receptors, their activities overlap substantially. IL-1 and TNF are unequivocally primary inflammatory cytokines, in that they set in motion the whole cascade of mediators. IL-6 tends to be more of a secondary mediator, fundamental for the acute-phase response in the liver, and has a regulatory function. The primary inflammatory cytokines IL-1 and TNF amplify leukocyte recruitment and survival of white cells in tissues, through secondary mediators produced or acting on the vessel wall. Therefore they amplify local mechanisms of innate resistance and set the stage for the activation of specific immunity. Systemically, production of IL-6 causes the production of acute-phase proteins in the liver, such as the pentraxins, serum amyloid P component and C-reactive protein, which opsonize microorganisms and debris and activate complement, thus amplifying systemic innate immunity.

The extravasation of leukocytes from the blood compartment and their accumulation in tissues represents an essential determinant of inflammatory and immune processes (Mantovani 1999, Mantovani et al 1997). Leukocyte recruitment and accumulation from the blood into tissues is a multistep process which involves a first step of rolling and adhesion to vascular endothelium followed by the transendothelial migration and passage through basement membrane. The process of extravasation from the blood is essentially controlled by the properties of vascular endothelium and chemotactic signals coming from tissues, able to induce directional migration of leukocytes. In general, the process of leukocyte recruitment can be considerably selective, causing the preferential accumulation of one or another white blood cell population. For instance, at sites of allergic reactions eosinophils are a prominent cell population. B cells associated to mucosal tissues are enriched in cells producing IgA. In general, the selectivity of leukocyte recruitment is not determined by a single molecule, selective for one or another cell population, but is the end result of a combination of molecular determinant (adhesion molecules and chemotactic factors). Thus, the current paradigm is that of an area code model where the address is generated by a combination of numbers and letters. Chemokines are a crucial component of the current paradigm. Here we will briefly discuss the IL-1 system, as a paradigm of the complexity of primary proinflammatory cytokines, and chemokines. Emphasis will be on negative pathways of regulation.

The IL-1 system

The IL-1 system includes two agonists, α and β, a specific converting enzyme, IL-1-converting enzyme (ICE), now called caspase 1, a receptor antagonist, of which three isoforms have been cloned, and three IL-1 'receptors'. The type I receptor and the accessory protein (type III) are signal transducing molecules essential for IL-1 activity (Fig. 1). The type II receptor represents a unique pathway of negative regulation acting as a decoy for IL-1. It is expressed on the cell surface and released in medium and biological fluids. Thus, the IL-1 system includes two unique negative pathways of regulation, the receptor antagonists and the decoy receptor, which have no counterparts for other cytokines or growth factors. The presence of these negative pathways of regulation highlights the need for tight control of the action of IL-1.

From a pharmacological point of view, it is of interest that certain anti-inflammatory agents coordinatedly regulate different elements in the system. For instance, anti-inflammatory cytokines such as IL-4 and IL-13 inhibit IL-1 production, increasing production of the receptor antagonist and the expression and release of the type II decoy receptor (Fig. 1).

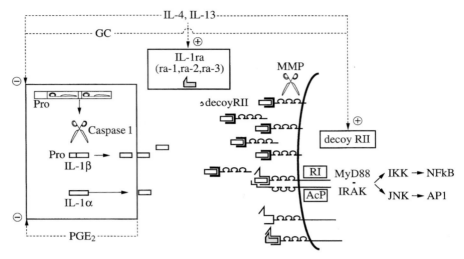

FIG. 1.   The IL-1 system. ICE, IL-1 converting enzyme; GC, glucocorticoid hormones; ra, receptor antagonist. The + and − signs indicate stimulation or inhibition of production.

Chemokines: an overview

Chemokine research is an area of intense effort, rapid progress and high expectations in terms of therapeutic results.

Several independent lines of work led to the identification of chemokines such as discussed here for monocyte chemotactic protein 1 (MCP-1). Already in the early 1970s it had been noted that supernatants of activated blood mononuclear cells contained attractants active on monocytes and neutrophils (Ward et al 1970). Subsequently a chemotactic factor active on monocytes was identified in culture supernatants of mouse (Meltzer et al 1977) and human (Bottazzi et al 1983a,b) tumour lines and called tumour-derived chemotactic factor (TDCF) (Bottazzi et al 1983a,b, 1985).

TDCF was at the time rather unique in that it was active on monocytes but not on neutrophils (Bottazzi et al 1983b) and had a low (12 kDa) molecular weight (Bottazzi et al 1983a,b). Moreover, correlative evidence suggested its involvement in the regulation of macrophage infiltration in murine and human tumours (Bottazzi et al 1983a,b, Mantovani et al 1992). A molecule with similar cellular specificity and physicochemical properties was independently identified in the culture supernatant of smooth muscle cells (SMDCF) (Valente et al 1984). The JE gene had been identified as an immediate-early PDGF-inducible gene in fibroblasts (Zullo et al 1985, Rollins et al 1988). Thus, in the mid 1980s a gene (JE) was in search of function and a monocyte-specific attractant was waiting for molecular definition. In 1989, MCP-1 was successfully purified from supernatants of a human glioma (Yoshimura et al 1989a) a human monocytic leukaemia (Matsushima et al 1989) and a human sarcoma (Zachariae et al 1990, Van Damme et al 1989, Graves et al 1989): sequencing and molecular cloning revealed its relationship with the long known JE gene (Furutani et al 1989, Yoshimura et al 1989b, Bottazzi et al 1990).

Chemokines are a superfamily of small proteins which play a crucial role in immune and inflammatory reactions and in viral infection (Baggiolini et al 1997, Hedrick & Zlotnik 1996, Rollins 1997). Most chemokines cause chemotactic migration of leukocytes, but these molecules also affect other functions such as angiogenesis, collagen production, the proliferation of haematopoietic precursors and the ontogeny of the CNS. Based on a cysteine motif, a CXC, CC, C and CX3C family have been identified. The chemokine scaffold consists of an N-terminal loop connected via Cys bonds to the more structured core of the molecule (three β sheets) with a C-terminal α helix. About 50 human chemokines have been identified.

Chemokines interact with seven-transmembrane domain, G protein-coupled receptors. Eight CC (CCR 1–8), 5 CXC (CXCR 1–5) and 1 CX3C (CX3CR1) receptors have been identified. Receptor expression is a crucial determinant of the spectrum of action of chemokines. For instance, recent results indicate that polarized T helper (Th) 1 and Th2 populations show differential receptor expression and responsiveness to chemokines (Bonecchi et al 1998a, Sallusto et al 1997, Loetscher et al 1998). Emerging evidence shows that regulation of receptor expression during activation or deactivation of monocytes is as important as regulation of chemokine production for tuning the chemokine system (Sica et al 1997, Sozzani et al 1998a).

The major eponymous function of chemokines is chemotaxis for leukocytes. Schematically, CXC (or α) chemokines are active on neutrophils (PMNs) and T lymphocytes, while CC (or β) chemokines exert their action on multiple leukocyte subtypes, including monocytes, basophils, eosinophils, T lymphocytes, dendritic cells and NK cells, but they are generally inactive on PMNs. Eotaxins (CC) represent the chemokines with the most restricted spectrum of action being selectively active on eosinophilic and basophilic granulocytes (Ben-Baruch et al 1995, Baggiolini et al 1997, Schall 1994, Sozzani et al 1996). Lymphotactin and fractalkine are the only proteins so far described with a C and CX3C motif, respectively (Kelner et al 1994, Bazan et al 1997, Pan et al 1997). They both act on lymphoid cells (T lymphocytes and NK cells) and fractalkine is also active on monocytes and PMN (Kelner et al 1994, Bazan et al 1997, Pan et al 1997, Bianchi et al 1996, S. Sozzani, P. Allavena & A. Mantovani, unpublished data).

Chemokines are redundant in their action on target cells (Table 1). No chemokine is active only on one leukocyte population and usually a given leukocyte population has receptors for and responds to different molecules

TABLE 1   Expression of chemokine receptors in leukocyte populations: a simplified view

Receptor[a]	Main ligands[b]	Main cells[c]
CCR1	MCP-3, RANTES, MIP-1α	Mo, T, NK, iDC, Neu
CCR2 B/A	MCPs (1–4)	Mo, T (act.) NK (act.)
CCR3	Eotaxin (1–3), MCP–3, RANTES	Eo, Ba, T (Th2)
CCR4	TARO, MDC	T (Th2, Tc2), NK, iDC
CCR5	MIP-1β, MIP-1α, RANTES	Mo, T (Th1, Tc1), iDC
CCR6	MIP-3α/LARC/Exodus	T, iDC (CD34)
CCR7	ELC/MIP-3β	T, Mo, mDC
CCR8	I309, TARC	T, (Th2/Tc2), Mo
CXCR1	IL-8, GCP-2	Neu
CXCR2	IL-8, GROs, NAP-2	Neu
CXCR3	IP10, MIG, ITAC	T (Th1, Tc1)
CXCR4	SDF-1	Widely expressed
CXCR5	BCA-1	B
CX3CR1	Fractalkine	Mo, NK, T

[a]CCR9 and CCR10 are identical and bind several CC chemokines. Since there is no clear evidence to date for signal transduction, this receptor should not be entitled to a CCR designation. For certain chemokines (e.g. thymus expressed chemokine, TECK, lymphotacin) the receptor(s) has (have) not been found yet.

[b]Only selected acronyms are presented. Abbreviations for chemokines: BCA-1, B cell-attracting chemokine 1; ELC, EBI (EBV-induced gene) 1-ligand chemokine (=MIP-3β/CKβ-11); GCP-2, granulocyte chemoattractant protein 2; IP-10, interferon γ inducible protein 10; GRO, growth-related oncogene; ITAC, interferon inducible T-cell a chemoattractant; LARC, liver and activation-regulated chemokine (MIP-3α/Exodus); MCP, monocyte chemotactic protein; MDC, macrophage-derived chemokine; MIG, monokine induced by IFN-γ; MIP, macrophage inflammatory protein; RANTES, regulated on activation normal T cell expressed and secreted; SDF-1, stromal cell-derived factor; SLC, secondary lymphoid tissue chemokine; TARC, thymus and activation regulated chemokine; TECK, thymus-expressed chemokine.

[c](act.), activated; Ba, basophils; DCs, dendritic cells; DC(CD34), DCs derived from CD34 cells *in vitro*, Eo, eosinophils; iDC and mDC, immature and mature dendritic cells; Mo, monocytes; Neu, neutrophils; Tc, T cytotoxic; Th, T helper.

(Table 1). Interestingly, mononuclear phagocytes, the most evolutionary ancient cell type of innate immunity, respond to the widest range of chemokines. These include most CC chemokines, fractalkine (CX3C) and certain CXC molecules (e.g. stromal cell-derived factor 1 (SDF-1) and, under certain conditions, IL-8).

The interaction of chemokines with their receptors is characterized by considerable promiscuity. Most known receptors have been reported to interact with multiple ligands and most ligands interact with more than one receptor. For instance, all four MCPs interact with CCR2, and at least MCP-2, MCP-3 and MCP-4 also recognize other receptors (CCR1 and CCR3).

Probably all cell types can produce chemokines under appropriate conditions. Two general modes of chemokine production can be defined. Molecules such as SDF-1 or macrophage-derived chemokine (MDC) are produced constitutively either by specialized cells and organs (macrophages, dendritic cells; thymus and lymphoid organs for MDC) or in a more diffuse way as for SDF-1. Most chemokines however are produced upon cell activation. Interestingly, chemokines are also produced in a redundant way (polyspeirism, πολυσ, many, οπειρω, make). Usually the same cell produces many chemokines concomitantly in response to the same stimulus. Again, polyspeirism is particularly striking for mononuclear phagocytes and endothelial cells exposed to bacterial products or primary inflammatory cytokines. In these cells, bacterial lipopolysaccharide (LPS), IL-1 and TNF elicit production of MCPs, macrophage inflammatory proteins [MIPs], RANTES (CC), fractalkine (CCX3C) and various CXC molecules (Baggiolini et al 1994, Schall 1994, Ben-Baruch et al 1995, Sozzani et al 1996, Mantovani et al 1997).

G protein-coupled receptors are a classic target in pharmacology. Given the role of chemokines in diverse human diseases, ranging from HIV infection to allergy, their pharmacology is a prime target for research. Although redundancy is a formidable stumbling block (see Table 1), available information suggests that blocking one agonist or one receptor can be beneficial in certain disease models as demonstrated for MCP-1 and its receptor CCR2 for monocytes and for eotaxins and their receptor CCR3 for eosinophils and basophils. Therefore simple chemicals with chemokine antagonistic properties or with selective inhibitory activity on chemokine production are a holy grail in present day cytokine pharmacology.

Chemokines in polarized type I and type II responses

Distinct sets of chemokines are known to control migration of T lymphocytes. Memory or activated T cells mainly express CCR1, CCR2, CXCR3 and CXCR5. On the contrary naïve T lymphocytes migrate in response to SDF1 (CXCR4) and MIP-3β (CCR7) (Sallusto et al 1998a, Baggiolini 1998), chemokines that are also active on mature dendritic cells (DCs). T cells can be subdivided in polarized type I and type II cells, depending on the spectrum of cytokines which they are able to produce. Th1 cells are characterized by production by of TNF and interferon (IFN)-γ and activate immunity based on macrophage activation and effector functions. At the other extreme of the spectrum, Th2 cells are characterized by IL-4 and IL-5 production and elicit immune responses based on the effector function of mast cells and eosinophils. The latter cell types are typically involved in allergic inflammation.

Recent results indicate that chemokines are part of the Th1 and Th2 paradigm. It was found that polarized Th1 and Th2 populations differentially express chemokine receptors. In particular, Th1 cells characteristically express high levels of CCR5 and CXCR3 whereas Th2 cells express CCR4, CCR8 and to a lesser degree, CCR3. Type I and type II CD8+ cells (Tc1 and Tc2, respectively) express a pattern of chemokine receptors very similar to that observed in Th1 and Th2 lymphocytes (D'Ambrosio et al 1998). In accordance with receptor expression, polarized Th1/Th2, as well as Tc1/Tc2 cells, differentially respond to appropriate agonists for these receptors, including, for Th1/Tc1 cells, MIP-1β and IFN-γ-induced protein (IP)-10, and for Th2/Tc2 cells MDC, 1309 and eotaxin (Bonecchi et al 1998a, Sallusto et al 1997, D'Ambrosio et al 1998). Production of IP-10 and similar CXCR3 agonists such as IFN-inducible T cell α chemoattractant (ITAC), is induced by IFN-γ. Conversely, production of eotaxin and/or MDC is induced by IL-4 and IL-13, typical Th2 cytokines (Mantovani et al 1998, Rollins 1997, Bonecchi et al 1998b). Thus, chemokines are an essential part of an amplification circuit of polarized type I and type II responses.

Regulation of chemokine receptors in tuning and shaping of the chemokine system

DCs are a heterogeneous system of leukocytes highly specialized in the priming of T cell-dependent immune responses. DCs are potent accessory cells and are believed to be indispensable to activate quiescent lymphocytes (Steinman 1991, Hart 1997, Banchereau & Steinman 1998). *In vivo* studies have demonstrated that DCs use specific pathways of tissue trafficking. Application of a fluorescent marker to skin induces migration of Langerhans cells (LCs) and a day later fluorescent LCs are found in the draining lymph nodes (Macatonia et al 1987). It has also been described that inhaled pathogens induce a very rapid mobilization of DCs in the airway epithelium (McWilliam et al 1994, 1996). The systemic treatment with LPS provokes a massive egress of DCs from several organs, such as skin, heart, kidney, intestine and cause DC migration from the marginal zone of the spleen to T cell areas (MacPherson et al 1995, Roake et al 1995, De Smedt et al 1996). After i.v. injection of inert particles, cells with DC characteristics are recruited to the hepatic sinusoid and then translocate to the hepatic lymph (Matsuno et al 1996).

Maturation of DCs can be induced *in vitro* by a variety of factors, in particular LPS and the inflammatory cytokines TNF and IL-1. Engagement of CD40 and of the newly described molecule TRANCE/RANK with their respective ligands expressed on activated T cells, also leads to maturation and activation of DCs (Banchereau & Steinman 1998, Cella et al 1997). *In vivo*, antigen uptake and the exposure of immature DCs to immune and inflammatory agonists cause the rapid mobilization of DCs from peripheral non-lymphoid tissues and activate the maturation process in these cells (Banchereau & Steinman 1998, Austyn 1996).

Previous studies performed with mononuclear phagocytes and lymphocytes have shown that cell activation is often associated with a rearrangement of the repertoire of expressed chemokine receptors. In particular, pro-inflammatory agonists (e.g. LPS, IL-1 and TNF) induce a down-modulation of certain CC chemokine receptors in monocytes, while anti-inflammatory signals, such as IL-10, have an opposite effect (Sica et al 1997, Sozzani et al 1998a, Penton-Rol et al 1998, Xu et al 1997, Tangirala et al 1997).

Recently it was shown that signals that induce maturation of DC also affect their migratory ability (Sozzani et al 1998b, Dieu et al 1998, Yanagihara et al 1998, Sallusto et al 1998b, Lin et al 1998). Exposure of DCs to LPS, IL-1 and TNF, or their culture in the presence of CD40 ligand, induces a rapid (< 1 h) inhibition of chemotactic response to MIP-1α, MIP-1β, MIP-3α, RANTES, MCP-3 and fMLP (Sozzani et al 1998b, Dieu et al 1998). As previously observed in phagocytes (Lloyd et al 1995, Sica et al 1997), inhibition of chemotaxis was followed, with a slower kinetics, by the reduction of membrane receptors and the down-regulation of mRNA receptor expression (Sozzani et al 1998b, Dieu et al 1998, Sallusto et al 1998b, Granelli-Pipemo et al 1998, Lin et al 1998). Concomitantly, in the same experimental conditions, the expression of CCR7 and the migration to its ligand MIP-3β, a chemokine constitutively expressed in lymphoid organs, were strongly up-regulated, with a maximal effect at 24 h. Up-regulation of CCR7 in DCs migrating to secondary lymphoid organs appears biologically relevant, since *in situ* hybridization analysis has shown that MIP-3β is specifically expressed in T cell-rich areas of tonsils and spleen, where mature DCs home, becoming interdigitating DCs (Dieu et al 1998, Ngo et al 1998). Receptor expression and chemotactic response to other constitutive chemokines, such as SDF-1 (CXCR4) and MDC (CCR4), were increased during maturation of DCs, in contrast to receptors for inducible chemokines, which were down-regulated (Sallusto et al 1998b, Lin et al 1998, Vecchi et al 1999).

Overall these findings provide a model for DC trafficking. Inflammatory chemokines acting through CCR1 and CCR5 may function as signals to localize DC precursors to non-lymphoid organs. On the contrary, the loss of responsiveness of DCs to the massive production of these chemokines during inflammation and immune reactions may play a permissive role for DCs to leave peripheral tissues. Meantime, the slower up-regulation of CCR7 prepares the cells to respond to MIP-3β and secondary lymphoid tissue chemokine (SLC; another ligand for CCR7) (Gunn et al 1998), expressed in lymphoid organs. The selective up-regulation of CCR7 may not be unique to this receptor inasmuch as similar results were observed with the platelet activating factor receptor (Sozzani et al

1997) and with a novel orphan receptor (S. Sozzami & A. Mantovani, unpublished results).

References

Austyn JM 1996 New insights into the mobilization and phagocytic activity of dendritic cells. J Exp Med 183:1287–1292

Baggiolini M 1998 Chemokines and leukocyte traffic. Nature 392:565–568

Baggiolini M, Dewald B, Moser B 1994 Interleukin-8 and related chemotactic cytokines CXC and CC chemokines. Adv Immunol 55:99–179

Baggiolini M, Dewald B, Moser B 1997 Human chemokines: an update. Annu Rev Immunol 15:675–705

Banchereau J, Steinman RM 1998 Dendritic cells and the control of immunity. Nature 392: 245–252

Bazan JF, Bacon KB, Hardiman G et al 1997 A new class of membrane-bound chemokine with a CX3C motif. Nature 385:640–644

Ben-Baruch A, Michiel DF, Oppenheim JJ 1995 Signals and receptors involved in recruitment of inflammatory cells. J Biol Chem 270:11703–11706

Bianchi G, Sozzani S, Zlotnik A, Mantovani A, Allavena P 1996 Migratory response of human natural killer cells to lymphotactin. Eur J Immunol 26:3238–3241

Bonecchi R, Bianchi G, Bordignon PP et al 1998a Differential expression of chemokine receptors and chemotactic responsiveness of type 1 T helper cells (Th1s) and Th2s. J Exp Med 187: 129–134

Bonecchi R, Sozzani S, Stine J et al 1998b Divergent effects of interleukin-4 and interferon-gamma on macrophage-derived chemokine production: an amplification circuit of polarazied T helper 2 responses. Blood 92:2668–2671

Bottazzi B, Polentarutti N, Acero R et al 1983a Regulation of the macrophage content of neoplasms by chemoattractants. Science 220:210–212

Bottazzi B, Polentarutti N, Balsari A et al 1983b Chemotactic activity for mononuclear phagocytes of culture supernatants from murine and human tumor cells: evidence for a role in the regulation of the macrophage content of neoplastic tissues. Int J Cancer 31:55–63

Bottazzi B, Ghezzi P, Taraboletti G et al 1985 Tumor-derived chemotactic factor(s) from human ovarian carcinoma: evidence for a role in the regulation of macrophage content of neoplastic tissues. Int J Cancer 36:167–173

Bottazzi B, Colotta F, Sica A, Nobili N, Mantovani A 1990 A chemoattractant expressed in human sarcoma cells (tumor-derived chemotactic factor, TDCF) is identical to monocyte hemoattractant protein-1/monocyte chemotactic and activating factor (MCP-1/MCAF). Int J Cancer 45:795–797

Cella M, Sallusto F, Lanzavecchia A 1997 Origin, maturation and antigen presenting function of dendritic cells. Curr Opin Immunol 9:10–16

Colotta F, Dower SK, Sims JE, Mantovani A 1994 The type II 'decoy' receptor: novel regulatory pathway for interleukin 1. Immunol Today 15:562–566

D'Ambrosio D, Iellem A, Bonecchi R et al 1998 Selective up-regulation of chemokine receptors CCR4 and CCR8 upon activation of polarized human type 2 Th cells. J Immunol 161: 5111–5115

De Smedt T, Pajak B, Muraille E et al 1996 Regulation of dendritic cell numbers and maturation by lipopolysaccharide in vivo. J Exp Med 184:1413–1424

Dieu MC, Vanbervliet B, Vicari A et al 1998 Selective recruitment of immature and mature dendritic cells by distinct chemokines expressed in different anatomic sites. J Exp Med 188:373–386

Dinarello CA 1996 Biologic basis for interleukin-1 in disease. Blood 87:2095–2147

Furutani Y, Nomura H, Notake M et al 1989 Cloning and sequencing of the cDNA for human monocyte chemotactic and activating factor (MCAF). Biochem Biophys Res Commun 159:249–255

Granelli-Piperno A, Delgado E, Finkel V, Paxton W, Steinman RM 1998 Immature dendritic cells selectively replicate macrophagetropic (M-tropic) human immunodeficiency virus type 1, while mature cells efficiently transmit both M- and T-tropic virus to T cells. J Virol 72:2733–2737

Graves DT, Jiang YL, Williamson MJ, Valente AJ 1989 Identification of monocyte chemotactic activity produced by malignant cells. Science 245:1490–1493

Gunn MD, Tangemann K, Tam C, Cyster JG, Rosen SD, Williams LT 1998 A chemokine expressed in lymphoid high endothelial venules promotes the adhesion and chemotaxis of naive T lymphocytes. Proc Natl Acad Sci USA 95:258–263

Hart DNJ 1997 Dendritic cells: unique leukocyte populations which control the primary immune response. Blood 90:3245–3287

Hedrick JA, Zlotnik A 1996 Chemokines and lymphocyte biology. Curr Opin Immunol 8: 343–347

Kelner GS, Kennedy J, Bacon KB et al 1994 Lymphotactin: a cytokine that represents a new class of chemokine. Science 266:1395–1399

Lin CL, Suri RM, Rahdon RA, Austyn JM, Roake JA 1998 Dendritic cell chemotaxis and transendothelial migration are induced by distinct chemokines and are regulated on maturation. Eur J Immunol 28:4114–4122

Lloyd AR, Biragyn A, Johnston JA et al 1995 Granulocyte-colony stimulating factor and lipopolysaccharide regulate the expression of interleukin 8 receptors on polymorphonuclear leukocytes. J Biol Chem 270:28188–28192

Loetscher P, Uguccioni M, Bordoli L et al 1998 CCR5 is characteristic of Th1 lymphocytes. Nature 391:344–345

Macatonia SE, Knight SC, Edwards AJ, Griffiths S, Fryers P 1987 Localization of antigen on lymph node dendritic cells after exposure to the contact sensitizer fluorescein isothiocyanate. Functional and morphological studies. J Exp Med 166:1654–1667

MacPherson GG, Jenkins CD, Stein MJ, Edwards C 1995 Endotoxin-mediated dendritic cell release from the intestine. Characterization of released dendritic cells and TNF dependence. J Immunol 154:1317–1322

Mantovani A 1999 The chemokine system: redundancy for robust outputs. Immunol Today 20:254–257

Mantovani A, Bottazzi B, Colotta F, Sozzani S, Ruco L 1992 The origin and function of tumor-associated macrophages. Immunol Today 13:265–270

Mantovani A, Bussolino F, Introna M 1997 Cytokine regulation of endothelial cell function: from molecular level to the bedside. Immunol Today 18:231–240

Mantovani A, Allavena P, Vecchi A, Sozzani S 1998 Chemokines and chemokine receptors during activation and deactivation of monocytes and dendritic cells and in amplification of Th1 versus Th2 responses. Int J Clin Lab Res 28:77–82

Mantovani A, Dinarello C, Ghezzi P 2000 Pharmacology of cytokines. Oxford University Press, Oxford

Matsuno K, Ezaki T, Kudo S, Uehara Y 1996 A life stage of particle-laden rat dendritic cells *in vivo*: their terminal division, active phagocytosis, and translocation from the liver to the draining lymph. J Exp Med 183:1865–1878

Matsushima K, Larsen CG, DuBois GC, Oppenheim JJ 1989 Purification and characterization of a novel monocyte chemotactic and activating factor produced by a human myelomonocytic cell line. J Exp Med 169:1485–1490

McWilliam AS, Nelson D, Thomas JA, Holt PG 1994 Rapid dendritic cell recruitment is a hallmark of the acute inflammatory response at mucosal surfaces. J Exp Med 179:1331–1336

McWilliam AS, Napoli S, Marsh AM et al 1996 Dendritic cells are recruited into the airway epithelium during the inflammatory response to a broad spectrum of stimuli. J Exp Med 184:2429–2432

Meltzer MS, Stevenson MM, Leonard EJ 1977 Characterization of macrophage chemotaxis in tumor cell cultures and comparison with lymphocyte-derived chemotactic factors. Cancer Res 37:721–725

Muzio M 1998 Signalling by proteolysis: death receptors induce apoptosis. Int J Clin Lab Res 28:141–147

Ngo VN, Tang HL, Cyster JG 1998 Epstein–Barr virus-induced molecule 1 ligand chemokine is expressed by dendritic cells in lymphoid tissues and strongly attracts naive T cells and activated B cells. J Exp Med 188:181–191

Pan Y, Lloyd C, Zhou H et al 1997 Neurotactin, a membrane-anchored chemokine upregulated in brain inflammation. Nature 387:611–617 (erratum: 1997 Nature 389:100)

Penton-Rol G, Polentarutti N, Luini W et al 1998 Selective inhibition of expression of the chemokine receptor CCR2 in human monocytes by IFN-γ. J Immunol 160: 3869–3873

Roake JA, Rao AS, Morris PJ, Larsen CP, Hankins DF, Austyn JM 1995 Dendritic cell loss from nonlymphoid tissues after systemic administration of lipopolysaccharide, tumor necrosis factor, and interleukin 1. J Exp Med 181:2237–2247

Rollins BJ 1997 Chemokines. Blood 90:909–928

Rollins BJ, Morrison ED, Stiles CD 1988 Cloning and expression of JE, a gene inducible by platelet-derived growth factor and whose product has cytokine-like properties. Proc Natl Acad Sci USA 85:3738–3742

Sallusto F, Mackay CR, Lanzavecchia A 1997 Selective expression of the eotaxin receptor CCR3 by human T helper 2 cells. Science 277:2005–2007

Sallusto F, Lanzavecchia A, Mackay CR 1998a Chemokines and chemokine receptors in T-cell priming and Th1/Th2-mediated responses. Immunol Today 19:568–574

Sallusto F, Schaerli P, Loetscher P et al 1998b Rapid and coordinated switch in chemokine receptor expression during dendritic cell maturation. Eur J Immunol 28:2760–2769

Schall TJ 1994 The chemokines. In: Thomson A (ed) The cytokine handbook, 2nd edn. Academic Press, London, p 419–460

Sica A, Saccam A, Borsatti A et al 1997 Bacterial lipopolysaccharide rapidly inhibits expression of C-C chemokine receptors in human monocytes. J Exp Med 185:969–974

Sozzani S, Locati M, Allavena P, Van Damme J, Mantovani A 1996 Chemokines: a superfamily of chemotactic cytokines. Int J Clin Lab Res 26:69–82

Sozzani S, Longoni D, Bonecchi R et al 1997 Human monocyte-derived and CD34+ cell-derived dendritic cells express functional receptors for platelet activating factor. FEBS Lett 418: 98–100

Sozzani S, Ghezzi S, Iannolo G et al 1998a Interleukin 10 increases CCR5 expression and HIV infection in human monocytes. J Exp Med 187:439–444

Sozzani S, Allavena P, D'Amico G et al 1998b Differential regulation of chemokine receptors during dendritic cell maturation: a model for their trafficking properties. J Immunol 161:1083–1086

Steinman RM 1991 The dendritic cell system and its role in immunogenicity. Annu Rev Immunol 9:271–296

Tangirala RK, Murao K, Quehenberger O 1997 Regulation of expression of the human monocyte chemotactic protein-1 receptor (hCCR2) by cytokines. J Biol Chem 272:8050–8056

Valente A J, Fowler SR, Sprague EA, Kelley JL, Suenram CA, Schwartz CJ 1984 Initial characterization of a peripheral blood mononuclear cell chemoattractant derived from cultured arterial smooth muscle cells. Am J Pathol 117:409–417

Van Damme J, Decock B, Lenaerts JP et al 1989 Identification by sequence analysis of chemotactic factors for monocytes produced by normal and transformed cells stimulated with virus, double-stranded RNA or cytokine. Eur J Immunol 19:2367–2373

Vecchi A, Massimiliano L, Ramponi S et al 1999 Differential responsiveness to constitutive versus inflammatory chemokines of immature and mature mouse dendritic cells. J Leukoc Biol 66:489–494

Ward PA, Remold HG, David JR 1970 The production by antigen-stimulated lymphocytes of a leukotactic factor distinct from migration inhibitory factor. Cell Immunol 1:162–174

Xu L, Rahimpour R, Ran L et al 1997 Regulation of CCR2 chemokine receptor mRNA stability. J Leukoc Biol 62:653–660

Yanagihara S, Komura E, Nagafune J, Watarai H, Yamaguchi Y 1998 EBI1/CCR7 is a new member of dendritic cell chemokine receptor that is up-regulated upon maturation. J Immunol 161:3096–3102

Yoshimura T, Robinson E A, Tanaka S, Appella E, Kuratsu J, Leonard EJ 1989a Purification and amino acid analysis of two human glioma-derived monocyte chemoattractants. J Exp Med 169:1449–1459

Yoshimura T, Yuhki N, Moore SK, Appella E, Lerman MI, Leonard EJ 1989b Human monocyte chemoattractant protein-1 (MCP-1). Full-length cDNA cloning, expression in mitogen-stimulation blood mononuclear leukocytes, and sequence similarity to mouse competence gene JE. FEBS Lett 244:487–493

Zachariae CO, Anderson AO, Thompson HL et al 1990 Properties of monocyte chemotactic and activating factor (MCAF) purified from a human fibrosarcoma cell line. J Exp Med 171: 2177–2182

Zullo JN, Cochran BH, Huang AS, Stiles CD 1985 Platelet-derived growth factor and double-stranded ribonucleic acids stimulate expression of the same genes in 3T3 cells. Cell 43: 793–800

DISCUSSION

MacNee: In which diseases is this processes dysregulated, and how could you then relate that to chronic obstructive pulmonary disease (COPD)? Does this occur in all inflammation?

Mantovani: I should have mentioned that for the MDC part of the work, which we expected to be relevant for type II responses such as allergy, there is *in vivo* evidence for overproduction of MDC. Recent work done at Millennium Pharmaceuticals suggests that anti-MDC antibodies block bronchial hyperreactivity (Lloyd et al 2000).

Williams: Apparently, in the early stages after sensitization and challenge, they found that eotaxin was the main recruiting agent for Th2 cells, but in the later stages it was MDC. Does this throw any light on the studies you have discussed with Christine Power about the CCR4 knockout animals? Do you know what they found?

Mantovani: I know of negative findings. They didn't find any protection against bronchial hyperreactivity. There are two possible explanations. One is that there is another receptor for MDC (we have evidence for this), the other is that a more appropriate genetic background has masked the role of CCR4. Finally, adaptation may have occurred in knockout mice. We should wait to see the actual data.

Williams: The findings you discussed at the end, where you can actually change the receptor expression on a particular leukocyte type and make it then respond to another chemokine towards which it was previously unresponsive, complicate the issue no end. A system has evolved where there are selective signals, i.e. chemokines, for particular cell types, but after particular stimuli this selectivity is lost and the cell now responds to a range of other chemokines. Why is this?

Mantovani: My explanation is that if you take an interferon-dominated type I inflammatory reaction, you will use eotaxin as an attractant for neutrophils. There is good evidence that neutrophils can play a role in activating and sustaining type I inflammation. This is a strategy where a non-canonical chemokine is utilized to do the job that is required under that particular condition. I would read the data on CXCR1/CXCR2 induction in monocytes in the same way.

Stolk: We have done a histological study on tissue from lung cancer patients with and without COPD. We found MCP-1 mRNA and protein were up-regulated as well as CCR2 on bronchial epithelial cells and macrophages. Furthermore, we found a correlation between IL-8 and COPD in the same tissue blocks. Would you predict that interfering with MCP-1 by using MCP-1 antagonists could be an option for clinical intervention?

Mantovani: Yes, it is generally thought that blocking MCP-1 might be an interesting strategy for various conditions. There is unpublished evidence that blocking MCP-1 may work in blocking type II inflammatory responses. It has been speculated that in coronary artery disease MCP-1 is a dominant mediator of the monocyte recruitment that is an initial step in arteriosclerosis. This is supported by animal models and pathological findings. There is evidence that statins are protective not only because they lower cholesterol, but because they are effective in blocking MCP-1 production.

Stolk: Do you also think that there is a chance that another mechanism may pop up to bypass these phenomena?

Mantovani: That is a question of general validity for chemokines. Many of the chemokines are functionally redundant. There are four MCPs, which all interact with CCR2. It was logical to expect that knocking out CCR2 would result in compromised mononuclear phagocyte recruitment. We were surprised to find that knocking out just MCP-1 gave mice with essentially the same phenotype. So in spite of this system being redundant, there is good reason to test the hypothesis that there may be therapeutic benefit by blocking one receptor and one agonist.

MacNee: What is the time course of release of MCP-1?

Mantovani: Hours.

MacNee: We studied the effects of acute and chronic cigarette smoking on various chemokines in bronchoalevolar lavage (BAL). Groα was the only one that turned out to be significantly higher in response to acute cigarette smoking. These subjects smoked two cigarettes, and one hour later we performed BAL. At this time point would one expect MCP-1 to be expressed and secreted?

Mantovani: The gene is certainly expressed after 30 minutes. Protein production will start shortly after.

Williams: Was that in BAL samples?

MacNee: Yes.

Williams: My colleagues have found that there is a lot of differential binding of different chemokines to components of sputum, so what you find in BAL is not necessarily what is generated in tissue.

MacNee: We always discuss the effects of IL-1 and TNF together. Are there different effects of IL-1 and TNF on cell activation?

Mantovani: Initially the two signalling pathways are distinct, but they both converge at the level of NIC. The two molecules are by and large synergistic. One can speculate that this reflects the fact that the two pathways are initially divergent.

MacNee: The reason I ask this is that we have an interest in the initiation of neutrophil sequestration in the pulmonary microcirculation, as a result of shape and deformability changes in the cells. It is interesting that IL-1 has no effect on neutrophil deformability, and yet TNF has a marked and fairly rapid effect on this neutrophil deformability which is a fairly good way of measuring cell activation. It is interesting that IL-1 is the one chemokine that has no effect on neutrophil deformability.

Mantovani: TNF has long been known to be an attractant for phagocytes, whereas IL-1 is completely inactive.

Williams: Are there any examples of decoy receptors for chemokines? You seem to have these decoy receptors to regulate the primary cytokines, but as far as I know the same thing does not occur at the level of the chemokines. I cannot see why one would be regulated one way and the other is not.

Mantovani: There are no examples. There is one receptor called DARC that is present on red cells, and it is recognized by *Plasmodium*. Nobody knows what it does, but apparently it does not signal.

Williams: That isn't quite the same, because we are talking about local regulation in tissues, which is extravascular, rather that in the circulation.

Mantovani: We think that there are functional decoys in the chemokine system. We are working on this.

Stolk: I would like to come back to the MCP-1 data. Professor Ziegler-Heitbrock in Munich states that there are certain subtypes of monocytes and macrophages involved in atherosclerosis, such as CD14- and CD16-positive cells that are important in the pathogenesis of atherosclerosis. Do you think that in COPD we should be looking at subsets as well as where these mechanisms that you describe are operating more aggressively?

Mantovani: The subset story that you are referring to may have a counterpart in COPD in terms of receptors and responsiveness towards MCP-1. There is evidence that the CD14/CD16-positive monocyte has higher levels of CCR2 and is more responsive to MCP-1. It would be logical to assume that this is the first monocyte to enter the tissue.

Nadel: I wanted to ask about the possible mechanism of interaction of IL-4 with monocyte-macrophages in the lungs. The reason for asking is that IL-4 induces 15-lipoxygenase in certain cells, and it induces the production and release of this enzyme in macrophages in coronary lesions and blood vessels in general. It is believed that the effects of this cascade are involved in the subsequent inflammatory responses in vascular disease. Is there an analogy to this in alveoli? Has anyone looked in the alveoli at the role of IL-4 and its interaction with macrophages?

Shapiro: All I know is that the thing we look at is when isolated macrophages are cultured with IL-4, this down-regulates macrophage metalloproteinases.

Dunnill: Dr Mantovani, you made an intriguing remark disagreeing about the lifespan of the leukocytes and macrophages. What did you mean by this?

Mantovani: One should never underestimate the lifespan of a neutrophil! The lifespan of a neutrophil *in vitro* (and I would assume the same is true *in vivo*) is dramatically affected by the context. Long ago, we and others found that cytokines (e.g. IL-1) and bacterial products dramatically prolong the survival of neutrophils.

Nadel: In the tissue, as far as I can tell from the literature and our own studies, neutrophils 'squirt out' from the postcapillary venules very close to the surface of the epithelium in airway tissue. They come from just under the basement membrane, and in a very short period of time they are through the epithelium and disappear into the lumen. When neutrophils are recruited into the airways by a stimulus, generally recruitment occurs within 1–4 h.

Jeffery: Isn't that time course modulated by whether or not ICAM-1 in the epithelium is up-regulated? I could imagine that tissue residence time would be extended by ICAM-1 up-regulation. This is supported by biopsy results particularly in patients with COPD where their neutrophils are apparently absent from subepithelial areas, but accumulate in consequence to recruitment, migration and retention of up-regulated epithelial ICAM-1.

Nadel: In animals, when ICAM-1 is up-regulated, the effects of neutrophils are changed.

Reference

Lloyd CM, Delaney T, Nguyen T et al 2000 CC chemokine receptor (CCR)3/eotaxin is followed by CCR4/monocyte-derived chemokine in mediating pulmonary T helper lymphocyte type 2 recruitment after serial antigen challenge *in vivo*. J Exp Med 191:265–274

Neutrophils in chronic obstructive pulmonary disease

Timothy J. Williams and Peter J. Jose

Leukocyte Biology Section, Biomedical Sciences Division, Sir Alexander Fleming Building, Imperial College School of Medicine, South Kensington, London SW7 2AZ, UK

Abstract. Neutrophil accumulation in the lung is a prominent feature of chronic obstructive pulmonary disease (COPD) and the activation of these cells, producing proteases and oxygen-derived free radicals, is thought to be important in the pathogenesis of the disease. An important step in recruitment is the local generation of a neutrophil chemoattractant signal which mediates the trapping and firm adhesion of rolling neutrophils on the microvascular endothelium, followed by migration via intercellular junctions. Two neutrophil chemoattractants are particularly important in this respect, C5a generated by cleavage of complement C5 in interstitial fluid, and interleukin (IL)-8 synthesized by cells in the lung, e.g. macrophages, epithelial cells, endothelial cells, smooth muscle cells and neutrophils themselves. Lipid mediators, such as leukotriene B4 (LTB4), are also potentially important. Several studies have been carried out to investigate the role of IL-8 in COPD. IL-8 has been detected in bronchoalveolar lavage fluid and sputum from such subjects and in the systemic circulation. The levels of IL-8 have been found to correlate with neutrophil numbers and markers of neutrophil activation, such as myeloperoxidase activity. Some studies have also found a correlation between IL-8 levels, neutrophil numbers and the degree of lung dysfunction. These parameters are insensitive to steroids. Thus, the mechanisms involved in neutrophil recruitment, i.e. chemoattractant secretion or action, adhesion and endothelial transmigration, are important potential targets for the development of novel therapy. The IL-8 receptors on neutrophils, CXCR1 and CXCR2, are of particular interest.

2001 Chronic obstructive pulmonary disease: pathogenesis to treatment. Wiley, Chichester (Novartis Foundation Symposium 234) p 136–148

The neutrophil is strongly implicated in the pathogenesis of chronic obstructive pulmonary disease (COPD) and thus there is considerable interest in mechanisms underlying its recruitment into the lung as these mechanisms can provide targets for the development of novel therapy. Neutrophils are found in the lung tissue and in sputum in COPD and it is thought that activation of these cells to release activated oxygen species and proteases such as elastase is important in tissue damage. Therapy aimed at suppressing neutrophil recruitment is, of course, potentially hazardous because of the importance of neutrophils in host defence to pathogens. This is evidenced by the difficulty of rare individuals with leukocyte adhesion deficiency to cope with infections.

In most microvascular beds, including the tracheobronchial microcirculation, a locally generated chemoattractant triggers the immobilization of rolling neutrophils on the venular endothelium. Rolling is mediated by interactions between selectins and their complementary receptors and firm adhesion by interactions between $\beta 2$ integrins on the neutrophil surface and ICAMs on the endothelium. This is then followed by migration through inter-endothelial cell junctions into the tissues. Neutrophil recruitment via the pulmonary microvasculature appears to be different. Here neutrophils are trapped and migrate from alveolar capillaries. Observed differences between mechanisms of recruitment in different animal models, e.g. differential dependence on $\beta 2$ integrins depending on the inflammatory stimulus (Doerschuk et al 1990), may reflect differences in the route of recruitment via these two microvascular beds.

Interleukin 8

There is currently considerable interest in interleukin (IL)-8 as an important mediator of neutrophil recruitment in COPD. IL-8 has been detected by ELISA in lung tissue and in the circulation of patients with COPD and the levels of IL-8 have been found to correlate with neutrophil accumulation (Riise et al 1995, Nocker et al 1996, Keman et al 1997, Yamamoto et al 1997, Mikami et al 1998, Pesci et al 1998).

IL-8 is a potent neutrophil chemoattractant CXC chemokine first detected as a secretory product of stimulated macrophages (Yoshimura et al 1987, Schroder et al 1987, Walz et al 1987). It is produced by many different cell types given appropriate stimulus and stimulates two different seven-transmembrane, G protein-coupled receptors on the neutrophil, CXCR1 and CXCR2. Several other structurally related CXC chemokines are also produced which act primarily on CXCR2. IL-8 also binds to a non-signalling seven-transmembrane receptor on red blood cells, the Duffy antigen. This may act as a sink to prevent the build up of chemokine in the circulation which could otherwise result in systemic activation of circulating neutrophils and thus compromise local recruitment to an inflammatory site.

C5a and its relationship to IL-8

C5a is a potent neutrophil chemoattractant protein with a rather longer history than IL-8. C5a is generated upon cleavage of the complement component C5 during complement activation. Thus, C5a can be produced far more rapidly than IL-8 which requires synthesis and secretion from cells. If tissue is sampled during on-going chronic inflammation it is not possible to determine the relationship between C5a and IL-8 generation. However, we have found in defined responses in animal inflammatory models that C5a is generated before IL-8 appears and that C5a production may often be a prerequisite for IL-8 production.

In a model of experimental peritonitis in rabbits induced by intraperitoneal injection of zymosan, we found that fluid exudate would induce increased microvascular permeability and neutrophil accumulation when assayed in rabbit skin *in vivo* (Forrest et al 1986). The critical mediator in the first 2 hours was found to be C5a (Williams & Jose, 1981, Jose et al 1981) which we discovered induces increased microvascular permeability by a neutrophil-dependent process (Wedmore & Williams 1981). All the permeability-increasing and neutrophil chemoattractant activity in exudate taken up to 2 hours could be neutralized by an antibody to C5a (Collins et al 1991). After 2 hours the exudate contained activity that was clearly not neutralized by anti-C5a. This activity was purified and found to be associated with two proteins, rabbit permeability factor (RPF)-1 and RPF-2. These proteins were sequenced, revealing that RPF-1 and RPF-2 were the rabbit homologues of human IL-8 and the related CXC chemokine, melanoma growth-stimulating activity (MGSA) (also known as Groα), respectively (Collins et al 1991, Beaubien et al 1990, Jose et al 1991).

A very similar relationship between C5a and IL-8 was found in a model of myocardial infarction in the rabbit. In this case a coronary artery was occluded for a fixed time and reperfusion allowed for different time periods. Myocardial tissue was extracted and measured for myeloperoxidase activity, as a marker of neutrophil accumulation, C5a and IL-8 using immunoassays. The neutrophil accumulation was again found to be associated with the immediate appearance of C5a and the delayed appearance of IL-8 (Ivey et al 1995). Interestingly, when circulating neutrophils were depleted using a pre-treatment of nitrogen mustard, this had no effect on the levels of C5a but virtually abolished the production of IL-8. Thus, C5a appears to be responsible for bringing in a first phase of neutrophils and these cells themselves seem to be the source of the IL-8 responsible for continuing neutrophil recruitment (Ivey et al 1995).

These results suggest that the neutrophil itself can be an important source of IL-8 *in vivo*. This is supported by the observation that neutrophils secrete IL-8 during phagocytosis (Bazzoni et al 1991, Au et al 1994). Evidence was obtained that interaction between the neutrophil and zymosan particles switches on the IL-8 gene and that this can be blocked by antibodies to β2 integrins and platelet activating factor (PAF) antagonists (Au et al 1994). Further experiments showed that co-incubation of neutrophils and isolated cardiac myocytes also resulted in IL-8 generation (Chivers 1999). In addition, a model has been established in the rabbit in which zymosan is used to stimulate an inflammatory response in the lung (Chivers 1999). IL-8 could be detected in lung extracts and bronchoalveolar lavage fluid, and again depletion of circulating neutrophils prevented the production of IL-8. This suggests that neutrophils are a major source of IL-8 in the lung in this model, or that a factor released by neutrophils can stimulate IL-8 production by other cells in the lung.

Conclusion

There is considerable evidence that IL-8 is important for neutrophil recruitment into the lung in COPD. Recent results show that antibodies to IL-8 neutralize only some of the neutrophil chemotactic activity detected in sputum from such patients (Mikami et al 1998). Although leukotriene B4 (LTB4) may account for some of the deficit (Mikami et al 1998) it may be that C5a is also an important contributor, as indicated by the results of the animal model studies described here.

These observations provide potential targets for therapeutic intervention in COPD. As pointed out earlier, such intervention is fraught with potential hazards as the neutrophil is essential for host defence against common pathogens. However, it may be possible to attenuate neutrophil accumulation and activation in order to limit neutrophil-mediated damage to the lung. One possibility is to block the IL-8 receptor(s) and one such study, using a small molecule CXCR2 antagonist, has been published by SmithKline Beecham (White et al 1998). Another possibility is to block the production of IL-8 and the results presented here suggest that a C5a antagonist, as well as providing one means of blocking the neutrophil accumulation, may secondarily block IL-8 production. Thus, this second approach may provide a more effective means of suppressing neutrophil accumulation. The observation showing the effectiveness of PAF antagonists on IL-8 generation also suggests that other pathways could be investigated to suppress neutrophil recruitment (Au et al 1994). Finally, it has been shown that phosphodiesterase IV (PDE IV) inhibitors are able to suppress IL-8 production and that this effect is markedly enhanced by agents that elevate cAMP such as prostaglandins or β-adrenoceptor agonists (Au et al 1998). This could be a contributing factor to the effectiveness of the PDE IV inhibitor as shown in the recent clinical trial using Ariflo (Torphy et al 1999).

References

Au BT, Williams TJ, Collins PD 1994 Zymosan-induced interleukin-8 release from human neutrophils involves activation via the CD11b/CD18 receptor and endogenous platelet activating factor as an autocrine modulator. J Immunol 152:5411–5419

Au BT, Teixeira MM, Collins PD, Williams TJ 1998 Effect of PDE4 inhibitors on zymosan-induced IL-8 release from human neutrophils: synergism with prostanoids and salbutamol. Br J Pharmacol 123:1260–1266

Bazzoni F, Cassatella MA, Rossi F, Ceska M, Dewald B, Baggiolini M 1991 Phagocytosing neutrophils produce and release high amounts of the neutrophil-activating peptide 1/ interleukin 8. J Exp Med 173:771–774

Beaubien BC, Collins PD, Jose PJ et al 1990 A novel neutrophil chemoattractant generated during an inflammatory reaction in the rabbit peritoneal cavity *in vivo*. Purification, partial amino acid sequence and structural relationship to interleukin 8. Biochem J 271:797–801

Chivers S 1999 Generation of the neutrophil chemoattractant interleukin-8 in inflammatory models of the rabbit heart and lung. PhD thesis, University of London, UK

Collins PD, Jose PJ, Williams TJ 1991 The sequential generation of neutrophil chemoattractant proteins in acute inflammation in the rabbit *in vivo*. Relationship between C5a and a protein with the characteristics of IL-8. J Immunol 146:677–684

Doerschuk CM, Winn RK, Coxson HO, Harlan JM 1990 CD18-dependent and -independent mechanisms of neutrophil emigration in the pulmonary and systemic microcirculation of rabbits. J Immunol 144:2327–2333

Forrest MJ, Jose PJ, Williams TJ 1986 Kinetics of the generation and action of chemical mediators in zymosan-induced inflammation of the rabbit peritoneal cavity. Br J Pharmacol 89:719–730

Ivey CL, Williams FM, Collins PD, Jose PJ, Williams TJ 1995 Neutrophil chemoattractants generated in two phases during reperfusion of ischemic myocardium in the rabbit. Evidence for a role for C5a and interleukin-8. J Clin Invest 95:2720–2728

Jose PJ, Forrest MJ, Williams TJ 1983 Detection of the complement fragment C5a in inflammatory exudates from the rabbit peritoneal cavity using radioimmunoassay. J Exp Med 158:2177–2182

Jose PJ, Collins PD, Perkins JA et al 1991 Identification of a second neutrophil-chemoattractant cytokine generated during an inflammatory reaction in the rabbit peritoneal cavity *in vivo*. Purification, partial amino acid sequence and structural relationship to melanoma-growth-stimulatory activity. Biochem J 278:493–497

Keman S, Willemse B, Tollerud DJ, Guevarra L, Schins RP, Borm PJ 1997 Blood interleukin-8 production is increased in chemical workers with bronchitic symptoms. Am J Ind Med 32:670–673

Mikami M, Llewellyn-Jones CG, Stockley RA 1998 The effect of interleukin-8 and granulocyte macrophage colony stimulating factor on the response of neutrophils to formyl methionyl leucyl phenylalanine. Biochim Biophys Acta 1407:146–154

Nocker RE, Schoonbrood DF, van de Graaf EA et al 1996 Interleukin-8 in airway inflammation in patients with asthma and chronic obstructive pulmonary disease. Int Arch Allergy Immunol 109:183–191

Pesci A, Balbi B, Majori M et al 1998 Inflammatory cells and mediators in bronchial lavage of patients with chronic obstructive pulmonary disease. Eur Respir J 12:380–386

Riise GC, Ahlstedt S, Larsson S et al 1995 Bronchial inflammation in chronic bronchitis assessed by measurement of cell products in bronchial lavage fluid. Thorax 50:360–365

Schroder JM, Mrowietz U, Morita E, Christophers E 1987 Purification and partial biochemical characterization of a human monocyte-derived, neutrophil-activating peptide that lacks interleukin 1 activity. J Immunol 139:3474–3483

Torphy TJ, Barnette MS, Underwood DC et al 1999 Ariflo (SB 207499), a second generation phosphodiesterase 4 inhibitor for the treatment of asthma and COPD: from concept to clinic. Pulm Pharmacol Ther 12:131–135

Walz A, Peveri P, Aschauer H, Baggiolini M 1987 Purification and amino acid sequencing of NAF, a novel neutrophil-activating factor produced by monocytes. Biochem Biophys Res Commun 149:755–761

Wedmore CV, Williams TJ 1981 Control of vascular permeability by polymorphonuclear leukocytes in inflammation. Nature 289:646–650

White JR, Lee JM, Young PR et al 1998 Identification of a potent, selective non-peptide CXCR2 antagonist that inhibits interleukin-8-induced neutrophil migration. J Biol Chem 273:10095–10098

Williams TJ, Jose PJ 1981 Mediation of increased vascular permeability after complement activation: histamine-independent action of rabbit C5a. J Exp Med 153:136–153

Yamamoto C, Yoneda T, Yoshikawa M et al 1997 Airway inflammation in COPD assessed by sputum levels of interleukin-8. Chest 112:505–510

Yoshimura T, Matsushima K, Tanaka S et al 1987 Purification of a human monocyte-derived neutrophil chemotactic factor that has peptide sequence similarity to other host defense cytokines. Proc Natl Acad Sci USA 84:9233–9237

DISCUSSION

MacNee: I would be interested in your comments on the relationship of the events you described to the stimulus, and then also on the differences that might occur between the bronchial circulation and the pulmonary circulation for these events. For example, if you give LPS or any other stimulus intratracheally rather than intravenously, this produces an influx of neutrophils into the airspaces and increased epithelial permeability. However, if the neutrophil influx is blocked by a neutrophil antibody, this doesn't prevent the increased epithelial permeability.

Williams: Clearly the mechanisms for the pulmonary circulation are a little different from those of the bronchial circulation. The bronchial circulation probably resembles the peripheral circulation more. Claire Doerschuk has done interesting work showing that there is a differential CD18 dependence of neutrophil accumulation in the lung depending on the stimulus. This may relate to the variable contribution of the pulmonary circulation and bronchial circulation in different situations. There are also clearly different routes for changing the barrier function of the microvascular endothelium. I gave one example of a neutrophil chemoattractant that has a dramatic effect on the permeability characteristics of the endothelium; an effect that is neutrophil-dependent, but there are also other situations. We looked at vascular permeability factor (VPF), which gives a potent permeability effect that is completely neutrophil-independent in our experiments. There may be a whole range of unidentified mediators that are involved in changing the barrier function of the microvascular endothelium. All these processes may be occurring together in response to complex inflammatory stimulus. There may also be situations where both the bronchial and pulmonary circulation are being affected. This will complicate the issue.

Hogg: I find the work on the marrow that you reported in the *Journal of Experimental Medicine* was very interesting (Palframan et al 1998). You use the guinea pig and we use the rabbit with a slightly different technique. We administer a bolus of BrdU to label the mitotic pool and then measure the appearance of labelled cells in the blood (Terashima et al 1996). We saw an effect with cigarette smoking which we think is relevant to the fact that smokers have higher white cell counts

(Terashima et al 1999). A similar effect is observed when alveolar macrophages are exposed to particles (PM10s) *in vitro* and the supernatants are placed in the lung (Terashima et al 1997). We are trying to figure out what is released by the alveolar macrophages that is having this effect. We have begun to look for responsible mediators and I wonder if you have considered IL-6?

Williams: We haven't looked at IL-6. As you give your IL-8 systemically, it may be that you have a lot of red cell binding. I don't know about IL-6 binding to red cells, but I have never heard it described. The big difference in potency could be due to this mopping up effect.

Hogg: Can you measure the IL-8 binding to red cells? Can you use flow cytometry for this?

Williams: The way we did this was to displace it. We added IL-8 to whole blood and then removed the plasma and saw how much was missing from the plasma.

Hogg: Can you actually measure it on red cells?

Williams: My colleague Peter Jose has a way of displacing it and measuring what is displaced. In your situation you are giving IL-6 or IL-8 down the airways. The amount that gets to the bone marrow is going to depend on how much gets bound in various places. If it goes down the airways there could be local binding, but once it gets into the circulation the IL-8 is going to bind to the red cells quite avidly.

Hogg: This may be why we didn't see an effect. I thought the IL-6 was interesting because of the work we heard about yesterday, concerning the exacerbations and the high IL-6 levels.

Wedzicha: We looked at 120 paired exacerbation plasma samples, and when we sampled early (within the first 48 h of exacerbation onset) we had a rise in IL-6 and then we had the rise in plasma fibrinogen. Certainly, IL-6 goes up and it was higher when we had viral infection and when we managed to isolate the virus.

MacNee: The other interesting observation is the ability of the neutrophil to release IL-8. We studied intratracheal instillation of LPS as a model of lung inflammation. We were able to obtain the cells that were sequestered in the pulmonary microcirculation by a pulmonary vascular lavage. We found that those cells that were sequestered in the microcirculation released more oxidants, and hence were more activated than the circulating neutrophils obtained from the same animal. I thought that these sequestered cells could also release IL-8 causing more cells to be sequestered.

Williams: Did you try incubating these cells to see whether the IL-8 gene was switched on?

MacNee: We didn't at the time, but it is something we might think about now. The hypothesis around that time was that cells within the microcirculation would cause injury without having to migrate. It is interesting to speculate that if these cells are more activated, they may release more IL-8.

Nadel: LPS induces the IL-8 gene in neutrophils (Inoue et al 1994).

Stockley: Endotoxin has its main effect by acting on epithelial cells in the airway. This causes the release of proinflammatory cytokines including IL-8, which will be another major source. If you find IL-8 within the neutrophil, this doesn't mean much because it is internalized when it binds to the receptor. Finding the source is going to be difficult unless you also detect the message.

Nadel: But orders of magnitude more IL-8 is released *in vivo* from neutrophils than from the epithelium.

Stockley: How do you know that?

Nadel: Because I measured it (see Inoue et al 1995, Fig. 1).

Stockley: You show correlations between IL-8 and neutrophils, but you are stuck on where the source is. Are the neutrophils there because of the IL-8, or is it the other way round? What Tim Williams has shown is that activating and recruiting neutrophils could in its own right not just generate IL-8 from the neutrophil, but also generates IL-8 locally.

Williams: I agree: we can't really distinguish those two mechanisms. We can say that the C5a is generated first, it recruits the neutrophils, and that the IL-8 generation is very much neutrophil dependent. Thus, if you deplete the neutrophils, there is very little IL-8 produced. We can show *in vitro* that the neutrophils can produce IL-8, but in the *in vivo* situation the neutrophils can be releasing other mediators that may stimulate IL-8 production from other cell types. So far I think we have not distinguished between these two possibilities.

Hogg: The fact that the neutrophil could be inducing other cells to release chemoattractants as well as releasing them itself is very interesting from a morphological viewpoint. Observations of neutrophil traffic through the epithelium show that it is not a random journey: they follow one another in lines as if they might be following a chemotactic gradient (Hulbert et al 1981).

Williams: If you look at a venule in mesentery, for example, and add a chemoattractant there is not an even sticking of neutrophils to the surface of the endothelium, or cells migrating through every junction. There is a conglomeration of cells and they go through one site which seems to be favoured for further migration. It could be that the tail of the leukocyte is used to stick on the following leukocyte, or once that junction is open it may be a favoured site for further recruitment.

Nadel: There could also be a difference in adhesion properties.

Hogg: David Walker and associates have shown that these cells migrate through the endothelium at the corners where three cells meet and produce little disruption in the tight junction (Walker et al 1994, 1998). They have reported clear evidence showing that neutrophils migrate along the surface developing both a europod and a lamellopodia as they move towards those corners. This suggests that there is some sort of interaction between neutrophils and endothelial cells that allow them to move along the endothelial surface.

Williams: There is also evidence that IL-8 generation from the endothelium tends to be in that area as well, i.e. where the three endothelial cells meet.

Stockley: I thought that this was a different form of IL-8: an extended peptide that is not a chemoattractant.

Rennard: There is another consequence of these sorts of channels of neutrophils or eosinophils going through specific areas. As the cells migrate, they can potentially cause damage or injury. If the migration of the cells is anatomically restricted to a specific site in the airway, the digestion or damage to the tissues through which they are migrating would be restricted. So my question for Ed Campbell would be, are there any histological ways of assessing this?

Campbell: It has always puzzled me how a tissue that experiences intense inflammatory cell traffic can remain relatively intact. I am attracted to the idea that a large number of cells move through small numbers of channels in favoured locations.

Rennard: I like the concept for the same reason. It may be that if all these cells are going to migrate along the same path, that they leave footprints. Could you see those footprints? Could you see things such as deposits of α_1-antitrypsin/elastase complexes on the neutrophil pathways in tissue?

Campbell: It is certainly possible that inflammatory cells leave things behind that could be detected.

Hogg: Dr Walker's group has also shown that interstitial fibroblasts extend pseudopods into holes in both the endothelial and epithelial basement membrane and that the neutrophils use the fibroblast surface as a guide as they migrate across the interstititum of the alveolar wall (Behzad et al 1996). The neutrophils must deform to get through tiny pre-existing holes in the basement membrane but don't seem to need to digest the membrane to get through (Walker et al 1995).

Senior: The question of how a neutrophil crosses a basement membrane is still unresolved. One of the enzymes that has been implicated is gelatinase B which is released rapidly from neutrophils upon stimulation by chemokines and other factors (Delclaux et al 1996). Using the gelatinase B-deficient mouse that we made with Steve Shapiro and Zena Werb's laboratory (Vu et al 1998) we tested the efficiency of neutrophil emigration in a setting where the neutrophil lacks gelatinase B. We tested emigration in three different tissues, the lung by instilling LPS via the trachea, the peritoneal cavity by putting in thioglycollate, and the skin using intracutaneous IL-8 (Betsuyaku et al 1999). We found that neutrophil emigration at all three sites was the same in the knockout mice as in wild-type mice. Other investigators using the neutrophil elastase-deficient mouse (Belaaouaj et al 1998) and proteinase inhibitors (Jill Mackarel et al 1999) did not detect disturbances in neutrophil emigration, further raising a question about the role of proteinases in this process.

Jeffery: The other aspect is that it may not be necessary for the neutrophil to migrate out of the vascular compartment to cause injury (however we define this). In studies that we did in collaboration with Professor Tim Williams' group, we looked at LPS given intravenously, and the measure of 'injury' in these experiments was the permeability to radioactive markers put into the intravascular compartment. There, LPS produced a permeability leak. When we looked at sections of the lung by light microscopy we had an 'apparently' marked neutrophil recruitment into the alveolar wall. However, at the electronmicroscopic level we saw that all the neutrophils were retained within the vascular compartment. I thought this might be an interesting comment in relation to what we talked about yesterday, that endothelial injury of itself might be a stimulus to the development of emphysema.

MacNee: Did you do the experiment to deplete the neutrophils?

Jeffery: No.

Hogg: Some years ago, Gie et al (1991) aerosolized fluorescein-labelled dextran into the lung and measured its appearance in the blood. They then injected zymozan-activated plasma that produced a tremendous accumulation of neutrophils in the pulmonary circulation and an immediate change in the rate of appearance of fluorescein-labelled dextran in the blood. When the neutrophils were depleted, the effect of the activated plasma was abolished and it reappeared when neutrophils from donors were reinfused. This happened very quickly, before any neutrophils had a chance to migrate and could be blocked with indomethacin.

Senior: Have you thought about testing the C5-deficient mouse in your system to see whether this mouse has altered kinetics of IL-8 expression?

Williams: No, we haven't.

Stockley: These mice don't get neutrophilic influx if you put bacteria in their lungs. This would fit with it being an early signal.

Nadel: Dr Williams, I would like to ask you about your model. The data that you showed were acute responses. If you infuse IL-8 into an animal model over a longer period, what happens to the neutrophil recruitment from the bone marrow?

Williams: This would be interesting, but I do not know what would happen.

Jackson: We keep coming to this question of why there aren't as many neutrophils as we would like in the airways. It is almost as if we are desperately trying to involve neutrophils in the response, rather than trying to find out whether they play a major role.

Jeffery: Your comment is applicable to monitoring cells in airway lumen, but neutrophils are there in the tissues and are compartmentalized. They are in the epithelium rather than the subepithelium. This might be a factor: initial rapid migration occurs through subepithelial tissues and then they are held up by ICAM-1 up-regulation in the epithelium.

Jackson: Taking mucus hypersecretion as an example, in a rat model of intra-tracheal LPS induced goblet cell metaplasia, there is a substantial increase in goblet cells at 48 h. This is preceded by an increase in neutrophils, and it is tempting to link the two. But vinblastine depletion of the neutrophils (at -72 h i.v.) demonstrates no effect on goblet cell metaplasia. So I would suggest that although they can be involved in the mucus hypersecretory response, neutrophils are not actually necessary for it.

Nadel: I wish to make a point about timing and I will use the following example: effects that occur *early* by free radical generation from neutrophils activate epidermal growth factor (EGF) receptors to produce mucins. The oxygen free radicals are generated within 4 h and the mucins are generated 48 h later. The time effect can be very misleading. There are many ways to induce mucins, and free radical release from neutrophils is used as an example because their recruitment, activation and release of oxygen free radicals occurs a day or more before goblet cells can be found in the epithelium.

Jeffery: Clearly neutrophils are not needed for the goblet cell hyperplastic response. In our early experiments with tobacco smoke in rats, which are obligate nasal breathers and presumably filter out most of the particles (which are important for neutrophil recruitment), we induced a goblet cell hyperplastic response without a significant increase in the number of tissue neutrophils. But in terms of the emphysematous reaction and the protease/antiprotease story, there is much evidence of neutrophils being present and excess of elastase in the lung. However, whether neutrophils are important or not to tissue destruction is difficult to conclude.

Shapiro: I would say that neutrophils aren't required for the emphysematous reaction, but if they are present, then they could certainly cause or accelerate emphysema. If one uses antibodies to deplete neutrophils, smoke-induced emphysema still occurs in rodents. So I'm not sure it is clear what they are doing in lung parenchyma either. I'm sure that Rob Stockley will have something to say about this blasphemy.

Stockley: No, I think mice are very useful for determining the pathogenesis of human diseases but caution needs to be expressed.

Nadel: No specific cell except an epithelial cell is required to form a goblet cell.

Jeffery: Is the neutrophil important to the pathogenesis of COPD?

Stockley: You will only answer that by making humans neutropenic and then allowing them to smoke freely. These discussions have gone on for 20–30 years.

MacNee: Drugs are being developed which have potent effects on most neutrophil functions. Unfortunately they have effects on many other processes.

Williams: This is an important issue. Pharmaceutical companies are busy producing antagonists for different chemokine receptors: soon we will have these reagents for blocking receptors and hence neutrophil recruitment. Where they are used will be the real issue. For people in industry this will be a major challenge.

Nadel: The fact that industry has them does not suggest that they will be used. We had a major protease meeting many years ago, and the one thing that everyone decided at that meeting was that neutrophil elastase inhibitors would be very important to study in disease. There was a single study by a single company, and this study was positive — I did it. The decision was that they would not go forward with it. The mere fact that an antagonist is produced doesn't mean that it will result in a drug!

Williams: John White's group have made a molecule that blocks CXCR2 very effectively (White et al 1998). All these things we are talking about can actually be translated into therapy. The industry people have a major challenge in deciding which cells are important and which receptors are important.

Stockley: But the more they hesitate and don't make that decision, the less we will ever know. There is a cry from the heart from people around this room who have been at this for years: we just go round in circles, and until the first definitive experiments with decent agents that block a pathway that we think is important are started, we won't know any more.

References

Behzad A, Chu F, Walker DC 1996 Fibroblasts are in a position to provide directional information to migrating neutrophils during pneumonia in rabbit lungs. Microvasc Res 51:303–316

Belaaouaj A, McCarthy R, Baumann M et al 1998 Mice lacking neutrophil elastase reveal impaired host defense against gram negative bacterial sepsis. Nature Med 4:615–618

Betsuyaku T, Shipley JM, Liu Z, Senior RM 1999 Neutrophil emigration in the lungs, peritoneum, and skin does not require gelatinase B. Am J Resp Cell Mol Biol 20:1303–1309

Delclaux C, Delacourt C, D'Ortho MP, Boyer V, Lafuma C, Harf A 1996 Role of gelatinase B and elastase in human polymorphonuclear neutrophil migration across basement membrane. Am J Resp Cell Mol Biol 14:288–295

Gie RP, Doerschuk CM, English D, Coxson HO, Hogg JC 1991 Neutrophil-associated lung injury after infusion of activated plasma. J Appl Physiol 70:2471–2478

Hulbert WC, Walker DC, Hogg JC 1981 The site of leukocyte migration through the tracheal mucosa in the guinea pig. Am Rev Respir Dis 124:310–316

Inoue H, Massion PP, Ueki IF et al *Pseudomonas* stimulates interleukin-8 mRNA expression selectively in airway epithelium, in gland ducts, and in recruited neutrophils. Am J Respir Cell Mol Biol 11:651–663

Inoue H, Hara M, Massion PP et al 1995 Role of recruited neutrophils in interleukin-8 production in dog trachea after stimulation with *Pseudomonas in vivo*. Am J Respir Cell Mol Biol 13:570–577

Jill Mackarel A, Cottell DC, Russell K J, FitzGerald MX, O'Connor CM 1999 Migration of neutrophils across human pulmonary endothelial cells is not blocked by matrix metalloproteinase or serine protease inhibitors. Am J Resp Cell Mol Biol 20:1209–1219

Palframan RT, Collins PD, Severs NJ, Rothery S, Williams TJ, Rankin SM 1998 Mechanisms of acute eosinophil mobilization from the bone marrow stimulated by interleukin 5: the role of specific adhesion molecules and phosphatidylinositol 3-kinase. J Exp Med 188:1621–1632

Terashima T, Wiggs B, English D, Hogg JC, van Eeden SF 1996 Polymorphonuclear leukocyte transit times in bone marrow during pneumococcal pneumonia. Am J Physiol 271:L587–L592

Terashima T, Wiggs B, English D, Hogg JC, van Eeden SF 1997 Phagocytosis of small carbon particles (PM10) by alveolar macrophages stimulates the release of polymorphonuclear leukocytes from the marrow. Am J Respir Crit Care Med 155:441–1444

Terashima T, Klut ME, English D, Hards J, Hogg JC, van Eeden SF 1999 Cigarette smoking causes sequestration of polymorphonuclear leukocytes released from the bone marrow in lung microvessels. Am J Respir Cell Mol Biol 20:171–177

Vu T, Shipley JM, Bergers G et al 1998 MMP-9/gelatinase B is a key regulator of growth plate angiogenesis and apoptosis of hypertrophic chondrocytes. Cell 93:411–422

Walker DC, MacKenzie A, Horsford S 1994 The structure of the tricellular region of endothelial tight junctions of pulmonary capillaries analyzed by freeze fracture. Microvasc Res 48:259–281

Walker DC, Behzad AR, Chu F 1995 Neutrophil migration through pre-existing holes in basal laminae of alveolar capillaries and epithelium during Streptococcal pneumonia. Microvasc Res 50:397–416

Walker DC, Brown LJ, MacDonell SD, Chu F, Burns AR 1998 Serial section reconstruction of neutrophils, endothelium and tight junction during diapedesis in capillaries of rabbit lung. FASEB J 12:A952

White JR, Lee JM, Young PR et al 1998 Identification of a potent, selective non-peptide CXCR2 antagonist that inhibits interleukin-8-induced neutrophil migration. J Biol Chem 273:10095–10098

Lymphocytes, chronic bronchitis and chronic obstructive pulmonary disease

Peter K. Jeffery

Lung Pathology Unit (Department of Gene Therapy), Imperial College School of Medicine at The Royal Brompton Hospital, Sydney Street, London SW3 6NP, UK

Abstract. Chronic obstructive pulmonary disease (COPD) is a cytotoxic T lymphocyte (CD8)- and macrophage (CD68)-predominant chronic inflammatory disorder of the conducting airways and alveoli. This is often associated with a neutrophilia, inflammation of small airways and destruction of tissue beyond the terminal bronchiolus, i.e. emphysema. In contrast, asthma is a helper T cell (CD4; type 2)-predominant chronic inflammatory disorder of the conducting airways in which there is T lymphocyte-derived gene expression for interleukin (IL)-4 and IL-5 but not interferon γ. There is fragility of airway surface epithelium, thickening of the reticular basement membrane, bronchial vessel congestion and (when severe) an increase in the mass of bronchial smooth muscle. This is usually (but not always) associated with tissue and peripheral blood eosinophilia rather than a neutrophilia and there is exudative plugging of the airways. These differences of inflammatory profile, remodelling and lung function are seen when smokers with COPD are compared with non-smoking mild asthmatics. However there may be important similarities and overlap, particularly in more severe asthma when neutrophils predominate and in the older and/or smoking asthmatic when reversibility of airflow is less obvious. We have recently demonstrated gene expression for IL-4 and IL-5 in and around the mucus-secreting glands of airways resected from smokers without a history of asthma. Also exacerbations of bronchitis may be associated with a tissue eosinophilia. On examination of bronchial biopsies from these patients we show surprisingly strong gene expression for IL-4, IL-5 and even human eotaxin and RANTES (regulated on activation normal T cell expressed and secreted). Whilst CD4 T lymphocytes of the Th2 phenotype might be expressing these cytokines in bronchitis, CD8 T lymphocytes are also capable of secreting IL-4 and IL-5. Viruses may modulate these changes in distinct lymphocyte functional phenotypes. The relevance and importance of CD4/CD8 T lymphocyte ratio to the development of COPD is discussed.

2001 Chronic obstructive pulmonary disease: pathogenesis to treatment. Wiley, Chichester (Novartis Foundation Symposium 234) p 149–168

The lung is a complex organ in both structure and function. Apart from its metabolic functions, the lung is designed to provide for the intimate approximation of blood and air. For the purpose of gas exchange an individual inhales and exhales some 10 000 litres of air a day at rest. In doing so, the lung is

149

exposed to a wide variety of noxious chemicals and particles, both inert and living, the last in the form of viruses, bacteria and fungal spores. It is not surprising that the lung has an immune system adapted to defend itself against such invasion and that it must do so in a highly specific and regulated fashion.

To carry out its many diverse functions the lung is composed of greater than 40 cell types and these include both 'fixed', structural cells and those which enter, migrate through and leave the lung (Jeffery 1997). Of the inflammatory cells some (e.g. neutrophils or macrophages) may react (e.g. to infection) in a non-specific way (innate). Others, such as lymphocytes, are programmed to recognize 'non-self' and respond in a highly specific fashion by initiating and orchestrating a carefully regulated inflammatory reaction (in respect of magnitude and type) which is designed to protect the host (referred to as acquired immunity). Lymphocytes are thus sentinel cells present normally in the airway wall and lung parenchyma, usually for relatively long periods of time. Subsets of lung lymphocytes migrate to and from mucosal sites in other organs (such as tonsils and Peyer's patches of the gut) to sample the external environment elsewhere, recognize and 'remember' foreign substances and return to the lung as 'memory' cells. These memory cells have increased awareness and the capacity to mount a more vigorous than usual response on subsequent exposure to the 'remembered' allergen. Such a response, whilst designed to protect the individual, if overly vigorous, may also damage or destroy the lung. The concept of an imbalance—an overly exuberant or inappropriate lymphocyte response to cigarette smoke exposure or viral/bacterial infection—is central to the hypotheses proposed finally herein to explain the inflammation and damage and destruction of tissue in the proportion of smokers who develop chronic obstructive pulmonary disease (COPD). There are many examples of such lymphocyte-driven inflammatory conditions and, the view taken here is that these include both COPD and asthma, two conditions of airflow obstruction, the first in which the defect in lung function is considered to be progressive and 'irreversible', the second reversible. The clinical expressions of COPD and asthma differ and the author's working hypothesis has been that this would be reflected as distinct profiles of inflammatory cell, regulatory cytokines/chemokines and tissue structure (Jeffery 2000).

Lymphocyte subsets

The early distinction between T and B lymphocytes was followed (in the late 1980s) by the introduction of a new functional terminology which split T lymphocytes (also referred to as T cells) into a number of subsets. An earlier Ciba Foundation symposium (1995) addressed this topic of T cell subsets (Mosmann et al 1995, Erard et al 1993, Croft et al 1994, Fong et al 1990, Maggi et al 1994, Seder et al 1992). T helper (Th) and T cytotoxic/suppressor (Tc) cells were phenotypically characterized by their cell surface expression of CD4 and CD8 epitopes. Mosmann and colleagues (1995) then found that most of the differences between 'help' and 'suppression' could be accounted for by the distinct pattern of cytokines synthesized and secreted. Th1 (and Tc1) cells tended to produce mainly interferon γ (IFNγ), and Th2 (and Tc2) cells characteristically produced interleukin (IL)-4 and IL-5. The results of these experiments, initially in mice, now appear to be largely true of humans also.

The relative proportion of these subsets varies in individuals normally and is influenced by both genetic background and environment (Amadori et al 1995). In investigating the inflammation of clinically distinct airway conditions we have discovered that the relative predominance of a T lymphocyte subset and its resultant cytokine profile appears to differ between asthma and COPD. These differences may be of fundamental importance and affect the consequences of the inflammatory reaction in respect of the capacity to rid the invader and the extent to which there is a 'bystander' effect and damage to host tissue.

Inflammation in asthma

From the point of view of the pathologist and immunologist, asthma is a CD4, Th2-type driven inflammatory condition of the conducting airways, associated with a thickening of the epithelial reticular basement membrane, which is usually but not always, associated with a tissue eosinophilia. Examination of bronchial biopsies demonstrates that, compared with healthy non-smoking subjects there is in non-smoking subjects with mild stable allergic (atopic) asthma, an approximately twofold increase in the numbers of (CD45+) leukocytes. Approximately half of the leukocytes are (CD3+) T cells and a quarter are (CD4+) T helper cells (Azzawi et al 1990). T helper cells predominate such that, in the subepithelial zone, that occupies the majority of the bronchial mucosa in biopsy specimens, the ratio of CD4 to CD8 cells is approximately 3:1 (Jeffery 1998).

Infiltration of the tissues by eosinophils is a characteristic of mild asthma but eosinophilia may not always predominate and neutrophils may outnumber them in severe intractable corticosteroid-dependent asthma (Wenzel et al 1997). Eosinophils may be difficult or impossible to detect in some cases even in fatal asthma (Azzawi et al 1992, Sur et al 1993, Gleich et al 1980). As expected the Th2 predominance is associated with gene expression (and the synthesis and secretion) of Th2-type regulatory cytokines (i.e. IL-4, I-L5, IL-10, IL-13) and also the pro-inflammatory cytokines granulocyte/macrophage colony stimulating factor (GM-CSF) and tumour necrosis factor (TNF)α (Hamid et al 1991, Robinson et al 1992, Humbert et al 1996). There is also up-regulation of eosinophil chemoattractants such as eotaxin by epithelium and also by endothelium and bronchial smooth muscle (Ying et al 1997). This provides the chemokine gradient necessary to the tissue recruitment of eosinophils (from the vascular compartment to the mucosa) and their eventual migration through the surface epithelium and into the airway lumen where they are cleared to the throat (Humbles et al 1997, Li et al 1997).

Inflammation in chronic bronchitis and COPD

Smoking tobacco induces an inflammatory response: it alters the immunoregulatory balance of T cell subsets found in blood, bronchoalveolar lavage (BAL), and tissues of the conducting airways and lung (Miller et al 1982, Costabel et al 1986). Smoking initiates a peripheral blood leukocytosis and a reversible decrease in the normally high CD4:CD8 cell ratio in blood of heavy smokers (i.e. > 50 pack-years). There is also a significant reduction of the CD4:CD8 cell ratio in BAL fluid but not blood of a group of milder smokers (14 pack-years). The increase in the number of BAL and tissue CD8 T cells is positively associated with pack-years smoked (Costabel et al 1986, Lams et al 1998, 2000).

Histological examination of airway tissues (taken at resection for tumour) from smokers demonstrates that inflammatory cells are present in and around submucosal glands. Scores of inflammation show a better association with the subjects who have symptoms of mucus-hypersecretion than does gland size *per se* (Mullen et al 1985). The safe use of the flexible fibre-optic bronchoscope as an investigative tool has allowed us the opportunity to investigate the changes that occur in relatively mild bronchitics and those which occur during the genesis of COPD. Of course this assumes that what is sampled proximately reflects the inflammatory changes seen in smaller, more peripheral airways and lung parenchyma and evidence is emerging to support this.

In bronchial biopsies of subjects with mild stable chronic bronchitis and COPD there is infiltration of the mucosa by inflammatory cells (Saetta et al 1993, Di Stefano et al 1996, O'Shaughnessy et al 1997 and Lams et al 2000) (Fig. 1A). This is associated with up-regulation of cell surface adhesion molecules of relevance to the inflammatory process (Di Stefano et al 1994). In the surface epithelium where, in contrast to the subepithelium, CD8 cells normally predominate, Fournier and colleagues have demonstrated by comparison with non-smokers, an increase in all inflammatory cell types in smokers with chronic bronchitis and mild COPD (Fournier et al 1989). We and others have shown that in the subepithelial zone, bronchial lymphomononuclear cells appear to form the predominant cell type with scanty neutrophils (in the absence of an exacerbation): the lymphomononuclear component comprises lymphocytes, plasma cells and macrophages. Significant increases are reported in the numbers of CD45 (total leukocytes), CD3 (T cells), CD25-activated and VLA-1 (late activation)-positive cells (presumed to be T cells) and of macrophages. The biopsy studies of

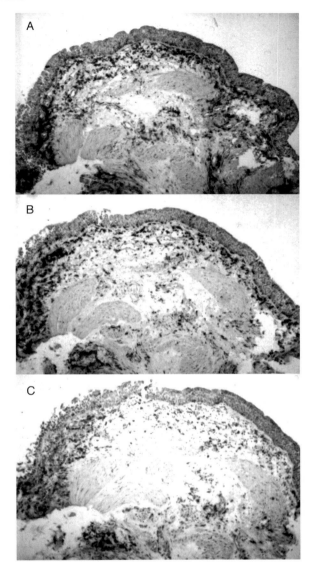

FIG. 1.   Tissue sections of bronchial biopsies from smokers with chronic bronchitis. Three adjacent sections have been cut and immunostained to identify three distinct inflammatory cell phenotypes: (A) a CD45 pan-leukocyte marker demonstrating extensive inflammatory cell infiltration; (B) a marker of CD8 T cells showing that a majority of the leukocytes present are of the cytotoxic/suppressor phenotype (C) a marker of CD4 T cells demonstrates that (unlike asthma) there are relatively few T helper cells. (The alkaline anti-alkaline phosphatase technique has been used and stains the immuno-positive cells.)

O'Shaughnessy and co-workers have demonstrated that in smokers with COPD, T-lymphocytes and neutrophils increase in the surface epithelium whilst T lymphocytes and macrophages increase in the subepithelium (O'Shaughnessy et al 1996, 1997). In contrast to asthma, we have shown that in COPD it is the CD8 T cell and not the CD4 T cell subset, which increases in number and proportion to become the predominant T cell subset (Figs 1B, C). Furthermore, the increase of CD8 cells shows a negative association with the forced expiratory volume in one second (FEV_1 expressed as a percentage of predicted) (Fig. 2A). This novel distinction between the relative proportions of T cell subsets of smokers with mild stable COPD and non-smoking mild asthmatics has received the support of subsequent studies of both resected tissues and bronchial biopsies (Lams et al 1998, Saetta et al 1997, Saetta 1998). The increase of the CD8 phenotype and of the CD8/CD4 ratio seen in the mucosa also occurs deeper in the airway wall in association with submucosal mucus-secreting glands in bronchitic smokers (Saetta et al 1997). The same profile of CD8-predominant T cell inflammation occurs deeper in the lung in both the 'small' airways (Lams et al 1998, Saetta 1998) and also the lung parenchyma (Finkelstein et al 1995). As with the findings in the large conducting airways there are significant negative associations of the numbers of CD8 cells and % predicted FEV_1 in both the small (peripheral) conducting airways (Fig. 2B) and lung parenchyma. However at these sites the negative correlations are stronger than in the large airways. The author's interpretation is that the patterns of inflammation are similar at both proximal and distal sites. However, the CD8 T cell predominance in the small airways and lung parenchyma is more closely associated with decreased lung function because they are causally related to the reduced values of FEV_1 obtained in these subjects with COPD.

Th2-type cytokines and eosinophils in chronic bronchitis

COPD and asthma seem to differ at the tissue level in a number of respects (Jeffery 2000). The author and co-workers have hypothesized that the CD8 T cell predominant profile of inflammation in smokers would not, therefore, be associated with a tissue eosinophilia nor the expression of IL-4, IL-5 or eosinophil chemoattractants. However, compared to normal healthy control tissue, there are a number of studies that report a small but significant increase in the number of tissue eosinophils in subjects with chronic bronchitis or COPD (Mullen et al 1985, O'Shaughnessy et al 1997, Lacoste et al 1993). Sputum eosinophilia is also reported in cases of 'eosinophilic bronchitis', i.e. patients without a history of asthma and without bronchial hyperresponsiveness (Gibson et al 1995, Pizzichini et al 1998). Furthermore, the numbers of tissue eosinophils are markedly and significantly increased when there is an exacerbation of bronchitis,

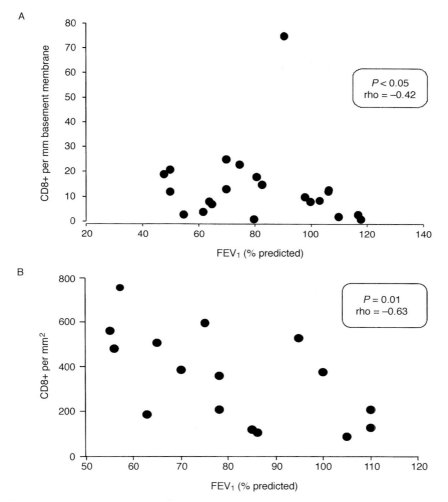

FIG. 2.   (A) Relationship of CD8+ T cells to FEV_1 in large airways (O'Shaughnessy et al 1997).
(B) Relationship of CD8+ T cells to FEV_1 in small airways (Saetta 1998).

(defined as a need by the patient to seek medical attention due to a sudden worsening of dyspnoea or an increase in sputum volume or purulence) (Saetta et al 1994, 1996).

IL-5 is a key molecule in the terminal differentiation and release of eosinophils from bone marrow. We have looked for IL-5 gene expression in tissues resected from asymptomatic and symptomatic (i.e. bronchitic) smokers and bronchial biopsies obtained from chronic bronchitic subjects who had recently experienced an exacerbation. To our surprise, we have found that in inflammatory cells associated with the mucus-secreting glands of bronchi resected from smokers' lungs demonstrate *gene* expression for both IL-4 and IL-5 and the numbers of these cells are significantly higher in subjects with chronic hypersecretion as compared with their asymptomatic controls (Zhu et al 1999a). As with CD4 cells, two types of cytotoxic CD8 (Tc) cells have been recently characterized on the basis of their distinctive cytokine profiles; Tc1 producing mainly IFNγ, and Tc2 secreting IL-4 and IL-5 (Erard et al 1993, Kemeny et al 1994). Due to the predominance of the CD8 phenotype in chronic bronchitis, we hypothesised that IL-4 and IL-5 would likely be secreted by Tc2 CD8 T cells. However, we have applied a double-labelling technique and demonstrate that whilst TNFα gene expression can be localized to CD8 cells, IL-4 and IL-5 cannot. Our data do not, therefore, support the hypothesis of a CD8 cellular origin and we will need to search further for the cellular origin of IL-4 and IL-5 in the mucus-secreting glands of patients with chronic bronchitis. Other candidate cells include CD4 cells and mast cells shown previously to be a source of IL-4 in asthma.

Thus, IL-4 and IL-5 gene expression is not restricted to asthma and, like the recent reports of fibrosing lung disease (Majumdar et al 1999), these regulatory cytokines are expressed also in chronic bronchitic smokers. Thus, whilst there are clinical and tissue-based differences reported between non-smoking asthmatics (with highly reversible disease) and smokers with chronic bronchitis and ('fixed' airflow obstruction), there are similarities and overlap also in respect of tissue eosinophilia and the expression of regulatory cytokines by lymphomononuclear cells. Recent data indicate that, in exacerbations of bronchitis, lymphomono-nuclear cells are also involved in the production of eosinophil chemoattractants including eotaxin and RANTES (Zhu et al 1999b). These data further emphasize the extent of similarity that may exist between acute exacerbations of bronchitis in smokers with chronic bronchitis and non-smokers with asthma.

Concluding comments

When stable, there is evidence of inflammation in both COPD and asthma but there are marked differences in terms of the predominant phenotype of T lymphocyte and the anatomic/mucosal site and functional consequences of such inflammation. Inflammation appears to be present throughout the bronchial tree and in the respiratory portion of the lung in COPD whereas the inflammation of asthma is centred usually on the bronchi and bronchioli. There are, however, reports of inflammation of the alveolar/bronchiolar attachments and an eosinophilia of the alveolar walls in asthma (Kraft et al 1996). The involvement of activated lymphocytes seems to be a common theme in both COPD and asthma yet the profound tissue eosinophilia of stable asthma does not appear in chronic bronchitic smokers until there is an exacerbation of bronchitis. Accordingly, the predominant lymphocyte subsets in COPD and asthma appear to be distinct, i.e. CD8 vs. CD4 cells respectively. There is a need to understand the pattern of cytokines produced by CD8 T cells in COPD. Identification of the lymphomononuclear cells producing IL-4 and IL-5 in chronic, symptomatic bronchitis remains to be elucidated.

Understanding the functional consequences of persistent inflammation and the associated injury and remodelling of airway and lung in COPD and asthma will be important. The mucus-hypersecretion that characterizes chronic bronchitis has traditionally been considered not to contribute to the accelerated rate of decline in FEV_1 and to the disability of COPD. However, even the role of this apparently innocuous feature of mucus hypersecretion has been re-questioned: relatively recent studies demonstrate that sputum volume is indeed associated with accelerated decline in FEV_1, increased hospital admission and increased mortality (Vestbo et al 1996). This is in addition to the undoubted detrimental effects of mucus on the stability of small airways in COPD. Tissue destruction and remodelling occurs primarily at the periphery of the lung in COPD. In contrast there is a trend towards involvement of relatively large proximal airways in asthma, particularly in respect of the early thickening of the reticular basement membrane and, in fatal asthma, enlargement of the mass of bronchial smooth muscle.

The clinical and tissue distinctions between COPD and asthma are by no means always clear: for example a 14 day course of oral prednisolone uncovers a sub-group of patients with COPD that show a degree of airways reversibility associated with histological features of asthma (Chanez et al 1997). There are also patients with sputum eosinophilia who do not show the characteristic clinical features of asthma (Gibson et al 1995). Smokers with combined structural and inflammatory features of both COPD and asthma are likely to be more frequent than we currently appreciate from the reported studies which compare highly selected groups, e.g. of smokers with mild COPD and non-smokers with asthma. We need now to investigate the inflammation of those who are asthmatic and who smoke.

Finally, many life-long smokers do not develop COPD, thus constitutional factors, which may influence the magnitude of the inflammatory response or the host tissue reaction to inflammation, are likely to be of importance also. Genetic deficiency of α_1-antitrypsin is well documented and smoking in this group clearly advances the onset of emphysema and accelerates its subsequent progression. Other genetic factors, such as variation in cellular response to cytotoxicity, phagocytosis and enzyme release by both neutrophils and macrophages and cytokine polymorphisms such as those recently reported for TNFα, may be important determinants of susceptibility to cigarette smoke. Studies of mice of distinct haplotypes demonstrate the importance of genetic make-up on the response, for example, to viral infection, a common cause of exacerbation in bronchitis. These studies clearly show the importance of the balance between CD4 and CD8 T lymphocyte responses. CD4 cell predominance encourages a tissue eosinophilia whereas CD8 cells producing IFNγ switch off the CD4-driven eosinophilic response (Hussell et al 1997).

I propose that airway (and lung) susceptibility to the effects of cigarette smoke will be greater in those individuals who already have a genetically determined low CD4:CD8 cell ratio in their peripheral blood (Amadori et al 1995). The last mentioned occurs in about 5% of the population and is a novel approach to explain why only a proportion (about 20%) of smokers succumb to the deleterious effects of tobacco smoke on the lung. It is probable that the tobacco smoke-induced predominance of CD8 T cells, particularly in individuals with a low CD4:CD8 ratio, will lead to an exaggerated CD8 response to viral infection which will result in damage of lung tissue by FAS-ligand and perforin/granzyme-induced apoptotic mechanisms. There is experimental evidence to support this proposal (Castleman et al 1985, Cannon et al 1988). This is a novel hypothesis based on a T lymphocyte-driven rather than a neutrophil-driven mechanism to explain the destruction of lung tissue in smokers who are predisposed to the development of COPD.

These hypotheses now require testing in the future by the application of immunopathological, molecular and epidemiological research.

References

Amadori A, Zamarchi R, De Silvestro G et al 1995 Genetic control of the CD4/CD8 T-cell ratio in humans. Nat Med 1:1279–1283

Azzawi M, Bradley B, Jeffery PK et al 1990 Identification of activated T lymphocytes and eosinophils in bronchial biopsies in stable atopic asthma. Am Rev Respir Dis 142:1407–1413

Azzawi M, Johnston PW, Majumdar S, Kay AB, Jeffery PK 1992 T lymphocytes and activated eosinophils in airway mucosa in fatal asthma and cystic fibrosis. Am Rev Respir Dis 145:1477–1482

Cannon MJ, Openshaw PJM, Askonas BA 1988 Cytotoxic T cells clear virus but augment lung pathology in mice infected with respiratory syncytial virus. J Exp Med 168:1163–1168

Castleman WL 1985 Bronchiolitis obliterans and pneumonia induced in young dogs by experimental adenovirus infection. Am J Pathol 119:495–504

Chanez P, Vignola AM, O'Shaughnessy T et al 1997 Corticosteroid reversibility in COPD is related to features of asthma. Am J Respir Crit Care Med 155:1529–1534

Ciba Foundation 1995 T cell subsets in infectious and autoimmune diseases. Wiley, Chichester (Ciba Found Symp 195)

Costabel U, Bross KJ, Reuter C, Ruhle K-H, Matthys H 1986 Alterations in immunoregulatory T-cell subsets in cigarette smokers. A phenotypic analysis of bronchoalveolar and blood lymphocytes. Chest 90:39–44 (erratum: 1986 Chest 92:1124)

Croft M, Carter L, Swain SL, Dutton RW 1994 Generation of polarized antigen-specific CD8 effector populations: reciprocal action of interleukin (IL)-4 and IL-12 in promoting type 2 versus type 1 cytokine profiles. J Exp Med 180:1715–1728

Di Stefano A, Maestrelli P, Roggeri A et al 1994 Upregulation of adhesion molecules in the bronchial mucosa of subjects with chronic obstructive bronchitis. Am J Respir Crit Care Med 149:803–810

Di Stefano A, Turato G, Maestrelli P et al 1996 Airflow limitation in chronic bronchitis is associated with T-lymphocyte and macrophage infiltration of the bronchial mucosa. Am J Respir Crit Care Med 153:629–632

Erard F, Wild MT, Garcia-Sanz JA, Le Gros G 1993 Switch of CD8 T cells to noncytolytic CD8-CD4- cells that make TH2 cytokines and help B cells. Science 260:1802–1805

Finkelstein R, Fraser RS, Ghezzo H, Cosio MG 1995 Alveolar inflammation and its relation to emphysema in smokers. Am J Respir Crit Care Med 152:1666–1672

Fong TA, Mosmann TR 1990 Alloreactive murine CD8+ T cell clones secrete the Th1 pattern of cytokines. J Immunol 144:1744–1752

Fournier M, Lebargy F, Le Roy Ladurie F, Lenormand E, Pariente R 1989 Intraepithelial T-lymphocyte subsets in the airways of normal subjects and of patients with chronic bronchitis. Am Rev Respir Dis 140:737–742

Gibson PG, Hargreaves FE, Girgis-Gabardo A, Morris M, Denburg JA, Dolovich J 1995 Chronic cough with eosinophilic bronchitis: examination for variable airflow obstruction and response to corticosteroid. Clin Exp Allergy 25:127–132

Gleich GJ, Motojima S, Frigas E, Kephart GM, Fujisawa T, Kravis LP 1980 The eosinophilic leucocyte and the pathology of fatal bronchial asthma: evidence for pathologic heterogeneity. J Allergy Clin Immunol 80:412–415

Hamid Q, Azzawi M, Ying S et al 1991 Expression of mRNA for interleukin-5 in mucosal bronchial biopsies from asthma. J Clin Invest 87:1541–1546

Humbert M, Durham SR, Ying S et al 1996 IL-4 and IL-5 mRNA and protein in bronchial biopsies from patients with atopic and nonatopic asthma: evidence against 'intrinsic' asthma being a distinct immunopathologic entity. Am J Respir Crit Care Med 154:1497–1504

Humbles AA, Conroy DM, Marleau S et al 1997 Kinetics of eotaxin generation and its relationship to eosinophil accumulation and the late reaction in allergic airway disease: analysis in a guinea pig model in vivo. J Exp Med 186:601–612

Hussell T, Baldwin CJ, O'Garra A, Openshaw PJ 1997 CD8+ T cells control Th2-driven pathology during pulmonary respiratory syncytial virus infection. Eur J Immunol 27:3341–3349

Jeffery PK 1997 Structural, immunologic and neural elements of the normal human airway wall. In: Holgate ST (ed) Asthma and rhinitis: implications for diagnosis and treatment. Blackwell Science, Oxford, p 80–108

Jeffery PK 1998 Structural and inflammatory changes in COPD: a comparison with asthma. Thorax 53:129–136

Jeffery PK 1999 Differences and similarities between chronic obstructive pulmonary disease and asthma. Clin Exp Allergy (suppl 2) 29:14–26

Jeffery PK 2000 Comparison of the structural and inflammatory features of COPD and asthma. Chest 117:251S-260S

Kemeny DM, Noble A, Holmes BJ, Diaz-Sanchez D 1994 Immune regulation: a new role for the CD8+ T cell. Immunol Today 15:107–110

Kraft M, Djukanovic R, Wilson S, Holgate ST, Martin RJ 1996 Alveolar tissue inflammation in asthma. Am J Respir Crit Care Med 154:1505–1510

Lacoste J-Y, Bousquet J, Chanez P et al 1993 Eosinophilic and neutrophilic inflammation in asthma, chronic bronchitis, and chronic obstructive pulmonary disease. J Allergy Clin Immunol 92:537–548

Lams BE, Sousa AR, Rees PJ, Lee TH 2000 Subepthelial immunopathology of large airways in smokers with and without chronic obstructive pulmonary disease. Eur Respir J 15:512–516

Lams BEA, Sousa AR, Rees PJ, Lee TH 1998 Immunopathology of the small-airway submucosa in smokers with and without chronic obstructive pulmonary disease. Am J Respir Crit Care Med 158:1518–1523

Li D, Wang D, Griffiths-Johnson DA et al 1997 Eotaxin protein gene expression in guinea-pigs: constitutive expression and upregulation after allergen challenge. Eur Respir J 10:1946–1954

Maggi E, Giudizi MG, Biagiotti R et al 1994 Th2-like CD8+ T cells showing B cell helper function and reduced cytolytic activity in human immunodeficiency virus type 1 infection. J Exp Med 180:489–495

Majumdar S, Li D, Ansari T et al 1999 Different cytokine profiles in cryptogenic fibrosing alveolitis and fibrosing alveolitis associated with systemic sclerosis: a quantitative study of open lung biopsies. Eur Respir J 14:251–257

Miller LG, Goldstein G, Murphy M, Ginns LC 1982 Reversible alterations in immunoregulatory T cells in smoking. Analysis by monoclonal antibodies and flow cytometry. Chest 82:526–529

Mosmann TR, Sad S, Krishnan L, Wegmann TG, Guilbert LJ, Belosevic M 1995 Differentiation of subsets of CD4+ and CD8+ T cells. In: T cell subsets in infectious and autoimmune diseases. Wiley, Chichester (Ciba Found Symp 195), p 42–54

Mullen JBM, Wright JL, Wiggs BR, Pare PD, Hogg JC 1985 Reassessment of inflammation of airways in chronic bronchitis. Br Med J (Clin Res Ed) 291:1235–1239

O'Shaughnessy TC, Ansari TW, Barnes NC, Jeffery PK 1996 Inflammatory cells in the airway surface epithelium of smokers with and without bronchitic airflow obstruction. Eur Respir J (suppl) 9:14s

O'Shaughnessy T, Ansari TW, Barnes NC, Jeffery PK 1997 Inflammation in bronchial biopsies of subjects with chronic bronchitis: inverse relationship of CD8+ T lymphocytes with FEV_1. Am J Resp Crit Care Med 155:852–857

Pizzichini E, Pizzichini MMM, Gibson P et al 1998 Sputum eosinophilia predicts benefit from prednisone in smokers with chronic obstructive bronchitis. Am J Respir Crit Care Med 158:1511–1517

Robinson DS, Hamid Q, Ying S et al 1992 Predominant TH2-type bronchoalveolar lavage T-lymphocyte popularity in atopic asthma. N Engl J Med 326:298–304

Saetta M 1998 CD8+ T-lymphocytes in peripheral airways of smokers with chronic obstructive pulmonary disease. Am J Respir Crit Care Med 157:822–826

Saetta M, Di Stefano A, Maestrelli P et al 1993 Activated T-lymphocytes and macrophages in bronchial mucosa of subjects with chronic bronchitis. Am Rev Respir Dis 147:301–306

Saetta M, Di Stefano A, Maetrelli P et al 1994 Airway eosinophilia in chronic bronchitis during exacerbations. Am J Respir Crit Care Med 150:1646–1652

Saetta M, Di Stefano A, Maestrelli P et al 1996 Airway eosinophilia and expression of interluekin-5 protein in asthma and in exacerbations of chronic bronchitis. Clin Exp Allergy 26:766–774

Saetta M, Turato G, Facchini FM et al 1997 Inflammatory cells in the bronchial glands of smokers with chronic bronchitis. Am J Respir Crit Care Med 156:1633–1639

Seder RA, Boulay JL, Finkelman F et al 1992 CD8+ T cells can be primed *in vitro* to produce IL-4. J Immunol 148:1652–1656

Sur S, Crotty TB, Kephart GM et al 1993 Sudden onset fatal asthma. A distinct entity with few eosinophils and relatively more neutrophils in the airway submucosa? Am Rev Respir Dis 148:713–719

Vestbo J, Prescott E, Lange P 1996 Association of chronic mucus hypersecretion with FEV_1 decline and chronic obstructive pulmonary disease morbidity. Am J Respir Crit Care Med 153:1530–1535

Wenzel SE, Szefler SJ, Leung DYM, Sloan SI, Rex MD, Martin RJ 1997 Bronchoscopic evaluation of severe asthma. Persistent inflammation associate with high dose glucocorticoids. Am J Respir Crit Care Med 156:737–743

Ying S, Robinson DS, Qiu M et al 1997 Enhanced expression of eotaxin and CCR3 mRNA and protein in atopic asthma. Association with airway hyperresponsiveness and predominant colocalization of eotaxin mRNA to bronchial epithelial and endothelial cells. Eur J Immunol 27:3507–3516

Zhu J, Majumdar S, Ansari T et al 1999 IL-4 and IL-5 mRNA in the bronchial wall of smokers. Am J Respir Crit Care Med 159:A450 (abstr)

Zhu J, Majumdar S, Turato G, Fabbri LM, Saetta M, Jeffery PK 1999 Airway eosinophilia in bronchitis and gene expression for IL-4, IL-5 and eotaxin in bronchial biopsies. Eur Respir J 14:360s (abstr)

DISCUSSION

Shapiro: You gave an elegant demonstration of the presence of the CD8 cells. It is intriguing what they are doing. In terms of the link to emphysema, the granzymes are pretty good at activating caspases which will cause nuclear damage, but they are not potent extracellular matrix degrading enzymes. I can understand how they may kill cells, but this still doesn't explain the damage to the matrix.

Jeffery: The granzymes and perforins are inducers of apoptosis, as you alluded to. Apoptosis has been linked to the development of emphysema, although I'm not clear as to the mechanisms. I agree that the direct connection to matrix destruction is tenuous.

Rogers: I was interested to see that IL-4-positive cells congregated around the epithelial glands. John Rankin's group has shown in mice that IL-4 is highly potent in inducing goblet cell hyperplasia in the surface epithelium (Temann et al 1996). From your data, if IL-4 is secreted around the glands, there may be IL-4 induction of submucosal gland hypertrophy.

Jeffery: IL-4 is an interesting molecule. It has a number of functions: as you say its induction of mucus may well be important here. IL-4 is also profibrotic, and has been shown to stimulate fibroblasts and collagen production, and therefore could also be involved in the fibrotic process, which is particularly important in the smaller airway in COPD.

Wedzicha: Listening to your paper, it seems to me that we have to start differentiating mild COPD from moderate and severe COPD. You showed data that CD8 T cells are related to lung function. We know from Keating et al (1996) that airway TNFα and IL-8 are also related to lung function. This could perhaps explain the differences we found in exacerbations from the data of Saetta et al (1994): she certainly studied a mild group of COPD patients. In fact, the falls in lung function at exacerbation in her study were larger than I would expect to see. This reinforces the fact that the airway will behave differently in COPD depending on the severity of the disease. The fact that we found virtually no eosinophils in exacerbation means that we got more neutrophils and lymphocytes, yet if we had looked at patients with milder disease we would have seen a different response.

Jeffery: It will also depend on the type of virus, bacterium or other factors initiating the exacerbation, how frequently that particular virus is invading, and how the T cells have been primed by prior viral infection. These and genetic factors control whether you will have the particular pattern of cytokines responsible for eosinophilia or not. Going back to the question about animal models, the advantage of the mouse is that you can look at a number of different haplotypes. It seems that the genetic background is terribly important in deciding the balance of CD4/CD8 response, and therefore the subsequent capacity to develop eosinophilia. This aspect could be investigated in humans in the future.

MacNee: In relation to this, studies indicate that there is a subgroup of COPD patients who have higher numbers of eosinophils in induced sputum, and these patients respond better to corticosteroids (Hargreave & Leigh 1999). This may appear to be a subgroup of COPD patients who may not be representative of the whole group.

Jeffery: This is the so-called 'eosinophilic bronchitic' group. I don't know of the role of the eosinophils there.

Rennard: Have you looked for HLA DR expression? There are several studies suggesting that epithelial cells can express HLA DR *in vivo* and *in vitro*, and then can actually function in terms of antigen presentation (Rossi et al 1990, Vignola et al 1993, Spurzem et al 1990). Is it possible that the epithelium is producing not only IL-4 but also is important in regulating lymphocytes in these tissues?

Jeffery: The epithelium does light up for HLA DR, but we haven't done a formal study to compare different groups.

Barnes: What do you think the mechanism might be for cigarette smoking to cause CD8 recruitment? There is much recent interest in IL-16 as an epithelial factor that will attract lymphocytes, so it may be interesting to look at IL-16 expression.

Jeffery: We haven't done that. I think RANTES might also be involved, and the up-regulation of RANTES expression in exacerbations is very marked. More specifically, it would be interesting to know what it is in the cigarette smoke that is acting perhaps at the bone marrow level or elsewhere that will trigger an influx of CD8 cells.

Barnes: It could be acting on epithelial cells. It is interesting that CD8 cells accumulate next to the epithelium, as if this may be the source of the signal.

Jeffery: I agree, the other signal with regard to epithelial-expressed adhesion molecules is CD103, which is a specific ligand for CD8 cells. This is up-regulated in smokers whether or not they have airflow obstruction. It might be this is a contributing mechanism.

Rennard: What is the location of RANTES in your patients?

Jeffery: The distribution of eotaxin and macrophage chemotactic protein (MCP)-4 is rather different in the human than it is in the guinea pig: we have had some experience working in collaboration with Tim Williams in the guinea pig. It is actually found in lymphomononuclear cells, and less so in the epithelium. In contrast, RANTES is markedly up-regulated in the epithelium. It is present in subepithelial lymphomononuclear cells, but there is a very different distribution than that of eotaxin.

Rennard: So there is a relationship between eosinophils in your patients and RANTES, and the eosinophils are in the deep tissue and RANTES is in the superficial tissue?

Jeffery: RANTES is particularly up-regulated in the epithelium, but is also found in lymphomononuclear cells beneath the epithelium. The eosinophils, however, are not particularly up-regulated in the epithelium.

Rennard: Do the deep eosinophils express TNFα? In the tissues that I have stained in COPD, they contain TNFα.

Jeffery: I have no idea. They could do.

Nadel: If they do and they are around glands, TNFα is an inducer of epidermal growth factor (EGF) receptors.

Silverman: I was interested in the relatively clean distinction you were able to make between asthmatics and COPD patients using the CD4:CD8 ratio. Is it known what happens to the CD4:CD8 ratio in the asthmatics as they age and also in asthmatics who smoke?

Jeffery: The age question hasn't been looked at. We know that in asthmatics who have a fatal attack, the CD8 levels are high, although I don't know the significance of this. One study that Peter Barnes and I have talked about which needs to be done, is to look at asthmatics who smoke. I have shown to you today two polarized ends of the spectrum, which we must first study, comparing non-smoking asthmatics with smoking COPD patients. Perhaps the behaviour of the smoking asthmatic will be very different, and this, I agree, is an important study to do.

Barnes: Smoking asthmatics appear to be less steroid responsive. Inflammation is most difficult to suppress with steroids at doses that are effective in non-smoking asthmatics. One would expect that cytokines would be suppressed by steroids, but this does not occur. Yet the same cytokines are suppressed by steroids in patients with asthma. This suggests that an active resistance mechanism is operating in COPD. I think that this may be induced by cigarette smoking.

Jeffery: I agree in general with what you say, but don't agree about the steroid non-responsiveness: it depends on how you define this. In biopsies of COPD we see a reductive response of mast cells to steroids.

Barnes: COPD is not as responsive as asthma.

Jeffery: Just to take this point one stage further, are you saying that the ISOLDE and Euroscope studies are showing that steroids have no effect in COPD whereas they have in asthma? The similar long-term study has not been done in asthma to my knowledge.

Calverley: No, and it is actually a study that probably ought to be done. However, I agree with Peter Barnes that the response is not as great in COPD as in asthma.

Stockley: Isn't this implicit in the definition of the disease? COPD is diagnosed as being a non-steroid-responsive condition.

Calverley: Almost certainly, yes. But the point is that there isn't a nice one-to-one correspondence of things. Steve Rennard started this morning talking about the complex, redundant mechanisms whereby epithelia generate mediators to sort out repair and healing. We are dealing with systems here that operate on several complex levels. We have already talked about the potential roles of hypersecretion. I don't think that there has to be a one-to-one correspondence of changes in hypersecretion and changes in lymphocytes in the airway wall. It is perfectly plausible to me that some parts—perhaps unimportant parts—of these mechanisms can be improved by an intervention. This can translate downstream into something that clinically matters at some stages of the illness. But in terms of how this process can be turned off—how we stop COPD progressing or, better still, make it regress—clearly, this is a very different issue when we treat with inhaled corticosteroids or oral corticosteroids in COPD than in asthma. Rob Stockley is right: we define these people as not responding to corticosteroids, so we shouldn't find it so surprising that they don't.

Jeffery: But when you come to a situation of exacerbation, and tell me about that, are you then dealing with a non-corticosteroid responsive situation? Corticosteroids are effective in the treatment of exacerbations, are they not?

Barnes: Corticosteroids are not very effective even in exacerbations of COPD, whereas they are almost always effective in asthma.

Nadel: This is a conference that is hypothesis driven, and I have another hypothesis to propose. What Peter Jeffery has described is that eosinophils occur deep in the airway tissue, not in the epithelium. There are two questions that could be answered. One is whether these are steroid sensitive. Intuitively, I would say that they will be. If they are steroid sensitive, is there some major function in that tissue that can be suppressed, and can you examine it? It is possible that gland hypersecretion in COPD could be related somehow to eosinophils. If this is correct, suppression of eosinophil function by steroids could suppress the hypersecretion and cough in COPD.

Barnes: But you are presupposing that steroids will inhibit eosinophils.

Nadel: No, I said that *if* they inhibit eosinophils, then you can test them on the tissue. People are giving steroids to patients with COPD. It wouldn't be difficult to find out whether this selectively suppresses eosinophils. You said that steroids do not suppress neutrophils.

Barnes: I don't think steroids suppress eosinophils. Although we don't see eosinophils, we have found very high levels of eosinophil cationic protein (ECP) and eosinophil peroxidase (EPO) in the sputum of COPD patients, which are unaffected by even high doses of steroids. In contrast, in asthmatics who have similar levels of these basic proteins, they are markedly suppressed by steroids. This is why I argue that there is an active resistance in COPD to the anti-inflammatory effects of steroids that we see so well in asthma.

MacNee: There seems to be a disagreement here. Does anyone agree with me that one of the problems we have at the moment is that we have studied relatively small numbers of patients? It is quite clear that when a large number of 'COPD' patients are examined, there is a subgroup that, for instance, has eosinophils in sputum. Is there a need to start to separate patients into subgroups with different responses to treatment. This seems to me to be more productive than blanket statements that steroids don't work in COPD.

Rennard: I think this has to be right. From what we have discussed at this meeting it is clear that COPD is a heterogeneous condition. There is not going to be one single entity called 'COPD'. A small study isn't going to be uniformly representative of these multiple sub-groups of patients. Results in one small study therefore may not agree well with results in another small study, and results in a big study may average out those differences. It is very likely that all the small studies are correct, and they are looking at subsets of patients from a heterogeneous patient population.

Agustí: I think we are discussing two separate issues. One is whether or not COPD responds better or worse to steroids. I think that most of the evidence suggests that the response is quite poor. The other issue is whether or not COPD patients have eosinophils in their bronchi. To my mind, the only paper showing eosinophils in the bronchi of COPD patients was by Marina Saetta, and these were patients with very mild COPD with an ambulatory type of exacerbation. I'm not sure that we have real evidence that COPD patients have eosinophils in their bronchi, particularly severe patients.

Jeffery: There are other papers, but there is some debate about whether the patients have COPD or asthma. Clearly there are some eosinophils present. However, what I showed in my paper today was the presence of eosinophil chemoattractants in exacerbations of bronchitis, using exactly the same biopsies that Marina Saetta studied.

Agustí: These were not COPD patients then.

Caverley: They were smokers with mild COPD.

Wedzicha: They had large falls in FEV_1 at exacerbation: greater than would be expected in COPD.

Caverley: Can I clarify this? Have you measured ECP in those paired BAL specimens? You have a larger collection of induced sputum specimens than anyone else; you have people with a very clearly defined exacerbation that is robust in a population that everyone would agree has COPD. One way to resolve some of the apparent conflict that we just heard would be if you had data about ECP, and it would also be interesting to know whether you had data from people who did and didn't have steroids.

Wedzicha: We do not have ECP data. Eighty per cent of our patients had inhaled steroids in fairly high doses. In our study, one of the inclusion criteria was that we did not want to change that in order not to disturb stability, even though the median exacerbation rate was high at 2.8. So if inhaled steroids are effective in COPD, I do not think they do a lot to exacerbations. They probably reduce them by a small amount. What I am rather confused about is the effect of inhaled steroids as pre-treatment versus treatment at the time of exacerbation. There is some confusion here. Long-term inhaled steroids probably do not work in the same way in COPD at exacerbation as when we treat the exacerbation with steroids. I think that steroids at exacerbation may work through reduction of oedema. Certainly, the time course is right for that.

Caverley: I think it would be nice to have a mechanism for the small effect of inhaled steroids in preventing exacerbations. It would be nice to know, because this might give us some idea of the mechanism that leads to the development of exacerbation.

Stockley: It looks like it is oedema. In the study that we did about one-third of our patients had non-infective exacerbations. We had about 40 patients that we followed through. Half went on to inhaled steroids and half were on placebo: the only difference we found biochemically was protein leakage. Those receiving inhaled steroids showed a reduction in protein leakage during their exacerbations.

Barnes: So they might be better helped by adrenaline, rather than steroids.

MacNee: I wonder whether we do now need some larger studies across a range of different patients.

Jeffery: We must look at the smoking asthmatic.

Barnes: The other thing I think is very interesting is the recent research showing persistence of inflammation in ex-smokers, some of whom stopped smoking more than 10 years ago. We see it in sputum induction studies of ex-smokers, who seem to have similar inflammatory patterns to active smokers. This raises important issues about the persistence of inflammation in COPD, independent of the causal mechanism. This is rather similar to what is seen in asthma, and we do not yet understand why inflammation persists in asthmatic airways.

Stockley: This could suggest to me that what we are looking at is exactly the wrong thing. If you believe the concept that a susceptible smoker progresses and cessation of smoking is a good thing, but what we are measuring as inflammation doesn't change, then we are measuring the wrong thing.

Barnes: That is exactly an issue that needs to be addressed. Or we are looking in the wrong place. How good is the evidence that when people stop smoking they actually go back to normal? The mean goes back to normal, but if you look at the individual patients, some continue to progress. We need to look at sub groups of patients after cessation of smoking.

Caverley: It may depend on the stage at which you stop smoking. In the Lung Health Study, the reduction in decline of FEV_1 is quite encouraging for people who are documented sustained quitters. If you look at the documented sustained quitters in the ISOLDE study, they had a lower rate of decline of FEV_1 than the continuing smokers, but that rate was still not normal.

Barnes: This is what I see in the clinic: people who stopped smoking 20 years ago still develop COPD.

Jeffery: But they are not non-symptomatic. They are still producing sputum.

Barnes: Not as much as continuing smokers.

MacNee: In the original data from the Fletcher & Peto (1977) I don't recall that we have data on the effects of smoking on severe disease.

Caverley: If you remember the original diagram, it used to be a dotted line, which means that it was made up.

MacNee: The suggestion was always that no matter what stage you stopped, the decline in FEV_1 slows, but I'm not sure we have the data to support this.

Caverley: The ISOLDE data, when fully analysed, should answer that question.

Campbell: Can you enlarge on the concept of genetic determinants of CD4:CD8 ratios? What is known about this in human beings?

Jeffery: The only study that I know has looked at this is by Amadori et al (1995) who looked at a group of about 400 blood donors over approximately three years. They found that their peripheral blood CD4:CD8 ratios were constant over that period. They also found that there was a relationship in their families that supported a genetic control of this ratio.

Campbell: Have we any idea of the mechanisms that control the CD4:CD8 ratio of lymphocytes that respond to an insult?

Jeffery: In terms of the genetic control, 5% of the population have a very low CD4:CD8 ratio (i.e. less than 1:1). The suggestion would be that if this 5% took up smoking, they would shift their ratio in favour of CD8. The day-to-day control in response to infection is the work of Hussell et al (1998) in mice. This CD4:CD8 response seems to be based on haplotype as well.

Nadel: There are clues from the early studies on leishmania in mice which can't mount a proper CD8 response. The animals die. These studied were definitive in terms of genotypes that do and do not mount CD4 and CD8 responses. These experimental conditions need to be tested for their relevance to humans.

Jeffery: We know a lot in the mouse regarding genetic associations, but not enough in humans.

Campbell: This might be important with regard to the genetic determinant of risk among cigarette smokers.

References

Amadori A, Zamarchi R, De Silvestro G et al 1995 Genetic control of the CD4/CD8 T-cell ratio in humans. Nat Med 1:1279–1283

Fletcher C, Peto R 1977 The natural history of chronic airflow obstruction. Br Med J 1:1645–1648

Hargreave FE, Leigh R 1999 Induced sputum, eosinophilic bronchitis, and chronic obstructive pulmonary disease. Am J Respir Crit Care Med 160: S53–S57

Hussell T, Georgiou A, Sparer TE, Matthews S, Pala P, Openshaw PJ 1998 Host genetic determinants of vaccine-induced eosinophilia during respiratory syncytial virus infection. J Immunol 161:6215–6222

Keatings VM, Collins PD, Scott DM, Barnes PJ 1996 Differences in interleukin-8 and tumor necrosis factor-alpha in induced sputum from patients with chronic obstructive pulmonary disease or asthma. Am J Respir Crit Care Med 153:530–534

Pizzichini E, Pizzichini M, Gibson P et al 1998 Sputum eosinophilia predicts benefit from prednisone in smokers with chronic obstructive bronchitis. Am J Respir Crit Care Med 158:1511–1517

Rossi GA, Sacco O, Balbi B et al 1990 Human ciliated bronchial epithelial cells: expression of the HLA-DR antigens and of the HLA-DR alpha gene, modulation of the HLA-DR antigens by gamma-interferon and antigen-presenting function in the mixed leukocyte reaction. Am J Respir Cell Mol Biol 3:431–439

Saetta M, Di Stefano A, Maestrelli P et al 1994 Airway eosinophilia in chronic bronchitis during exacerbations. Am J Respir Crit Care Med 150:1646–1652

Spurzem JR, Sacco O, Rossi GA, Rennard SI 1990 MHC class II expression by bronchial epithelial cells is modulated by lymphokines and corticosteroids. Am Rev Respir Dis 141:A681

Temann U-A, Prasad B, Gallup MW et al 1996 A novel role for murine IL-4 *in vivo*: induction of MUC5AC gene expression and mucin hypersecretion. Am J Respir Cell Mol Biol 16:471–478

Vignola AM, Campbell AM, Chanez P et al 1993 HLA-DR and ICAM-1 expression on bronchial epithelial cells in asthma and chronic bronchitis. Am Rev Respir Dis 148:689–694

Oxidants/antioxidants and chronic obstructive pulmonary disease: pathogenesis to therapy

William MacNee

ELEGI/Colt Laboratories, Department of Medical and Radiological Sciences, Wilkie Building, The University of Edinburgh, Medical School, Teviot Place, Edinburgh EH8 9AG, UK

Abstract. There is now considerable evidence for an increased oxidant burden in smokers, particularly in those smokers who develop chronic obstructive pulmonary disease (COPD), as shown by increased markers of oxidative stress in the airspaces, breath, blood and urine. The presence of increased oxidative stress is a critical feature in the pathogenesis of COPD, since it results in inactivation of antiproteinases, airspace epithelial injury, mucus hypersecretion, increased sequestration of neutrophils in the pulmonary microvasculature, and gene expression of pro-inflammatory mediators. The sources of the increased oxidative stress in patients with COPD derive from the increased burden of oxidants present in cigarette smoke, or from the increased amounts of reactive oxygen species released from leukocytes, both in the airspaces and in the blood. Antioxidant depletion or deficiency in antioxidants also contributes to oxidative stress. The development of airflow limitation is related to dietary deficiency of antioxidants and hence dietary supplementation may be a beneficial therapeutic intervention in this condition. Oxidative stress also has a role in enhancing the airspace inflammation, which occurs in smokers and patients with COPD through the activation of redox-sensitive transcriptions factors such as NF-κB and AP-1, which regulate the genes for pro-inflammatory mediators and protective antioxidant gene expression. Antioxidants that have good bioavailability or molecules that have antioxidant enzyme activity are therefore therapies that not only protect against the direct injurious effects of oxidants, but also may fundamentally alter the inflammatory events which have a central role in the pathogenesis of COPD.

2001 Chronic obstructive pulmonary disease: pathogenesis to treatment. Wiley, Chichester (Novartis Foundation Symposium 234) p 169–188

The lungs are continuously exposed to oxidants generated either endogenously (e.g. released from phagocytes) or exogenously (e.g. air pollutants or cigarette smoke). Furthermore, intracellular oxidants (e.g. from mitochondrial electron transport) are involved in cellular signalling pathways. Cells are protected against this oxidative challenge by well-developed enzymatic and non-enzymatic antioxidant systems (Halliwell 1996).

When the balance between oxidants and antioxidants shifts in favour of oxidants, from either an excess of oxidants and/or depletion of antioxidants, *oxidative stress* is said to occur. Oxidative stress not only produces direct injurious effects, but also initiates the molecular mechanisms that control lung inflammation.

Oxidative stress is thought to play an important role in the pathogenesis of a number of lung diseases, including chronic obstructive pulmonary disease (COPD). Smoking is the main aetiological factor in this condition. Since cigarette smoke contains 10^{17} molecules per puff, this together with other evidence of increased oxidative stress in smokers and in patients with COPD, led to the proposal that an oxidant/antioxidant imbalance is important in the pathogenesis of this condition (Rahman & MacNee 1996).

Oxidants in cigarette smoke

Cigarette smoke is a complex mixture of over 4700 chemical compounds of which free radicals and other oxidants are present in high concentrations (Pryor & Stone 1993). The gas-phase of cigarette smoke contains approximately 10^{15} radicals per puff, primarily of the alkyl and peroxyl types.

The tar phase of cigarette smoke also contains a high concentration of radicals (10^{17} spins/gram, measured by electron spin resonance), which are more stable, such as the semiquinone radical, which can reduce oxygen to produce O_2^-, the hydroxyl radical (OH) and hydrogen peroxide (H_2O_2). Since both cigarette tar and lung epithelial lining fluid contain metal ions, such as iron, Fenton chemistry will occur to produce the hydroxyl radical a very reactive and potent oxidant.

Cell-derived oxidants

The increase in the oxidative burden produced by inhaling cigarette smoke can be enhanced in smokers' lungs by the release of oxygen radicals from inflammatory leukocytes, which are known to migrate into the lungs in increased numbers in cigarette smokers (Repine et al 1997).

Alveolar macrophages from the lungs of smokers release increased amounts of oxidants such as O_2^- and H_2O_2. Superoxide anion and H_2O_2 can be generated by xanthine/xanthine oxidase (XO). XO activity is increased in cell free bronchoalveolar lavage fluid from COPD patients, compared with normal subjects and is associated with increased O_2^- production.

The generation of oxidants in epithelial lining fluid in smokers may be further enhanced by the presence of increased amounts of free iron in the airspaces of patients with COPD (Halliwell & Gutteridge 1990). Free iron in the ferrous form catalyses Fenton's reaction and the Haber–Weiss reaction, which generate the hydroxyl radical, a free radical which is extremely damaging to all tissues.

Reactive oxygen species (ROS) and reactive nitrogen species (RNS) can also be generated intracellularly from several sources, such as mitochondrial respiration, the NADPH oxidase system, xanthine/xanthine oxidase and, in the case of RNS, from arginine by the action of nitric oxide synthetase in response to tumour necrosis factor (TNFα) and lipopolysaccharide (LPS), which are relevant inflammatory stimuli in COPD. Lipid peroxidation following the reaction of free radicals with polyunsaturated fatty acid side chains in membrane or lipoproteins is a further reaction that can result in cell damage and is self-perpetuating process which continues as a chain reaction (Gutteridge 1995).

Increased oxidative stress in smokers and patients with COPD

The evidence for the presence of increased oxidative stress in smokers and patients with COPD is now overwhelming (Repine et al 1997, Rahman & MacNee 1996).

Evidence of local oxidative stress in the lungs

The epithelial lining fluid (ELF) forms an interface between the airspace epithelium and the external environment and therefore forms a critical defence mechanism against inhaled oxidants or those produced by cells in the airspaces (Cross et al 1994).

The antioxidant properties of mucus derive from the abundance of sulfydryl and disulfide moieties in its structure, which effectively scavenge oxidants (Cross et al 1984), and its metal binding properties. Oxidants have been shown to cause the release of mucous. Antioxidant species in ELF comprise low molecular weight antioxidants, metal binding proteins, antioxidant enzymes and sacrificial reactive proteins and unsaturated lipids. The concentrations of non-enzymatic antioxidants vary in ELF and some are concentrated in ELF compared to plasma, such as glutathione (GSH) and ascorbate which may indicate their relative importance (Cross et al 1994) (Table 1).

There is limited information on the respiratory epithelial antioxidant defences in smokers, and even less for patients with COPD. GSH is elevated in bronchoalveolar lavage (BAL) fluid in chronic smokers (Morrison et al 1994). Rahman and colleagues (Rahman et al 1995, 1996a) studied the acute effects of cigarette smoke condensate (CSC) on GSH metabolism in a human alveolar epithelial cell line *in vitro*, and in rat lungs *in vivo* after intra-tracheal CSC instillation. They found a dose and time-dependent depletion of intracellular GSH, concomitant with the formation of GSH-conjugates.

TABLE 1   Antioxidant constituents of epithelial lining fluid

Antioxidant	Plasma, μM	ELF, μM
Ascorbic acid	40	100
Glutathione	1.5	100
Uric acid	300	90
Albumin-SH	500	70
α-tocopherol	25	2.5
β-carotene	0.4	—

Studies have shown variable changes in antioxidants in BAL fluid (Rahman et al 1996b). There appears to be no consistent change in antioxidant defences in ELF in smokers. The apparent inconsistencies between these studies in the levels of the different antioxidants in ELF and alveolar macrophages may be due to differences in the recent smoking histories in chronic smokers.

Markers of oxidative stress

Oxidative stress has been assessed by measuring markers of the effects of radicals on lung biomolecules such as lipid protein or DNA, or by measuring the stress responses to an increased oxidant burden. Patients with COPD have higher levels of exhaled H_2O_2 in breath condensate, a direct measurement of airspace oxidant burden, than normal non-smokers, and levels are even higher during exacerbations of COPD (Nowak et al 1996, 1998). However there are concerns over its reproducibility as a marker on oxidative stress.

Nitric oxide (NO) has been used as a marker of airway inflammation and indirectly as a measure of oxidative stress. Increased levels of NO occur in exhaled breath in some but not all studies of patients with COPD, but are not at the levels reported in asthmatics (Corradi et al 1999, Rutgers et al 1999). Smoking increases NO levels in breath which limits the usefulness of this marker in COPD.

Urine isoprostane $F_2\alpha$-III, which is an isomer of prostaglandin, formed by free radical peroxidation of arachidonic acid, is elevated in patients with COPD, compared with healthy controls, and is even more elevated in exacerbations of the condition (Pratico et al 1997).

Systemic oxidative stress

Patients with acute exacerbations of COPD show increased production of superoxide anion from their peripheral blood neutrophils compared with measurements in stable patients (Rahman et al 1997). Activation may be even more pronounced in neutrophils which are sequestered in the pulmonary microcirculation in smokers and in patients with COPD (Brown et al 1995).

Polyunsaturated fats and fatty acids in cell membranes are a major target of free radical attack, resulting in lipid peroxidation. Products of lipid peroxidation are significantly increased in plasma or in BAL fluid in healthy smokers and patients with acute exacerbations of COPD, compared with healthy non-smokers (Morrison et al 1999, Rahman et al 1996b). In addition increased circulating levels of F_2-isoprostane, which is a more direct measurement of lipid peroxidation have been found in smokers (Morrow et al 1995).

The plasma antioxidant capacity is significantly decreased both after smoking (Fig. 1) and also in patients with acute exacerbations of COPD, when compared with plasma from age-matched non-smoking controls. The levels of antioxidant capacity in the plasma have a negative correlation with the increased release of oxygen radicals from circulating neutrophils in patients with exacerbations of COPD (Rahman et al 1996b).

Oxidative stress and airspace epithelial injury

Injury to the airway epithelium is an important early event following exposure to cigarette smoke, and is shown by an increase in airspace epithelial permeability (Morrison et al 1994). The injurious effects of cigarette smoke on human alveolar epithelial cell monolayers, can be shown by increased epithelial cell detachment, decreased cell adherence and increased cell lysis (Lannan et al 1994). These effects are in part oxidant-mediated since they were partially prevented by the antioxidant GSH in concentrations $(500\,\mu M)$ which are present in the epithelial lining fluid. Extra- and intracellular GSH appears to be critical to the maintenance of epithelial integrity following exposure to cigarette smoke. This was shown in studies which demonstrate that the increased epithelial permeability of epithelial cell monolayers *in vitro*, and in rat lungs *in vivo* following exposure to cigarette smoke condensate, was associated with profound changes in the homeostasis of the antioxidant GSH (Li et al 1994, Rahman et al 1995, 1996c, Li et al 1996). Depletion of lung GSH alone can induce increased airspace epithelial permeability both *in vitro* and *in vivo* (Li et al 1994, 1995, 1996). Human studies have also shown increased epithelial permeability in chronic smokers compared with non-smokers, with a further increase in ^{99m}Tc-DTPA clearance following acute smoking (Morrison et al 1994).

Oxidative stress and neutrophil influx in the lungs

The lungs contain a large pool of non-circulating neutrophils, which are either retained or slowly moving within the pulmonary microcirculation (MacNee & Selby 1993). This results from the size differential between neutrophils (average

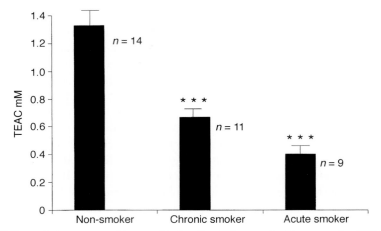

FIG. 1.   Effects of cigarette smoking on Trolox Equivalent antioxidant capacity (TEAC).
(Data shown are mean with SEM, ***P <0.001, compared with non-smokers.)

diameter 7 μM) and pulmonary capillary segments (average diameter 5 μM), which
causes a proportion of the circulating neutrophils to deform, and thus move
slowly, in order to negotiate the smaller capillary segments. In normal subjects
there is a correlation between neutrophil deformability measured *in vitro* and the
subsequent sequestration of these cells in the pulmonary microcirculation
following their re-injection — the less deformable the cells the more
sequestration of these cells occurs in the pulmonary circulation (Selby et al 1991).
The sequestration of neutrophils in the pulmonary capillaries allows time for the
neutrophils to interact and adhere to the pulmonary capillary endothelium, and
thereafter their transmigration across the alveolar capillary membrane to the
interstitium and airspaces of the lungs. There is a transient increase in neutrophil
sequestration in the lungs during smoking (MacNee & Selby 1993), which is due to
a decrease in the deformability of circulating neutrophils (Fig. 2) (Drost et al
1993). *In vitro* studies show that the decreased neutrophil deformability induced
by cigarette smoke exposure is abolished by antioxidants, such as GSH,
suggesting this event is oxidant-mediated (Drost et al 1992). Oxidative stress
reaches the circulation during cigarette smoking (Fig. 1), which could decrease
the deformability of neutrophils by polymerizing actin (Drost et al 1992).

Inhalation of cigarette smoke by hamsters increases neutrophil adhesion to the
endothelium of both arterioles and venules (Lehr et al 1993), which is thought to be
mediated by superoxide anion derived from cigarette smoke, since it was inhibited
by pre-treatment with Cu/Zn superoxide dismutase (SOD). Neutrophils
sequestered in the pulmonary circulation of the rabbit following cigarette smoke
inhalation also show increased expression of CD18 integrins (Klut et al 1993).

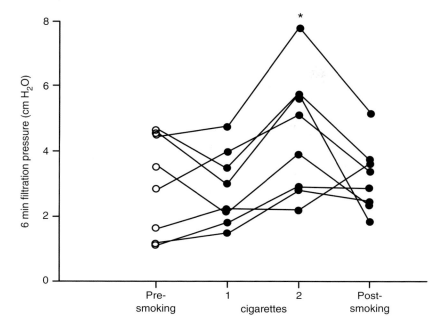

FIG. 2.   Neutrophil deformability measured *ex vivo* as 6 min filtration pressure (increasing pressure, decreased deformability) in eight healthy smokers, before, during and after smoking 1–2 cigarettes. There is a significant (*$P < 0.01$) increase in filtration pressure during smoking the 2nd cigarette which decreases 10 min post-smoking.

Activation of neutrophils sequestered in the pulmonary microvasculature could also induce the release of reactive oxygen intermediates and proteases within the microenvironment with limited access for free radical scavengers and antiproteases. Thus destruction of the alveolar wall, as occurs in emphysema, could result from a proteolytic insult derived from the intra-vascular space, without the need for the neutrophils to migrate into the airspaces.

Neutrophil sequestration in the microcirculation, allows chemotaxis to occur. Nicotine itself is chemotactic. Smoke exposure results in increased levels of chemotactic factors in the airspaces (Morrison et al 1998).

Oxidative stress and proteinase/antiproteinase imbalance

An increased elastase burden in the lungs and a functional 'deficiency' of α_1-antitrypsin due to its inactivation by oxidants is the basis of the protease/anti-protease theory of pathogenesis. This 'functional α_1-antitrypsin deficiency' is thought to be due to inactivation of the α_1-antitrypsin by oxidation of the methionine residue at its active site by oxidants in cigarette smoke. This is an over-simplification, because other proteinases and other antiproteinases are likely to have a role.

The acute effects of cigarette smoking on the functional activity of α_1-antitrypsin in BAL fluid have been studied, and have shown a transient, but non-significant fall in the anti-protease activity of BAL fluid 1 hour after smoking (Abboud et al 1985). Studies assessing the oxidative inactivation of α_1-antitrypsin in chronic smokers have also failed to produce clear support for this hypothesis.

Oxidant/antioxidant imbalance and the development of airways obstruction

Epidemiological studies have shown a relationship between circulating neutrophil numbers and the forced expiratory volume in one second (FEV_1) (Chan-Yeung & Dybuncio 1984) and between the change in peripheral blood neutrophil count and the change in airflow limitation over time (Chan-Yeung et al 1988). There is also a relationship between peripheral blood neutrophil oxidant release and measures of airflow limitation in young cigarette smokers (Richards et al 1989). Lipid peroxidation products in plasma has also been shown to correlate inversely with the percentage predicted FEV_1 in a population study (Schunemann et al 1997).

In the general population there is an association between dietary intake of the antioxidant vitamin E and lung function, supporting the hypothesis that this antioxidant may have a role in protecting against the development of COPD (Britton et al 1995, Grievink et al 1998). Antioxidant vitamin supplementation has been shown to reduce oxidant stress, measured as a decrease in pentane levels in breath as an assessment of lipid peroxides (Steinberg & Chait 1998).

Oxidative stress and gene expression

Pro-inflammatory genes

Numerous markers of inflammation have been shown to be elevated in the sputum of patients with COPD, including interleukin (IL)-8 and tumour necrosis factor (TNF)α. Genes for many of these inflammatory mediators are regulated by transcription factors such as nuclear factor kappa B (NF-κB). NF-κB is present in the cytosol in an inactive form linked to its inhibitory protein IκB. Many stimuli, including cytokines and oxidants, activate NF-κB, resulting in ubiquitination, cleaving of IκB from NF-κB and the destruction of IκB in the proteozome (Rahman & MacNee 1998). This critical event in the inflammatory response is redox sensitive. In vitro studies using both macrophage cell lines and alveolar and bronchial epithelial cells, show that oxidants cause the release of inflammatory mediators such as IL-8, IL-1 and NO and that these events are associated with increased expression of the genes for these inflammatory mediators and increased nuclear binding or activation of NF-κB (Parmentier et al 2000). Furthermore, stimuli relevant to the development of exacerbations of COPD, such as particulate air pollution that has oxidant properties also activates NF-κB in alveolar epithelial cells (Jimenez et al 2000).

Thiol antioxidants such as N-acetylcysteine and N-acystelin, which have potential as therapies in COPD, have been shown in *in vitro* experiments to block the release of these inflammatory mediators from epithelial cells and macrophages, by a mechanism involving increasing intracellular GSH and decreasing NF-κB activation (Parmentier et al 2000).

Antioxidant genes

An important response to oxidative stress is the up-regulation of protective antioxidant genes. The antioxidant GSH is concentrated in epithelial lining fluid compared with plasma (Rahman et al 1999) and appears to have an important protective role, together with its redox enzymes in the airspaces and intracellularly in epithelial cells. Human studies have shown that GSH is elevated in epithelial lining fluid in chronic cigarette smokers, compared with non-smokers, an increase which does not occur during acute cigarette smoking (Fig. 3) (Morrison et al 1999). The effects of acute and chronic cigarette smoking can be mimicked by exposure of airspace epithelial cells to cigarette smoke condensate *in vitro*. This produces an initial decrease in intracellular GSH with a rebound increase when the cells are washed and culture is continued for 24 h (Fig. 4) (Rahman et al 1995, 1996a,c). This effect *in vitro* was mimicked by a similar change in GSH in rat lungs *in vivo* following intratracheal instillation of cigarette smoke condensate (Rahman et al 1996a), associated with an increase in the oxidized form (GSSG). The increase in GSH following cigarette smoke exposure is due to transcriptional up-regulation of mRNA for γ-glutamyl cysteine synthetase (γGCS), the rate-limiting enzyme in GSH synthesis (Fig. 4) (Rahman et al 1996a,c). The mechanism of the up-regulation of γGCS mRNA is by the activation by cigarette smoke of the redox-sensitive transcription factor activator protein 1 (AP-1) (Rahman et al 1998, 1999). A proximal AP-1 site in the promoter region of the gene is critical for the regulation of γGCS gene expression in response to various oxidants including cigarette smoke (Rahman et al 1999). These events are likely to account for the increased GSH levels seen in the epithelial lining fluid in chronic cigarette smokers, which acts as a protective mechanism, whereas the more injurious effects of cigarette smoke may occur repeatedly during and immediately after cigarette smoking when the lung is depleted of antioxidants including GSH. The cytokine TNF, which is thought to have a role in the lung inflammation in COPD also decreases intracellular GSH levels initially in epithelial

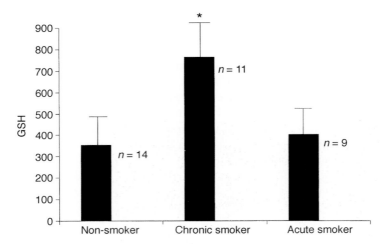

FIG. 3.   Effect of smoking on glutathione (GSH) lung epithelial lining fluid in smokers, non-smokers, smokers who have not smoked for 12 h (chronic smokers) and those who have smoked 1–2 cigarettes 1 h before the measurement (acute smokers). Data shown are mean plus SEM; there is a significant (*$P < 0.05$) increase in glutathione in epithelial lining fluid in chronic cigarette smokers compared with non-smokers which is not present following acute smoking.

cells by a mechanism involving intracellular oxidative stress, which is followed 12–24 h thereafter by a rebound increase in intracellular GSH as a result of AP-1 activation and increased γGCS expression (Rahman et al 1999). Corticosteroids have been used as anti-inflammatory agents in COPD, but there is still doubt over their effectiveness in reducing airway inflammation in COPD. Dexamethasone also

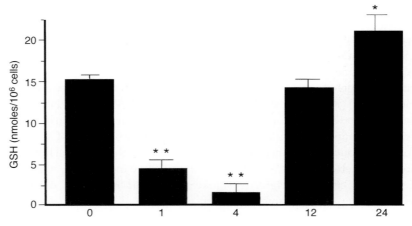

FIG. 4.   (*Caption opposite*)

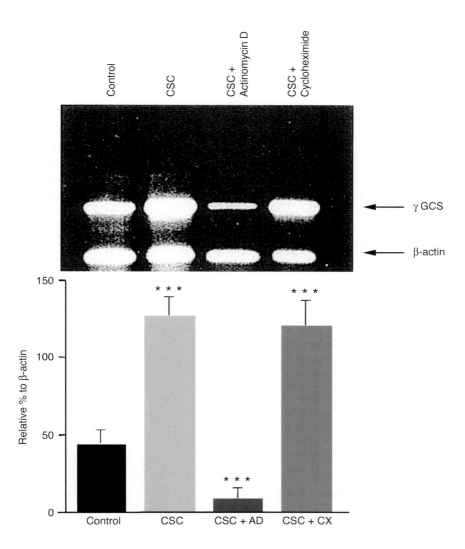

FIG. 4.   The effects of cigarette smoke condensate on glutathione (*facing page*) and γ-glutamylcysteine synthetase (γGCS) mRNA (*this page*) by RT-PCR in alveolar epithelial cells (A549). Cigarette smoke condensate produces an initial decrease in glutathione up to 4 h with a rebound increase in glutathione at 24 h. The rebound increase in intracellular glutathione is associated with up-regulation of mRNA for γGCS. This up regulation is at the transcriptional levels since it is blocked by actinomycin D but not by cyclohexamide. (*$P < 0.05$, **$P < 0.01$, ***$P < 0.001$).

causes a decrease in intracellular GSH in airspace epithelial cells, but no rebound increase compared with the effects of TNF (Rahman et al 1999). Moreover, the rebound increase in GSH produced by TNF in epithelial cells is prevented by co-treatment with dexamethasone (Rahman et al 1999). These effects may have relevance for the treatment of COPD patients with corticosteroids.

Rats exposed to whole cigarette smoke for up to 14 d show increased expression of a number of antioxidant genes in bronchial epithelial cells, including manganese superoxide dismutase (MnSOD), metallothionein (MT) and GSH peroxidase (Gilks et al 1998).

The c-*fos* gene belongs to a family of growth and differentiation-related immediate early genes, the expression of which generally represents the first measurable response to a variety of chemical and physical stimuli. Studies in various cell lines have shown enhanced gene expression of the c-*fos* in response to cigarette smoke condensate (Muller & Gebel 1998). These effects of cigarette smoke condensate can be mimicked by peroxynitrite and smoke-related aldehydes in concentrations that are present in cigarette smoke condensate. The effects of cigarette smoke condensate can be enhanced by pre-treatment of the cells with buthionine sulfoxamine to decrease intracellular GSH and can be prevented by treatment with N-acetylcysteine, a thiol antioxidant (Muller & Gebel 1998). These studies emphasize the importance, of intracellular levels of the antioxidant GSH in gene expression.

Thus oxidative stress, including that produced by cigarette smoke, causes increased gene expression of both injurious pro-inflammatory genes by oxidant-mediated activation of transcription factors such as NF-κB, but also activation of protective genes such as γGCS. A balance may therefore exist between pro and 'anti-inflammatory' gene expression in response to cigarette smoke, which may be critical to whether cell injury is induced by cigarette smoking (Fig. 5). Knowledge of the molecular mechanisms that regulate these events may open new therapeutic avenues in the treatment of COPD.

Oxidative stress and susceptibility to COPD

Since only a proportion — 15–20% — of cigarette smokers appear to be susceptible to its effects, showing a rapid decline in FEV_1 and development of the disease, there has been considerable interest in identifying those who are most susceptible.

Polymorphisms of various genes have been shown to be more prevalent in smokers who develop COPD than in smokers with a similar smoking history who have not developed COPD (Sandford et al 1997). Relevant to the effects of cigarette smoke is a polymorphism in the gene for microsomal epoxide hydrolase, which is an enzyme involved in the metabolism of highly reactive epoxide intermediates which are present in cigarette smoke (Smith & Harrison 1997).

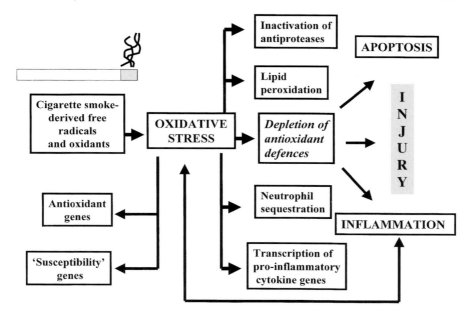

FIG. 5.   Effects of oxidative stress in the lungs of smokers.

The proportion of individuals with a slow microsomal epoxide hydrolase activity (homozygotes) was significantly higher in patients with COPD and a subgroup of patients shown pathologically to have emphysema (COPD 22%; emphysema 19%) compared with control subjects (6%).

Therapy to redress the oxidant/antioxidant imbalance in COPD

There are various options to enhance the lung antioxidant screen (Rahman & MacNee 1996, Repine et al 1997). One approach would be the molecular manipulation of antioxidant genes, such as GSH peroxidase or genes involved in the synthesis of GSH, such as γGCS or by developing molecules with activity similar to those of antioxidant enzymes, such as catalase and SOD.

Another approach would simply be to administer antioxidant therapy. This has been attempted in cigarette smokers using various antioxidants such as vitamin C and vitamin E (Kondo et al 1994, York et al 1976, Toth et al 1986, Nowak et al 1998). The results have been rather disappointing. Attempts to supplement lung GSH have been tried using GSH or its precursors (MacNee et al 1991). Nebulised GSH has also been used therapeutically but this has been shown to induce bronchial hyper-reactivity. The cysteine-donating compound N-acetylcysteine acts as a cellular precursor of GSH and becomes de-acetylated in the gut to cysteine following oral administration. It reduces disulfide bonds and has the potential to interact directly with oxidants. N-acetylcysteine given 600 mg 3 times daily for 5 days produced a significant increase in plasma GSH levels, but no associated rise in GSH in BAL or in lung tissue (Bridgeman et al 1991, 1994). These data seem to imply that producing a sustained increase in lung GSH is difficult using N-acetylcysteine in subjects who are not already depleted of GSH. In spite of this continental European studies have shown that N-acetylcysteine reduces the number of exacerbation days in patients with COPD (Bowman et al 1983). This was not confirmed in a British Thoracic Society study of N-acetylcysteine (British Thoracic Society Research Committee 1985). Nacystelyn is a lysine salt of N-acetylinecysteine. It is also a mucolytic and oxidant thiol compound, which in contrast to N-acetylcysteine, which is acid, has a neutral pH. Nacystelyn can be aerosolized into the lung without causing significant side effects. Studies comparing the effects of Nacystelyn and N-acetylcysteine found that both drugs enhanced intracellular GSH in alveolar epithelial cells and inhibited hydrogen peroxide and superoxide anion release from neutrophils harvested from peripheral blood from smokers and patients with COPD (Nagy et al 1997). There are as yet no clinical data on this drug.

In summary, there is now very good evidence for an oxidant/antioxidant imbalance in COPD and increasing evidence that this imbalance is important in the pathogenesis of this condition. There are a number of important effects of oxidative stress in smokers that are relevant to the development of COPD. Oxidative stress may also be critical to the inflammatory response to cigarette smoke, through the up-regulation of redox-sensitive transcription factors and hence pro-inflammatory gene expression; but is also involved in the protective mechanisms against the effects of cigarette smoke by the induction of antioxidant genes. Inflammation itself induces oxidative stress in the lungs and polymorphisms on genes for inflammatory mediators or antioxidant genes may have a role in the susceptibility to the effects of cigarette smoke. Knowledge of the mechanisms of the effects of oxidative stress should in future allow the development of potent antioxidant therapies which test the hypothesis that oxidative stress is involved in the pathogenesis of COPD, not only by direct injury to cells, but also as a fundamental factor in inflammation in smoking-related lung disease.

Acknowledgements

Funded by the Medical Research Society, The British Lung Foundation and The Colt Foundation.

References

Abboud RT, Fera T, Richter A, Tabona MZ, Johal S 1985 Acute effect of smoking on the functional activity of α-1-protease inhibitor in bronchoalveolar lavage fluid. Am Rev Respir Dis 131:79–85

Bridgeman MME, Marsden M, MacNee W, Flenley DC, Ryle AP 1991 Cysteine and glutathione concentrations in plasma and bronchoalveolar lavage fluid after treatment with N-acetylcysteine. Thorax 46:39–42

Bridgeman MME, Marsden M, Selby C Morrison D, MacNee W 1994 Effect of N-acetyl cysteine on the concentrations of thiols in plasma, bronchoalveolar lavage fluid, and lining tissue. Thorax 49:670–675

Bowman G, Backer U, Larsson S et al 1983 Oral acetylcysteine reduces exacerbation rate in chronic bronchitis. Eur J Respir Dis 64:405–415

British Thoracic Society Research Committee 1985 Oral N-acetylcysteine and exacerbation rates in patients with chronic bronchitis and severe airways obstruction. Thorax 40:823–835

Britton JR, Pavord ID, Richards KA et al 1995 Dietary antioxidant vitamin intake and lung function in the general population. Am J Respir Crit Care Med 151:1383–1387

Brown DM, Drost E, Donaldson K, MacNee W 1995 Deformability and CD11/CD18 expression of sequestered neutrophils in normal and inflamed lungs. Am J Respir Cell Mol Biol 13:531–539

Chan-Yeung M, Dybuncio A 1984 Leucocyte count, smoking and lung function. Am J Med 76:31–37

Chan-Yeung M, Abboud R, Dybuncio A, Vedal S 1988 Peripheral leucocyte count and longitudinal decline in lung function. Thorax 43:462–468

Corradi M, Majori M, Cacciani GC, Consigli GF, de'Munari E, Pesci A 1999 Increased exhaled nitric oxide in patients with stable chronic obstructive pulmonary disease. Thorax 54:576–580

Cross CE, Halliwell B, Allen A 1984 Antioxidant protection: a function of tracheobronchial and gastrointestinal mucus. Lancet 1:1328–1330

Cross CE, van der Vliet A, O'Neill CA, Louie S, Halliwell B 1994 Oxidants, antioxidants, and respiratory tract lining fluids. Env Health Perspect (suppl) 102:185–191

Drost EM, Selby C, Lannan S, Lowe GDO, MacNee W 1992 Changes in neutrophil deformability following *in vitro* smoke exposure: mechanism and protection. Am J Respir Cell Mol Biol 6:287–95

Drost E, Selby C, Bridgeman MME, MacNee W 1993 Decreased leukocyte deformability following acute cigarette smoking in smokers. Am Rev Respir Dis 148:1277–1283

Gilks CB, Price K, Wright JL, Churg A 1998 Antioxidant gene expression in rat lung after exposure to cigarette smoke. Am J Path 152:269–278

Grievink L, Smit HA, Ocke MC, van't Veer P, Kromhout D 1998 Dietary intake of antioxidant (pro)-vitamins, respiratory symptoms and pulmonary function: the MORGEN study. Thorax 53:166–171

Gutteridge JMC 1995 Lipid peroxidation and antioxidants as biomarkers of tissue damage. Clin Chem 41:1819–1828

Halliwell B 1996 Antioxidants in human health and disease. Annu Rev Nutr 16:33–50

Halliwell B, Gutteridge JMC 1990 Role of free radicals and catalytic metal ions in human disease: an overview. Methods Enzymol 186:1–85

Jimenez LA, Thomson J, Brown D, Rahman I, Hay RT, Donaldson K, MacNee W 2000 PM10 particles activate NF-κB in alveolar epithelial cells. Toxicol Appl Pharmacol 166:101–110

Klut DE, Doerschuk CM, Van Eeden JF, Burns AF, Hogg JC 1993 Activation of neutrophils within the pulmonary microvasculature of rabbits exposed to cigarette smoke. Am J Respir Cell Mol Biol 39:82–90

Kondo T, Tagami S, Yoshioka A, Nishumura M, Kawakami Y 1994 Current smoking of elderly men reduces antioxidants in alveolar macrophages. Am J Respir Crit Care Med 149:178–182

Lannan S, Donaldson K, Brown D, MacNee W 1994 Effects of cigarette smoke and its condensates on alveolar cell injury *in vitro*. Am J Physiol 266:L92–L100

Lehr HA, Kress E, Menger MD et al 1993 Cigarette smoke elicits leukocyte adhesion to endothelium in hamsters: inhibition by CuZnSOD. Free Radic Biol Med 14:573–581

Li XY, Donaldson K, Rahman I, MacNee W 1994 An investigation of the role of glutathione in the increased epithelial permeability induced by cigarette smoke *in vivo* and *in vitro*. Am J Respir Crit Care Med 149:1518–1525

Li XY, Donaldson K, Brown D, MacNee W 1995 The role of tumour necrosis factor in increased airspace epithelial permeability in acute lung inflammation. Am J Resp Cell Mol Biol 13:185–195

Li XY, Rahman I, Donaldson K, MacNee W 1996 Mechanisms of cigarette smoke induced increased airspace permeability. Thorax 51:465–471

MacNee W, Selby C 1993 New perspectives on basic mechanisms in lung disease. 2. Neutrophil traffic in the lungs: role of haemodynamics, cell adhesion, and deformability. Thorax 48:79–88

MacNee W, Bridgeman MME, Marsden M et al 1991 The effects of N-acetylcysteine and glutathione on smoke-induced changes in lung phagocytes and epithelial cells. Am J Med 90:60s–66s

Morrison D, Lannan S, Langridge A, Rahman I, MacNee W 1994 Effect of acute cigarette smoking on epithelial permeability, inflammation and oxidant status in the airspaces of chronic smokers. Thorax 49:1077 (abstr)

Morrison D, Strieter RM, Donnelly SC, Burdick M, Dunkel SL, MacNee W 1998 Neutrophil chemokines in bronchoalveolar lavage fluid and leukocyte-conditioned medium from nonsmokers and smokers. Eur Respir J 12:1067–1072

Morrison D, Rahman I, Lannan S, MacNee W 1999 Epithelial permeability, inflammation and oxidant stress in the air spaces of smokers. Smoking as a cause of oxidative damage. Am J Respir Crit Care Med 159:473–479

Morrow JD, Frei B, Longmire AW et al 1995 Increase in circulating products of lipid peroxidation (F_2-isoprostanes) in smokers. N Engl J Med 332:1198–1203

Muller T, Gebel S 1998 The cellular stress response induced by aqueous extracts of cigarette smoke is critically dependent on the intracellular glutathione concentration. Cardiogenesis 19:797–801

Nagy AM, Vanderbist F, Parij N, Maes P, Fondu P, Neve J 1997 Effect of the mucoactive drug nacystelyn on the respiratory burst of human blood polymorphonuclear neutrophils. Pulm Pharmacol Ther 10:287–292

Nowak D, Antczak A, Krol M et al 1996 Increased content of hydrogen peroxide in expired breath of cigarette smokers. Eur Respir J 9:652–657

Nowak D, Kasielski M, Pietras T, Bialasiewicz P, Antczak A 1998 Cigarette smoking does not increase hydrogen peroxide levels in expired breath condensate of patients with stable COPD. Monaldi Arch Chest Dis 53:268–273

Parmentier M, Drost ME, Hirani N et al 1999 Thiol antioxidants inhibit neutrophil chemotaxis by decreasing release of IL-8 from macrophages and pulmonary epithelial cells. Am J Respir Crit Care Med 159:A286

Pratico D, Basili S, Vieri M, Cordova C, Violi F, Fitzgerald GA 1997 Chronic obstructive pulmonary disease associated with an increase in urinary levels of isoprostane $F_2\alpha$-III, an index of oxidant stress. Am J Resp Crit Care Med 158:1709–1714

Pryor WA, Stone K 1993 Oxidants in cigarette smoke: radicals hydrogen peroxides peroxynitrate and peroxynitrite. Ann NY Acad Sci 686:12–28

Rahman I, MacNee W 1996 Role of oxidants/antioxidants in smoking-induced lung diseases. Free Radic Biol Med 21:669–681

Rahman I, MacNee W 1998 Role of transcription factors in inflammatory lung diseases. Thorax 53:601–612

Rahman I, Li XY, Donaldson K, Harrison DJ, MacNee W 1995 Glutathione homeostasis in alveolar epithelial cells *in vitro* and lung *in vivo* under oxidative stress. Am J Physiol 269:L285–L292

Rahman I, Smith CAD, Lawson M, Harrison DJ, MacNee W 1996a Induction of γ-glutamylcysteine synthetase by cigarette smoke condensate is associated with AP-1 in human alveolar epithelial cells. FEBS Letters 396:21–25

Rahman I, Morrison D, Donaldson K, MacNee W 1996b Systemic oxidative stress in asthma, COPD, and smokers. Am J Respir Crit Care Med 154:1055–1060

Rahman I, Bel A, Mulier B et al 1996c Transcriptional regulation of γ-glutamylcysteine synthetase-heavy subunit by oxidants in human alveolar epithelial cells. Biochem Biophys Res Commun 229:832–837

Rahman I, Skwarska E, MacNee W 1997 Attenuation of oxidant/antioxidant imbalance during treatment of exacerbations of chronic obstructive pulmonary disease. Thorax 52:565–568

Rahman I, Smith CAD, Antonicelli F, MacNee W 1998 Characterisation of γ-glutamylcysteine synthetase-heavy subunit promoter: a critical role for AP-1. FEBS Lett 427:129–133

Rahman I, Antonicelli F, MacNee W 1999 Molecular mechanisms of the regulation of glutathione synthesis by tumour necrosis factor-α and dexamethasone in human alveolar epithelial cells. J Biol Chem 274:5088–5096

Repine JE, Bast A, Lankhorst I 1997 Oxidative stress in chronic obstructive pulmonary disease. Oxidative Stress Study Group. Am J Resp Crit Care Med 156:341–357

Richards GA, Theron AJ, van der Merwe CA, Anderson R 1989 Spirometric abnormalities in young smokers correlate with increased chemiluminescence responses of activated blood phagocytes. Am Rev Respir Dis 139:181–187

Rutgers SR, van der Mark TW, Coers W et al 1999 Markers of nitric oxide metabolism in sputum and exhaled air are not increased in chronic obstructive pulmonary disease. Thorax 54:576–680

Sandford AJ, Weir TD, Pare PD 1997 Genetic risk factors for chronic obstructive pulmonary disease. Eur Respir J 10:1380–1391

Schunemann HJ, Muti P, Freudenheim JL et al 1997 Oxidative stress and lung function. Am J Epidemiol 146:939–948

Selby C, Drost E, Wraith PK, MacNee W 1991 *In vivo* neutrophil sequestration within the lungs of man is determined by *in vitro* 'filterability'. J Appl Physiol 71:1996–2003

Smith CAD, Harrison DJ 1997 Association between polymorphism in gene for microsomal epoxide hydrolase and susceptibility to emphysema. Lancet 350:630–633

Steinberg FM, Chait A 1998 Antioxidant vitamin supplementation and lipid peroxidation in smokers. Am J Clin Nutr 68:319–327

Toth KM, Berger EM, Buhler CJ, Repine JE 1986 Erythrocytes from cigarette smokers contain more glutathione and catalase and protect endothelial cells from hydrogen peroxide better than do erythrocytes from non-smokers. Am Rev Respir Dis 134:281–284

York GK, Pierce TH, Schwartz LS, Cross CE 1976 Stimulation by cigarette smoke of glutathione peroxidase system enzyme activities in rat lung. Arch Environ Health 31:286–290

DISCUSSION

Calverley: There is a problem, and it worries me a little bit. Although these are quite large studies and uniformly positive, there is a strong suspicion that some studies that were not positive were not reported. The limit of meta-analysis is that it depends on the data that go into it. These are very interesting data, and one would hope that they are going to be tested.

MacNee: I haven't seen the full report of this meta-analysis.

Calverley: There is also a meta-analysis that is in the Cochrane database by Peter Black and Phillipa Poole from Auckland, which draws the same kind of conclusion (Poole & Black 2000). For a long time people have looked at end points such as breathlessness, FEV_1 and exercise performance, and almost all the studies are negative. The exacerbation story, which has been going on mainly in mainland Europe and not in the transatlantic community, is worth thinking about.

Barnes: But you have to distinguish between mucolytics and antioxidants. *N*-acetylcysteine is classified as a mucolytic, which is wrong. It has this effect, but it is also an antioxidant.

MacNee: *N*-acetylcysteine is deacetylated in the gastrointestinal tract to cysteine, which is then synthesized into GSH in the liver. If *N*-acetylcysteine has an effect *in vivo* as an antioxidant, it has this property by acting as a precursor of GSH. We have measured the increase in GSH in BAL after *N*-acetylcysteine and found a small transient increase. In a parallel group study we measured GSH levels in lung tissue in subjects who took *N*-acetylcysteine before lung resection, and we showed a trend to higher GSH levels compared with subjects who did not receive this treatment, but this did not achieve statistical significance following a large dose of *N*-acetylcysteine. Perhaps one of the problems with the variable effects of *N*-acetylcysteine in clinical studies is the relatively small dose of *N*-acetylcysteine in some studies.

Barnes: We have studied oral *N*-acetylcysteine, and were not able to show any effect on exhaled markers of oxidative stress. As we have discussed before, the best way to give these drugs may be by inhalation. Oxidative stress will be highest in the lumenal compartment of the lung.

MacNee: *N*-acetylcysteine is acidic, and has been shown to produce bronchial hyper-reactivity when it is inhaled. A similar compound called Nacystelin is entering clinical trials and can be given in a dry powder since it has a neutral pH. We have used this drug in experimental studies, and it produced the same protective antioxidant effects as *N*-acetylcysteine. It has not yet been licensed for COPD, but is being used on clinical trials in cystic fibrosis.

Barnes: In Europe, nebulized *N*-acetylcysteine is available and is widely used.

MacNee: There are preliminary data which show that Nacystelin can reduce markers of inflammation in sputum in cystic fibrotic patients. One other interesting feature is that it has been shown to alter the proteolytic activity in sputum in patients with cystic fibrosis. This is thought to be due to the provisions of reducing groups which affect elastase activity.

Stockley: We showed that in 1984.

MacNee: There you are: there is nothing new under the sun.

Barnes: The other thing that you briefly alluded to that I think would be interesting to further study is the role of peroxynitrite in COPD. If there is increased oxidative stress and NO is also produced, peroxynitrite can form, which will nitrate proteins at tyrosine residues — probably for the life of the protein. Neutrophils are also effective at causing the nitration of proteins as well, through the release of myeloperoxidase. Some of these effects might be important in COPD. For example, tyrosine nitration can affect the signalling of epidermal growth factor (EGF) receptors and other signalling proteins that use tyrosine phosphorylation. This is something that needs to be explored, and it is a further reason for considering antioxidants as therapy.

MacNee: One of the important features that is becoming apparent is that it is not necessarily only the direct oxidative stress that is important: there are also several products of the initial oxidative stress which may also be important to the effects of oxidative stress downstream from this which may be vital. With respect to peroxynitrite, we are not sure what this does exactly *in vivo*.

Barnes: It nitrates tyrosine residues, which can be demonstrated immunocytochemically. The issue is whether this has functional consequences.

MacNee: Also I think products of the lipid peroxidation are important. We have been interested in a particular lipid peroxidation product which is likely to be highly relevant in activating of signal transduction events. The other point to remember is that inflammation itself causes oxidative stress. For example, TNF causes increased reactive oxygen species released from electron transport in mitochondria, producing intracellular oxidative stress, which can activate gene expression.

Rennard: I want to pick up on Peter Barnes' comment about the misclassification of the drugs that we use. Some antibiotics have antioxidant functions. To what degree is that responsible for some of their potential therapeutic benefits during COPD exacerbations? Would they be useful in COPD even outside of exacerbations?

MacNee: I don't know.

Nadel: Macrolides prevent the surface expression of elastase.

Stockley: They can act in two different ways — there is almost a biphasic response. In some instances they can actually enhance superoxide production by neutrophils, particularly where there are neutrophils that are not particularly good at this. In other instances, they actually scavenge. There are also data showing that some of these agents reduce leukotriene (LT)B4 or interleukin (IL)-1β production by inflammatory cells. It is not a stupid suggestion as there are patients who respond to antibiotics who shouldn't, and the issue has to be whether this is just a non-antimicrobial effect that these agents are having.

Nadel: They prevent the secretagogue effect of neutrophils *in vivo* and *in vitro*.

Calverley: One of the things that comes across is that oxidative stress is a very important mechanism in inflammation. It is also clear that there is a systemic component in inflammation in COPD. What I am not quite sure about is the size of the effect we are seeing in COPD given the amount of inflammation we get in the airways and how this compares with other inflammatory conditions such as interstitial lung disease, where there is a lot of inflammation going on. If you look at the same range of markers, are you getting the same kind of signal, or is it bigger in COPD than some of these other conditions? Is it possible even to make that kind of cross-disease comparison?

MacNee: Peter Barnes can talk about the surrogate markers of oxidative stress because he has measured these in idiopathic pulmonary fibrosis (IPF). We have looked at antioxidant capacity in plasma in IPF and shown depletion indicating oxidative stress in IPF similar to COPD. What is even more interesting related to cigarette smoking is that IPF is one of the conditions in which GSH is depleted in epithelial lining fluid. However, in patients that smoke, GSH is not depleted: it is not raised, as it is in normal chronic smokers, but it is not depleted. Therefore the mechanisms which activate AP-1 and γGCS in smokers also occur in cigarette smokers who have IPF.

Calverley: What happens to the stress responses in ex-smokers?

Barnes: We have looked at patients with rhinitis, and they do not have increased oxidative stress in their exhaled air. We have looked at a range of lung diseases, and all show increased oxidative stress. The highest of all is in cystic fibrosis, then COPD is next, then severe asthma. Mild asthma has less oxidative stress, and smokers without COPD have more than mild asthma patients. Oxidative stress is markedly increased in exacerbations of COPD and asthma, and may account for some of the inflammatory changes that occur. It is interesting that the studies with N-acetylcysteine have shown reduction of exacerbations and this might be a situation where these drugs are particularly indicated.

Massaro: Is there a change in evidence of inflammation if you put COPD patients on chronic oxygen therapy?

MacNee: Interestingly, I recently reviewed a paper which measured plasma antioxidant capacity and showed evidence of systemic oxidative stress by a decrease in antioxidant capacity in patients who were having long-term home oxygen therapy.

Agustí: To follow on that point, perhaps it is worth remembering that reactive oxygen species are being produced constantly in the mitochondria, for example, under physiological conditions. In an isolated preparation in our lab we showed that if you induce hypoxia, all of the immediate early response genes are up-regulated, and this can be blocked by N-acetylcysteine.

Reference

Poole PJ, Black PN 2000 Mucolytic agents for chronic bronchitis or chronic obstructive pulmonary disease. In: The Cochrane Library, issue 2. Update Software, Oxford

Proteases and antiproteases

Robert A. Stockley

Department of Medicine, Queen Elizabeth Hospital, Edgbaston, Birmingham B15 2TH, UK

Abstract. Serine proteases have been implicated in the pathogenesis of chronic obstructive pulmonary disease (COPD) since the identification of α_1-antitrypsin deficiency in 1963. This inhibitor efficiently inactivates several enzymes released by activated neutrophils including neutrophil elastase, cathepsin G and proteinase 3, all of which have been shown to generate features of COPD in animal models. Recent studies have identified the mechanisms of enzyme release from activated neutrophils and indicate that the concentrations are usually two orders of magnitude above that of normal α_1-antitrypsin. This results in an area of obligate proteolysis in the immediate vicinity of a migrating neutrophil. The area is greatly enlarged in α_1-antitrypsin deficiency explaining the increased susceptibility of such patients to develop lung damage. The migration into and activation of neutrophils in the lung is likely to be a major determinant of the development of COPD. Understanding the processes has important implications for the design of new therapeutic strategies.

2001 Chronic obstructive pulmonary disease: pathogenesis to treatment. Wiley, Chichester (Novartis Foundation Symposium 234) p 189–204

The protease–antiprotease theory of the pathogenesis of chronic obstructive pulmonary disease (COPD) and, in particular, the features of emphysema and bronchial disease has dominated respiratory research into these conditions for some 36 years. The research programme was instigated in 1963 following the discovery of the strong association between the development of early onset emphysema and the inherited deficiency of the serine proteinase inhibitor α_1-antitrypsin (Laurell & Eriksson 1963). Initial studies in this genetic condition indicated not only the early onset of disease but also the presence of chronic bronchitis and predominantly lower zone emphysema which were linked to rapid progression and a reduction in life expectancy particularly if individuals continued to smoke (Larsson 1978).

Because α_1-antitrypsin is the major serum inhibitor of serine proteinases this observation suggested that a serine proteinase normally controlled by this inhibitor was responsible for the pathogenic features of the lung disease. Studies subsequently indicated that α_1-antitrypsin was also the major inhibitor of serine proteinases present in the alveolar region of the lung (Gadek et al 1981a) adding further support to the concept that a serine proteinase was responsible for the disease.

Initial studies in the early 1960s demonstrated that proteinases were capable of producing lung lesions similar to human emphysema in experimental animals. These initial studies included the use of the metalloproteinase papain and subsequently the serine enzyme, porcine pancreatic elastase. However, eventually serine proteinases more relevant to humans were shown to produce emphysema. Initially, Senior and his colleagues demonstrated that the human enzyme neutrophil elastase could produce these pathological changes (Senior et al 1977) and subsequently Kao et al (1988) demonstrated that a further serine proteinase from the neutrophil (proteinase 3) could also produce emphysema. Both of these enzymes are inhibited specifically and strongly by α_1-antitrypsin, providing indirect evidence that they may be the critical proteinases involved in the development of emphysema in α_1-antitrypsin deficiency.

Processes involved

These serine proteinases are capable of damaging many components of the extracellular matrix, including, collagen, laminin, fibrillin and elastin. However, the majority of evidence suggests that it is destruction of lung elastin that is of importance in the development of emphysema. As cells migrate from the vascular space into the airway they release neutrophil elastase producing a destructive pathway to facilitate the migration. Although lung elastin can then repair itself, it results in the deposition of amorphous elastin and a loss of the fibrillary structure which is probably critical for maintaining peripheral airway integrity.

Evidence in humans

Although all the forgoing observations and experiments provide compelling evidence that serine proteinases and perhaps most specifically neutrophil elastase is a major mediator of the development of emphysema, evidence in humans is at best indirect. Undoubtedly, elastase and proteinase 3 will be delivered to the airway and interstitium as neutrophils migrate. In addition, a small sub-population of monocytes have also been shown to contain neutrophil elastase, and these cells represent a subset that respond to chemotactic agents and may migrate into the lung releasing their elastase in a similar way to the neutrophil (Owen et al 1994). In addition, it is known that neutrophils are increased in the lower airways in cigarette smokers (Hunninghake & Crystal 1983) which is a major risk factor in the development of emphysema. Elegant studies by Damiano et al (1986) using immunohistochemistry demonstrated a strong relationship between the severity of emphysematous change in the human lung and the amount of neutrophil elastase present in the interstitium, although it remains unproven whether this represents a cause or effect.

Quantum proteolysis

Activated neutrophils adherent to connective tissue substrate have been shown to cause its degradation. In studies using fibronectin as the substrate, evidence would suggest that the majority of the digestion of this connective tissue is caused by neutrophil elastase (Chamba et al 1991) rather than cathepsin G and proteinase 3, both of which are also present within the same azurophil granule. The process of tissue degradation occurs in very close proximity to the neutrophil and cannot be completely abrogated even in the presence of a supranormal physiological concentration of inhibitor.

Early studies suggested that the tight adherence between the neutrophil and the substrate largely excluded proteinase inhibitors from this site thereby facilitating the activity of the released enzymes (Campbell et al 1982). Indeed, data had shown that the process could be more successfully prevented by pre-incubation of the substrate with proteinase inhibitors such as secretory leukoproteinase inhibitor which became adherent to the substrate thereby providing a local protective shield (Llewellyn-Jones et al 1994). Nevertheless, more recent elegant studies have confirmed that the ability of neutrophils to destroy connective tissue even in the presence of high concentrations of the appropriate proteinase inhibitor is largely based on a major mismatch between the concentration of enzyme being released from the cell and that of the inhibitors surrounding it (Liou & Campbell 1995).

Neutrophil elastase is stored pre-formed within the azurophil granule where the concentrations have been shown to be approximately 5 mM, which is several orders of magnitude higher even than the serum concentration of α_1-antitrypsin, its major inhibitor (30 μM). As the azurophil granules are released from an activated neutrophil the neutrophil elastase starts to diffuse away from the granule and the concentration gradually drops until it equals that of the surrounding inhibitor whereupon it is completely inactivated and no further connective tissue degradation takes place.

Mathematical modelling of this process has indicated that in the immediate vicinity of the granule concentration drops very rapidly. However, as the enzyme diffuses further away the fall in concentration gradually slows down producing an exponential concentration/distance relationship. Observation of the area of connective destruction occurring in close proximity to the azurophil granule has shown that the relationship between the area of enzyme activity and the concentration of inhibitor in the system follows the same theoretical curve predicting the concentration of enzyme (Liou & Campbell 1996). Of importance to note in this relationship is that the exponential part of the curve occurs at concentrations below $10\,\mu M$. In α_1-antitrypsin deficiency the serum concentrations are typically of the order of $5\,\mu M$, and thus on theoretical grounds it would be predicted that every migrating neutrophil would produce a much larger area of damage in the lung in α_1-antitrypsin deficiency than in subjects without deficiency. This may explain the susceptibility of such individuals not only to develop clinically important lung disease but to do so at a younger age and to progress more rapidly (see Fig. 1).

Implications for non-deficient subjects

Quantum proteolysis and the physiological role of the neutrophil indicate that connective tissue destruction will be a constant and necessary feature of cell migration. Thus, it would be expected that the greater the number of neutrophils that migrate into the lung and the longer the period of time over which they do so, the more likely it will be that the individual will develop clinically significant disease even in the presence of normal α_1-antitrypsin. These possibilities have not been explored in great detail in patients with emphysema although there is some evidence in the literature to support the processes that may be of importance. For instance, McCrea and colleagues studied healthy smokers and non-smokers, and demonstrated that in the majority of smokers the concentration of the neutrophil

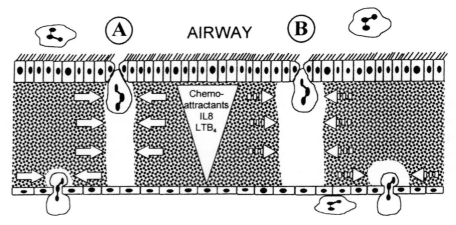

FIG. 1.   (A) Migration of neutrophils into the airway in response to chemoattractants such as IL-8 and LTB4 result in an area of degradation of connective tissue which is tightly limited by the surrounding proteinase inhibitors. (B) In α_1-antitrypsin deficiency the critically low concentration of the inhibitor results in a much larger area of connective tissue destruction as the neutrophil migrates.

chemoattractant, interleukin (IL)-8 in alveolar lavage fluid was the same as in healthy non-smokers (McCrea et al 1994). However, a small proportion of their smokers (approximately 10%) had concentrations of IL-8 that were above the healthy control subjects and demonstrated that these samples had greater chemotactic activity for neutrophils than normal. Since it is generally accepted that only a proportion of cigarette smokers are susceptible to the development of emphysema this observation may be of central importance. It is possible that individual subjects exposed to cigarette smoke respond differently in their production and release of cytokines that are chemotactic for the neutrophil. Indeed cigarette smoke can stimulate epithelial cells to release IL-8 (Mio et al 1997) and if the release is excessive in susceptible subjects then neutrophil recruitment would also be expected to be excessive resulting in increased tissue damage with time. Whether the increased release of IL-8 is genetic or represents a secondary phenomenon has yet to be explored in depth. Studies have suggested that latent adenovirus infections may increase the susceptibility of epithelial cells to release inflammatory cytokines like IL-8 when stimulated (Keicho et al 1997). Indeed, latent adenovirus infections have been shown to be a feature of patients with COPD (Matsuse et al 1992) but this possibility of a secondary effect leading to the development of COPD remains to be clarified. Alternatively it is possible that excessive inflammation may also arise because of other gene polymorphisms such as that seen for the tumour necrosis factor (TNF) receptor which is more common in COPD (Huang et al 1997). TNF is a cytokine which stimulates IL-8 production (Kwon et al 1994) and thus the receptor polymorphism could account for increased IL-8 production in susceptible smokers but studies have to be undertaken to determine whether this is a possible mechanism.

Neutrophil defects

It is possible that the neutrophils themselves respond differently in the susceptible individual. Studies in smokers with emphysema have shown that the neutrophils demonstrate an increased chemotactic response to a standard chemoattractant when compared to age and smoking related control subjects (Burnett et al 1987). The implication of this finding is that even if the amount of neutrophil chemoattractant in the lung was being released in normal quantities the cell response would be excessive, again resulting in greater neutrophil traffic with time.

Finally, studies have also shown that the neutrophils from patients with emphysema have an increased potential to digest connective tissue both in the resting state and when stimulated (Burnett et al 1987); again the cause is unknown. However in this instance the implication would be that the number of neutrophils migrating could remain normal but for each cell a greater area of destruction would be produced as it migrates. Again the net result would be a greater than 'normal' amount of connective tissue damage with time.

Treatment strategy

As argued above, the major problem in α_1-antitrypsin deficiency is the excessive area of connective tissue damage that occurs as cells migrate which is related to the critically low concentration of α_1-antitrypsin. Thus, theoretically, replacement therapy with α_1-antitrypsin should result in a marked reduction in this destructive process. Replacement therapy for α_1-antitrypsin has been present for many years in the USA (and some European countries) although unfortunately no controlled clinical trial has been carried out. However, scientific data have shown that infusions of α_1-antitrypsin result in a rapid increase in circulating levels and that with appropriate strategy the concentrations at the end prior to the next infusion can be maintained above the critical level of $10\,\mu M$ (Wewers et al 1987). As indicated above, this level is likely to be appropriate to markedly limit the area of connective tissue destruction produced around an activated neutrophil. It is not possible directly to determine the α_1-antitrypsin concentration in the interstitium, however, replacement therapy does normalize the α_1-antitrypsin concentration in the epithelial lining fluid of the lung (Gadek et al 1981b) suggesting it is also 'normal' in the interstitium. Also the major restriction to protein movement is the epithelial rather than the endothelial cell layer. Concentrations of proteins in the serum and interstitium are thought to be similar (Gorin & Stewart 1979) whereas concentrations in airway lining fluid are markedly reduced (approximately 10% of that present in the serum). Thus it is likely that the interstitial concentration of α_1-antitrypsin in patients on replacement therapy remains close to the protective threshold of $10\,\mu M$ even when levels in the plasma are lowest ($\sim 10\,\mu M$) just prior to the next infusion.

Patients with α_1-antitrypsin deficiency also have increased neutrophil recruitment to the lung which would still produce greater damage than normal to the interstitial connective tissue. The exact mechanism is unproven but Hubbard at al (1991) demonstrated that the concentration of neutrophil chemoattractant leukotriene B4 (LTB4) is increased in lung lavage fluid in deficient subjects.

These authors thought that the source was the alveolar macrophage and that the α_1-antitrypsin deficiency was responsible. They argued that the low α_1-antitrypsin concentration in the lung would fail to inactivate any neutrophil elastase released at this site and the active enzyme then stimulated the macrophages to release LTB4 (Hubbard et al 1991). Replacement therapy should thus reverse this second process as well. Normalization of lung α_1-antitrypsin would restore the ability to inactivate elastase in the lung, removing free elastase activity and hence the stimulus to release

LTB4 from the macrophages. The net result would be a reduction in neutrophil traffic and restoration of the α_1-antitrypsin above the critical protective threshold. To date however, the effects of replacement therapy on neutrophil traffic has not been determined.

Non-deficient patients

Patients without α_1-antitrypsin deficiency should already have adequate protective levels of this proteinase inhibitor in the interstitium. However, as argued above, it is likely to be the neutrophil traffic and state of activation of these cells that are critical in the development of emphysema. Thus, it may be that the appropriate strategy for such patients is to either reduce the concentration of elastase within the azurophil granule or alter the cellular response to activation or the chemoattractants. In this respect agents like non-steroidal anti-inflammatory drugs have been shown to alter the nature of the neutrophil resulting in both reduced chemotactic response and degranulation as shown in Figs 2 and 3 (Barton et al 2000). Therapeutic trials of such a strategy have yet to be undertaken although one study investigating the role of colchicine (which

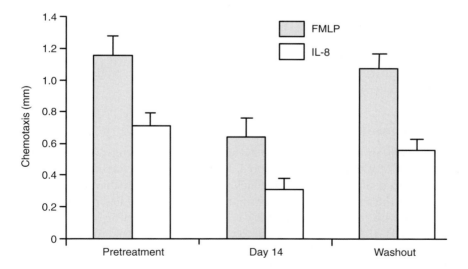

FIG. 2. The chemotactic response of neutrophils is indicated on the vertical axis using the distance migrated under agarose in response to FMLP (formyl-methionyl-leucyl-phenylalanine; hatched histogram) and IL-8 (open histogram). Results are shown from neutrophils from healthy individuals prior to treatment at the end of 14 days therapy with indomethacin and 2 weeks after cessation of therapy in the washout phase. Indomethacin caused a highly significant reduction in response by the end of therapy.

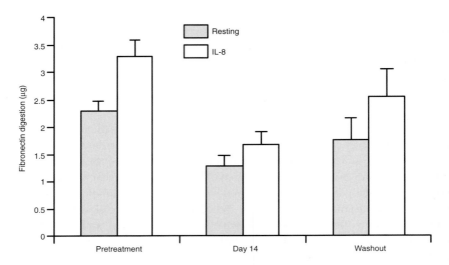

FIG. 3.   Fibronectin degradation by neutrophils is shown on the vertical axis for cells at rest (hatched histogram) and following stimulation with IL-8 (open histogram) results are shown for neutrophils obtained prior to therapy at the end of 14 days therapy with indomethacin and the washout 2 weeks after cessation of therapy. Indomethacin caused highly significant reduction in fibronectin degradation both by unstimulated and stimulated cells.

reduces cell migration) was not shown to have an effect on inflammation as monitored in the airway (Cohen et al 1990). Nevertheless, studies modulating neutrophil function are ongoing.

Other features of COPD

Because serine proteinases have been shown to cause emphysema their role in other features of COPD has also been explored. Neutrophil elastase has been shown to be a potent cause of airway damage which can be observed within 20 minutes of infusion of the enzyme into the airway (Suzuki et al 1996). Pathological studies show that with time ciliated epithelia is lost and goblet cell hyperplasia occurs (Snider et al 1985), the airways become lined with mucus and elastase and cathepsin G have been shown to be major secretagogues (Sommerhoff et al 1990). In addition, elastase will reduce ciliary beat frequency (Smallman et al 1984) and can damage epithelial cells (Amitani et al 1991) which would facilitate bacterial colonization, another important feature of COPD. Finally, elastase has also been shown to reduce the secretion of secretory leukproteinase inhibitor (Sallenave et al 1994), an elastase inhibitor which also has antibacterial activity (Hiemstra et al 1996) and this effect would also facilitate bacterial colonization.

Free elastase activity has been detected in the major airways of patients with α_1-antitrypsin deficiency (Hill et al 1999) and in patients without deficiency particularly during acute bacterial exacerbations (Stockley & Burnett 1979). Furthermore, the amount of free elastase activity detected in patients with α_1-antitrypsin deficiency during exacerbations is even greater than in non-deficient subjects (Hill et al 1999). Thus, the balance between serine proteinases and their inhibitors in the airway may also be important in determining the clinical features of COPD. Recent studies have suggested that chronic bronchial disease is an important risk factor in the prognosis of lung disease and if so the presence of elastase activity may be a central feature. Thus bacterial colonization and in particular the degree of bacterial colonization and the number and severity of bacterial exacerbations may be of major importance, and alternative strategies may be required to stabilize COPD. These may include inhalation of appropriate serine proteinase inhibitors or a more aggressive antimicrobial therapy approach to the management of these patients.

Summary

In summary, serine proteinases and their inhibitors have long been implicated in the pathogenesis of COPD. Our present understanding means that it should be possible to develop strategies that would interfere with the processes thought to be responsible. Such strategies would help confirm or refute the role of enzymes such as neutrophil elastase and may prove highly successful in stabilization of the disease and prevention of progression.

References

Amitani R, Wilson R, Rutman A et al 1991 Effects of human neutrophil elastase and *Pseudomonas aeruginosa* proteinases on human respiratory epithelium. Am J Respir Cell Mol Biol 4:26–32

Barton AE, Bayley DL, McCarmie M et al 2000 Phenotypic changes in neutrophils related to anti-inflammatory therapy. Biochem Biophys Acta 1500:108–118

Burnett D, Chamba A, Hill SL, Stockley RA 1987 Neutrophils from subjects with chronic obstructive lung disease show enhanced chemotaxis and extracellular proteolysis. Lancet 8567:1043–1046

Campbell EJ, Senior RM, McDonald JA, Cox DL 1982 Proteolysis by neutrophils. Relative importance of cell-substrate contact and oxidative inactivation of protease inhibitors *in vitro*. J Clin Invest 70:845–852

Chamba A, Afford SC, Stockley RA, Burnett D 1991 Extracellular proteolysis of fibronectin by neutrophils: characterisation and the effects of recombinant cytokines. Am J Respir Cell Mol Biol 4:330–337

Cohen AB, Girard W, McLarty J et al 1990 A controlled trial of colchicine to reduced the elastase load in the lungs of cigarette smokers with chronic obstructive pulmonary disease. Am Rev Respir Dis 142:63–72

Damiano VV, Tsang A, Kucich U et al 1986 Immunolocalization of elastase in human emphysematous lung. J Clin Invest 78:482–493

Gadek JE, Fells GA, Zimmerman RL, Rennard SI, Crystal RG 1981a Antielastases of human alveolar structures. Implications for the protease–antiprotease theory of emphysema. J Clin Invest 68:889–898

Gadek JE, Klein HG, Holland PV, Crystal RG 1981b Replacement therapy of alpha 1-antitrypsin deficiency. Reversal of protease–antiprotease balance within alveolar structures of PiZ subjects. J Clin Invest 68:1158–1165

Gorin AB, Stewart PA 1979 Differential permeability of endothelial and epithelial barriers to albumin flux. J Appl Physiol 47:1315–1324

Hiemstra PS, Maassen RJ, Stolk J, Heinzel-Wieland R, Steffens GJ, Dijkman JH 1996 Antibacterial activity of antileukoprotease. Infect Immun 64:4520–4524

Hill AT, Campbell EJ, Bayley DL et al 1999 Evidence for excessive bronchial inflammation during an acute exacerbation of COPD in patients with alpha 1-antitrypsin deficiency. Am J Respir Crit Care Med 160:1968–1975

Huang SL, Su CH, Chang SC 1997 Tumor necrosis factor-α gene polymorphism in chronic bronchitis. Am J Respir Crit Care Med 156:1436–1439

Hubbard RC, Fells G, Gadek J, Pacholok S, Humes J, Crystal RG 1991 Neutrophil accumulation in the lung in alpha 1-antitrypsin deficiency. Spontaneous release of leukotriene B4 from alveolar macrophages. J Clin Invest 88:891–897

Hunninghake GW, Crystal RG 1983 Cigarette smoking and lung destruction. Accumulation of neutrophils in the lungs of cigarette smokers. Am Rev Respir Dis 128:833–838

Kao RC, Wehner NG, Skubitz KM, Hoidal JR 1988 Proteinase 3: a distinct human polymorphonuclear leukocyte proteinase that produces emphysema in hamsters. J Clin Invest 82:1963–1973

Keicho N, Elliott WM, Hogg JC, Hayashi S 1997 Adenovirus E1A regulates IL-8 expression induced by endotoxin in pulmonary epithelial cells. Am J Physiol (Lung Cell Mol Physiol) 272:L1046–L1052

Kwon OJ, Au BT, Collins PD et al 1994 Tumor necrosis factor-induced interleukin-8 expression in cultured human airway epithelial cells. Am J Physiol 267:L398–L405

Larsson C 1978 Natural history and life expectancy in severe α_1-antitrypsin deficiency, Pi Z. Acta Med Scand 204:345–351

Laurell C, Eriksson S 1963 The electophoretical1-globulin pattern of serum in a1-antitrypsin deficiency. Scand J Clin Lab Invest 15:132–140

Liou TG, Campbell EJ 1995 Nonisotropic enzyme–inhibitor interactions: a novel nonoxidative mechanism for quantum proteolysis by human neutrophils. Biochemistry 34:16171–16177

Liou TG, Campbell EJ 1996 Quantum proteolysis resulting from release of single granules by human neutrophils: a novel, nonoxidative mechanism of extracellular proteolytic activity. J Immunol 157:2624–2631

Llewellyn-Jones CG, Lomas DA, Stockley RA 1994 Potential role of recombinant secretory leucoprotease inhibitor in the prevention of neutrophil mediated matrix degradation. Thorax 49:567–572

McCrea KA, Ensor JE, Nall K, Bleecker ER, Hasday JD 1994 Altered cytokine regulation in the lungs of cigarette smokers. Am J Respir Crit Care Med 150:696–703

Matsuse T, Hayashi S, Kuwano K, Keunecke H, Jefferies WA, Hogg JC 1992 Latent adenoviral infection in the pathogenesis of chronic airways obstruction. Am Rev Respir Dis 146:177–184

Mio T, Tadashi M, Romberger DJ et al 1997 Cigarette smoke induces interleukin-8 release from human bronchial epithelial cells. Am J Respir Crit Care Med 155:1770–1776

Owen CA, Campbell MA, Boukedes SS, Stockley RA, Campbell EJ 1994 A discrete subpopulation of human monocytes expresses a neutrophil-like proinflammatory (P) phenotype. Am J Physiol 267:L775–L785

Sallenave JM, Shulmann J, Crossley J, Jordana M, Gauldie J 1994 Regulation of secretory leukocyte proteinase inhibitor (SLPI) and elastase specific inhibitor (ESI-elafin) in human airway epithelial cells by cytokines and neutrophilic enzymes. Am J Respir Cell Mol Biol 11:733–741

Senior RM, Tegner H, Kuhn C et al 1977 Induction of pulmonary emphysema with human leukocyte elastase. Am Rev Respir Dis 116:469–475

Smallman LA, Hill SL, Stockley RA 1984 Reduction of ciliary beat frequency *in vitro* by sputum from patients with bronchiectasis: a serine protenaise effect. Thorax 39:663–667

Snider GL, Stone PJ, Lucey EC et al 1985 Eglin-c, a polypeptide derived from the medicinal leech, prevents human neutrophil elastase induced emphysema and bronchial secretory cell metaplasia in the hamster. Am Rev Respir Dis 132:1155–1161

Sommerhoff CP, Nadel JA, Basbaum CB, Caughey GH 1990 Neutrophil elastase and cathepsin G stimulate secretion from culture bovine airway glands serus cells. J Clin Invest 85:682–689

Stockley RA, Burnett D 1979 Alpha-1-antitrypsin and leukocyte elastase in infected and non-infected sputum. Am Rev Respir Dis 120:1081–1086

Suzuki T, Wang W, Linn J-T, Shirato K, Mitsuhashi H, Inoue H 1996 Aerosolised human neutrophil elastase induces airways constriction and hyper-responsiveness with protection by intravenous pre-treatment with half-length secretory leukoprotease inhibitor. Am J Respir Crit Care Med 153:1405–1411

Wewers MD, Casolaro MA, Sellers S et al 1987 Replacement therapy for alpha 1-antitrypsin deficiency associated with emphysema. N Eng J Med 316:1055–1062

DISCUSSION

MacNee: What is the latest on elastase inhibitors? Earlier Jay Nadel mentioned that he had used an elastase inhibitor that had never made it to the clinic.

Stockley: That was a great inhibitor, and it looked very promising. To this day I don't know why that programme was shelved. There were various other companies who also had elastase inhibitors. I was talking to GlaxoWellcome the other day, and they had a well developed programme, and all of a sudden their inhibitor went down because it was toxic. This may partly be because people have got cold feet, for two reasons. First, there are expanding concepts about other proteinases causing emphysema, which has muddied the waters. Second, there is the issue of how to produce and fund a study that shows prevention of progression. This would take several hundred patients and take five years if your outcome is FEV_1. I think — and I am sure that Jan Stolk will support this — that if we are really looking at something that prevents progression of emphysema, we can actually see emphysema using high resolution computed tomography (CT) scans. What Jan has shown and we are also showing is that you can see the disease progress by the year in small numbers of patients. If that ever became accepted as a primary outcome measure — or at least as the next phase of the development of a drug — this could be done much more economically.

MacNee: So there are no elastase inhibitors going into clinical trials.

Stockley: For emphysema that is true, but there are studies of elastase inhibitors coming on for other aspects of COPD where elastase may play a role.

Nadel: Specifically, in terms of the ICI anti-elastase, the decision was made not to proceed with drug development. As far as can be determined, it was not an issue of drug toxicity. It may have been an issue of cost of manufacture, but there is no one I know who knows. The only study that was done was positive.

Barnes: The drug was a peptide and there were pharmacokinetic problems. This is the likely reason why it was not developed commercially.

Calverley: I am trying to model what goes on in emphysema, and I am trying to make the leap from what Jim Hogg was saying about neutrophils migrating through post-capillary venules to Rob Stockley's fibronectin fluorescein-labelled gel digested area. If I understand the logic of what is being said, the problem is that these neutrophils are crawling around. What is actually happening in the process of alveolar destruction? Have we any ideas at all, or is it just too slow moving a process for us to even second-guess? Are we talking about cells that just go straight through, or are we talking about cells that are trapped in the interstitium, going round chewing holes in it to get out in the way that we are hearing described? It is quite important, because it will dictate how we intervene.

Campbell: It is clear that the lung can tolerate a lot of neutrophil traffic. There may be situations in which the amount of traffic is so excessive that there is permanent injury resulting simply from the inflammatory cells passing through. There may also be situations in which the inflammatory cells are activated in the interstitium as they pass through, and finally there is the well-explored situation in which there is a low concentration of protease inhibitors that makes the interstitial space vulnerable when inflammatory cells traffic through. The real challenge is understanding why some individuals who have an apparently normal complement of proteinase inhibitors can still suffer permanent loss of alveolar structures during the process of inflammation. Some of the most interesting information on this subject came from the studies that Dr Vic Damiano performed did before he retired (Damiano et al 1986). He showed that leukocyte elastase was present on elastic fibres in the interstitial space. The responsible neutrophils were nowhere to be seen. Presumably they had transmigrated, and during the process of migrating through the interstitial space they had released their granule contents close enough to the elastic fibres that the leukocyte elastase was able to bind to the elastic fibres. It has been shown that once elastase binds an elastic fibre, it tends to be persistent on the fibre and digest it over a prolonged period of time. Once it binds to the elastic fibre, naturally occurring inhibitors can't inhibit the elastase, even though it goes from site to site on the fibre (Morrison et al 1999, Padrines & Bieth 1991). Perhaps very occasionally when an inflammatory cell traverses the interstitial space, something goes wrong such that some of the enzyme gets in the wrong place. At least in the case of leukocyte elastase, this can result in an ongoing injury until the elastic fibre breaks. The dynamics of this process are totally unexplored *in vivo*.

Hogg: Older work (Boren 1962) and a more recent report (Nagai & Thurlbeck 1991) suggest that the first thing that happens is that the pores enlarge and become fenestrae. This suggests there might be more neutrophil margination and more opportunity to migrate near the pores of Kohn. A fascinating thing that is present in experimental emphysema (Massaro & Massaro 1997) and in human emphysema (Coxson et al 1999) is that the volume goes up first and surface area is relatively well maintained. Something must happen to the elastic fibres that make the pressure–volume characteristics of those units different so that they expand at functional residual capacity (FRC). This can be appreciated in direct measurements of the centrilobular emphysematous (CLE) space where the CLE spaces behave like paper bags (Hogg et al 1969). That is, they are easy to inflate until they reach their limit. I think it is possible that the initial damage to the elastin allows the volume to increase and that this decreases the surface area to volume ratio with the destruction of the lung surface coming at a later stage of the disease.

Nadel: It is possible to make neutrophils migrate into the airway and then out to the goblet cells. When a neutrophil is recruited and approaches a goblet cell, neutrophil elastase comes to the surface. Then an interaction between the neutrophil and the goblet cell occurs, which is Mac1-dependent and ICAM-1 dependent. This interaction causes a second messenger in the neutrophil to change the structure of the elastase that is on the surface and allow it to free itself from the neutrophil and cause goblet cell degranulation. This can be blocked with certain inhibitors. There is a close contact interaction between the neutrophil and the target cell, the goblet cell, which allows elastase-induced degranulation to occur in the goblet cell. Antibodies to the adhesion molecules block goblet cell degranulation. I would make an alternative suggestion, therefore, that as neutrophils are moving through tissue, although the elastase is on the surface, it won't necessarily be released. If it is not released onto that substrate, perhaps you won't get tissue breakdown. It is possible that the adhesion-interaction occurs in the degradation of lung tissue, just as it does in the degranulation of goblet cells.

Jeffery: A word of caution. I remember the Damiano paper (Damiano et al 1986), but this work was repeated again later by Professor Bernard Fox and Dr Terry Tetley using a more rigorous immunoelectron microscopic methodology, where they wished to control for the spurious effects of tissue charge attraction. They found no evidence of long term retention of elastase on the fibres. I don't know if Damiano's work has been confirmed.

Stockley: It has been repeated by a Chinese group who got the same results (Ge et al 1990).

Campbell: It has also been studied by the Italian group in several murine models (de Santi et al 1995, Cavarra et al 1996, 1999).

Jeffery: Obviously neutrophils can move through tissues without apparently destroying alveolar architecture. It seems to me that neutrophils can leave the vessels without damaging them, and yet in some way in emphysema they destroy the tissue. Often what happens in emphysema is that you are left with a lot of vessels, whereas the rest of the tissue has gone. There seems to be a two-phase process: whatever the neutrophil uses in order to migrate through the basement membranes surrounding the vessels is different to what it is doing when it is destroying tissue.

Stockley: I think that is right: I don't think neutrophils use elastase to go through the basement membrane.

MacNee: I was under the impression that loss of pulmonary capillary bed was an early feature of emphysema.

Dunnill: That is true. In the immunoelectron microscopy study, did they not show some remodelling taking place of the elastic tissue?

Campbell: The local distribution of elastase as interstitial elastin correlated with local airspace enlargement in Damiano's study (Damiano et al 1986). In de Santi's studies, the burden of elastase correlated with lung elastin loss (de Santi et al 1995).

Dunnill: With regard to the neutrophils, a lot of early work showed that cigarette smoke induced degranulation in neutrophils. Is that still valid?

Stockley: Probably.

MacNee: There were also studies which assessed whether neutrophils had lost elastase during migration, and they hadn't. The question I would ask is, does the neutrophil have to migrate at all?

Hogg: Dr Robert Wright reported some excellent morphological studies of the elastin network in normal and emphysematous lungs in the 1960s (Wright 1961). I think it might be useful to do more studies of this type and develop a model that would allow us to think about how cells come in contact with the network and damage it. A good model could also be used to predict how these changes would affect the mechanical properties of the tissue.

Calverley: There is another thing that these data raise: you have calculated volume and density, and presumably we know something about the elastin content of the tissue. We have always assumed that we are dealing with a single process that moves uniformly throughout the evolution of this disease. It would be nice to know whether the relative loss of elastin per gram of content is the same throughout the disease evolution: if it is not, this suggests that different processes may be more relevant at different disease stages.

Stockley: There aren't many data suggesting that total lung elastin is lost in this condition. The actual amount of elastin is increased. The same is true of the animal model. What we are talking about is lysis of elastin which is then re-laid in an amorphous fashion rather than as fibrils.

Nadel: Is the elastin breakdown dependent on adhesion of neutrophils?

Hogg: I don't know whether there is adhesion, but I do know that in David Walker's study of neutrophil migration, he felt that the neutrophils were often in contact with elastin.

Paré: What is the 250 g loss in lung weight seen in lung volume reduction surgery patients if it is not loss of elastin along with other lung connective tissue?

Hogg: Lung tissue has disappeared, and I don't know how the elastin gets reorganized around the edge of that hole, and how much of that weight the elastin would actually contribute.

Paré: I was fascinated by the data on neutrophils in bronchoalveolar lavage from α_1-antitrypsin-deficient patients. Were those non-smoking α_1-antitrypsin patients?

Stockley: Those were patients who smoked, but there are data from Ron Crystal's group looking at subjects with no abnormal lung function who were non-smoking α_1-antitrypsin patients (Hubbard et al 1991). They had an increased neutrophil load and increased LTB4 in the lavage fluid.

Paré: Does that predict which patients will go on to develop emphysema?

Stockley: I don't think it does for the simple reason that if you are α_1-antitrypsin deficient and a non-smoker your prognosis is quite good. Only about half of those patients go on to develop progressive airflow obstruction, but at a later age than the smokers.

References

Boren HG 1962 Alveolar fenestrae. Relationship to the pathology and pathogenesis of pulmonary emphysema. Am Rev Respir Dis 85:328–344

Cavarra E, Martorana PA, Gambelli F, de Santi M, van Even P, Lungarella G 1996 Neutrophil recruitment into the lungs is associated with increased lung elastase burden, decreased lung elastin, and emphysema in alpha 1 proteinase inhibitor-deficient mice. Lab Invest 75:273–280

Cavarra E, Martorana PA, de Santi M et al 1999 Neutrophil influx into the lungs of beige mice is followed by elastolytic damage and emphysema. Am J Respir Cell Mol Biol 20:264–269

Coxson HO, Rogers RM, Whittall KP et al 1999 A quantification of lung surface area in emphysema using computed tomography. Am J Respir Crit Care Med 159:151–156

Damiano VV, Tsang A, Kucich U et al 1986 Immunolocalization of elastase in human emphysematous lungs. J Clin Invest 78:482–493

de Santi MM, Martorana PA, Cavarra E, Lungarella G 1995 Pallid mice with genetic emphysema. Neutrophil elastase burden and elastin loss occur without alteration in the bronchoalveolar lavage cell population. Lab Invest 73:40–47

Ge YM, Zhu Y J, Luo WC, Gong YH, Zhang XQ 1990 Damaging role of neutrophil elastase in the elastic fiber and basement membrane in human emphysematous lung. Chin Med J (Engl) 103:588–594

Hogg JC, Nepszy S, Macklem PT, Thurlbeck WM 1969 The elastic properties of the centrilobular emphysematous space. J Clin Invest 48:421–431

Hubbard RC, Fells G, Gadek J, Pacholok S, Humes J, Crystal RG 1991 Neutrophil accumulation in the lung in alpha 1-antitrypsin deficiency. Spontaneous release of leukotriene B4 by alveolar macrophages. J Clin Invest 88:891–897

Massaro GD, Massaro D 1997 Retinoic acid treatment abrogates elastase-induced pulmonary emphysema. Nat Med 3:675–677

Morrison HM, Welgus HG, Owen CA, Stockley RA, Campbell EJ 1999 Interaction between leukocyte elastase and elastin: quantitative and catalytic analyses. Biochim Biophys Acta 1430:179–190

Nagai A, Thurlbeck WM 1991 Scanning electron microscopic observations of emphysema in humans. A descriptive study. Am Rev Respir Dis 144:901–908

Padrines M, Bieth JG 1991. Elastin decreases the efficiency of neutrophil elastase inhibitors. Am J Respir Cell Mol Biol 4:187–193

Wright RR 1961 Elastic tissue in normal and emphysematous lungs. Am J Pathol 39:355–367

Matrix metalloproteinases and TIMPs: properties and implications for the treatment of chronic obstructive pulmonary disease

Tim Cawston, Severine Carrere, Jon Catterall, Richard Duggleby, Sarah Elliott, Bill Shingleton and Andrew Rowan

Department of Rheumatology, Department of Medicine, University of Newcastle, Framlington Place, Newcastle upon Tyne NE2 4HH, UK

Abstract. The matrix metalloproteinases (MMPs) are a unique family of metalloenzymes that, once activated, can destroy connective tissue. The active enzymes are all inhibited by tissue inhibitors of metalloproteinases (TIMPs). The relative amounts of active MMPs and TIMPs are important in determining whether tissues are broken down in disease. Although elastase is often regarded as the target enzyme in chronic obstructive pulmonary disease (COPD), both the neutrophils and macrophages in the lung contain metalloproteinases and both collagen and elastin are degraded in disease. Transgenic studies have shown that when MMP1 is over-expressed, pulmonary emphysema develops in mice, while MMP12 knockout mice do not develop pulmonary emphysema when exposed to cigarette smoke. New drugs that can specifically block active MMPs are now available. These potent inhibitors are effective *in vitro* and prevent the destruction of tissue in animal models. Future patient trials will test the effectiveness of these compounds in preventing tissue destruction.

2001 Chronic obstructive pulmonary disease: pathogenesis to treatment. Wiley, Chichester (Novartis Foundation Symposium 234) p 205–228

Connective tissue is made up of a variety of proteins that comprise the extracellular matrix and include collagens, elastin and proteoglycans. In normal tissue a steady state exists where the turnover of these molecules is in equilibrium. Any change in this steady-state situation will rapidly affect the healthy function of the tissue. During connective tissue turnover, active proteinases degrade the extracellular matrix. In many tissues these proteinases are made by the same connective tissue fibroblasts that synthesise matrix components, but various inflammatory cells such as neutrophils or macrophages also synthesize and secrete such enzymes.

The extracellular matrix — structure and function

Collagen is the most common protein of the vertebrate body and has a unique structure (Prockop & Kivirikko 1995). Three coils of polypeptide form a triple helical molecule with globular regions at either end of a rod-shaped triple helix. On leaving the cell the globular regions are removed and the collagen molecules align to produce the characteristic staggered arrangement. Cross-links form between the collagen molecules, which increases stability. These self-aligned collagen molecules form the collagen fibres that give connective tissues their strength and rigidity (Fig. 1). Type I collagen is present in the lung but other collagen types such as type II can also be found in the cartilage surrounding larger airways.

In the lung, elastic fibres are present in the blood vessel walls and in the interstitium of the whole respiratory tree, where they help to restore the lung to

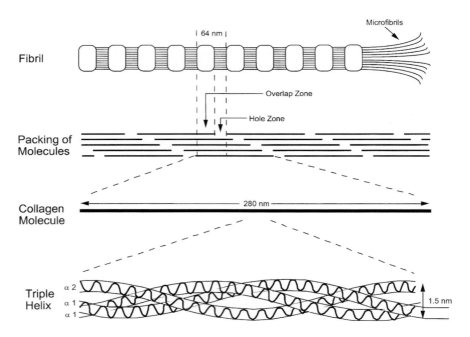

FIG. 1.   Structure and assembly of collagen. The three chains of the triple helical precursor molecule, procollagen, are synthesized within the cell where these chains wind about each other to form the triple helix. After secretion, globular regions are cleaved from each end of the procollagen molecule to form collagen and the individual molecules assemble in a staggered array. Cross-links form between the individual molecules and these form the collagen fibres which give tissues strength and rigidity.

its original shape and volume during expiration. Lung elastic fibres account for 1–2.5% of the dry weight of the lung, they are highly branched and the majority are less than 0.1 mm in diameter. The elastic fibres, composed of elastin and microfibrillar proteins, are responsible for the resilient properties of the respiratory structures at low levels of stress. At higher levels of stress collagenous fibres provide the resistance to further stretch and generate tissue recoil. Thus both of these components of the extracellular matrix are important to the healthy functioning of the tissue.

How is connective tissue broken down?

There are four main classes of proteinase that are classified according to the amino acid or chemical group at the catalytic centre of the enzyme (Barrett 1994). Proteolytic pathways can be divided into (1) intracellular pathways, catalysed by cysteine and aspartate proteinases, which cleave proteins at low pH inside the cell and (2) extracellular pathways, catalysed by serine and metalloproteinases, which act at neutral pH outside the cell. These pathways all play a part in the turnover of connective tissue and the pathway that predominates varies with different resorptive situations.

Matrix metalloproteinases — a unique family of proteinases

A family of closely related proteinases called matrix metalloproteinases (MMPs) has been characterized that can destroy all the proteins of the extracellular matrix (Woessner 1991, Birkedal-Hansen et al 1993, Cawston 1996). MMPs have a number of common properties (see Fig. 2): (1) they contain common sequences of amino acids; (2) they are secreted as inactive proenzymes; (3) activation can be achieved proteolytically; (4) they contain zinc at the active centre; and (5) the active enzymes are all inhibited by tissue inhibitors of metalloproteinases (TIMPs), act at neutral pH and cleave extracellular matrix proteins.

The MMP family can be divided into four main groups (Murphy & Knäuper 1997), which differ in size and substrate specificity (Table 1). These are called stromelysins, collagenases, gelatinases and membrane-type metalloproteinases. The different MMPs (with some previously used names) are listed in Table 1. Two of these proteinases (MMP8 and MMP9) are found stored within the specific granules of the neutrophil while the others are produced by a variety of cells after stimulation by different cytokines and other mediators (Birkedal-Hansen et al 1993, Goldring 1993). There are some enzymes (e.g. stromelysins and collagenases) with very similar substrate specificities but their pattern of expression is often distinct. This allows different enzymes to be used to digest the

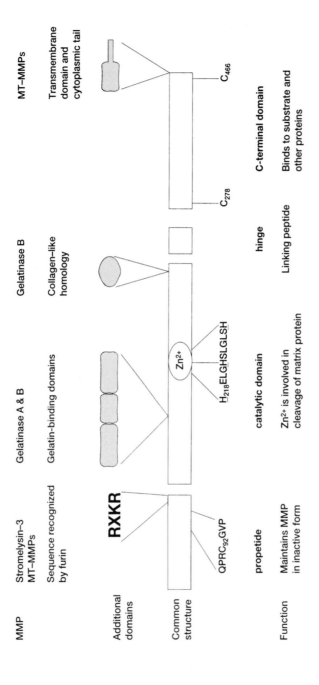

FIG. 2. Domain structure of MMPs. The MMPs have a domain structure that is common to all members of the family. A propeptide domain contains a conserved cysteine that binds to the active site zinc that is bound to three highly conserved histidines in the N-terminal domain. A hinge peptide with little secondary structure links to the C-terminal domain. This domain is important for binding to substrate and, in the progelatinases, also binds to the tissue inhibitors of metalloproteinases TIMP-1 and TIMP-2. The gelatinases have collagen-like and gelatin binding domains inserted into the N-terminal domain. In the membrane-type metalloproteinases, a membrane-spanning domain is found at the C-terminus of the protein and in these enzymes (and stromelysin 3) a sequence of 10–12 amino acids is found between the propeptide and the N-terminal domain. This is recognised by furin-like enzymes so allowing activation prior to secretion.

same matrix component at different times or locations thus providing a high degree of precise control.

How are the MMPs controlled?

As the MMPs are so potent they are carefully controlled at a number of key points: (1) various cytokines, growth factors and other agents can stimulate the synthesis and secretion of proMMPs (Goldring 1993); (2) the proMMPs require activation (Springman et al 1990); and (3) the active enzymes can be inhibited by TIMPs, which are also regulated by growth factors (Cawston 1996). These control points are illustrated in Fig. 3, which uses collagenase as an example.

Control of the MMPs — synthesis and secretion

Interleukin (IL)-1 and tumour necrosis factor (TNF)α both stimulate cells to produce proinflammatory and degradative effects, and these cytokines stimulate the synthesis and secretion of MMPs. Other growth factors and cytokines such as platelet-derived growth factor (PDGF), fibroblast growth factor (FGF)-2 and epidermal growth factor (EGF) can up-regulate some MMPs (Goldring 1993) and other agents such as IL-4, IL-13, IL-10 and transforming growth factor (TGF)β can often down-regulate these enzymes (van Roon et al 1996, Cawston et al 1996). Oncostatin M, which is present in macrophages within the rheumatoid joint and other inflammatory sites, can synergize with IL-1 to markedly up-regulate MMP1 and MMP3 (Korzus et al 1997, Cawston et al 1998). Thus the control of the MMPs is complex and cell–cell interactions and cell–matrix interactions can also modify the responses of individual cell types (Fig. 3).

Control of MMPs — how are the proMMPs activated?

Activation of the MMPs is an important control point in connective tissue breakdown. All the metalloproteinases are produced in a proenzyme form that requires the removal of a 10 kDa N-terminal fragment to activate the enzyme. Many active MMPs activate other proMMPs suggesting that a carefully controlled activation cascade occurs *in vivo* involving different members of the MMP family. Some MMPs (e.g. MT-MMPs, MMP11) have a conserved sequence of some 10–12 amino acids between the propeptide and the N-terminal domain, which is recognized by the furin family of serine proteinases (Pei & Weiss 1995). This allows the activation of the enzyme within the Golgi and these enzymes arrive at the cell surface in an active form. They are thought to be responsible for the activation of other members of the MMPs. There is strong evidence for a role for the plasminogen–plasmin system in the activation of proMMPs in some

TABLE 1   The matrix metalloproteinase family (MMPs) — names, numbers and substrates

Enzyme name (and previously used names)	Substrate
The collagenases	collagen type I,II,III,VII,VIII,X
interstitial collagenase	gelatins, aggrecan, tenascin, entactin
MMP1	link protein, α_1-proteinase inhibitor
vertebrate collagenase	
fibroblast collagenase	
neutrophil collagenase	collagen type I,II,III,VII,X,
MMP8	aggrecan, gelatin, fibronectin
collagenase 2	α_1-proteinase inhibitor
collagenase 3	collagen I,II,III,VII, X
MMP13	aggrecan, gelatin
collagenase 4	collagen I, gelatin
MMP18	
The gelatinases	
gelatinase A	gelatin collagens I,IV,V,VII,X, XI, XIV
MMP2	fibronectin, elastin
72 kDa gelatinase	laminin, aggrecan
type IV collagenase	vitronectin, α1-proteinase inhibitor
gelatinase B	gelatin
MMP9	collagens I,IV,V,VII, X,XI, X1V
92 kDa-gelatinase	vitronectin, elastin, aggrecan
type V collagenase	α1-proteinase inhibitor
The stromelysins	
stromelysin 1	aggrecan, link protein fibronectin, laminin,
MMP3	elastin gelatin, α1-proteinase inhibitor
transin	collagens I,III, IV, V,VIII, IX, X,XI
proteoglycanase	vitronectin, tenascin, decorin procollagen
procollagenase activator	peptides, activates procollagenase
stromelysin 2	gelatin
MMP10	collagen Types I,III,IV,V,VIII,IX
transin 2	fibronectin, laminin
	elastin, aggrecan
	activates procollagenase
stromelysin 3	N-terminal domain cleaves casein
MMP11	α1-proteinase inhibitor
matrilysin	gelatins, elastin
MMP7	aggrecan, α1-proteinase inhibitor
pump 1	fibronectin, link protein
uterine metalloproteinase	vitronectin, tenascin-C
	entactin, laminin
	activates procollagenase

(*continued*)

TABLE 1   (*continued*)

Enzyme name (and previously used names)	Substrate
The membrane bound MMPs	
MT-1 MMP	collagen I,II,III, proteoglycan
MMP14	laminin B chain, fibronectin, gelatin
	vitronectin activates progelatinase-A
MT-2 MMP	collagens I, II, fibronectin, proteoglycan
MMP15	tenascin, laminin
MT-3 MMP	collagen III, fibronectin, gelatin, laminin
MMP16	activates progelatinase-A and possibly other
	MMPs
MT-4 MMP	gelatin
MMP17	
MT-5-MMP	gelatin, fibronectin, proteoglycan
MMP24	
Other enzymes	
Metalloelastase	elastin, fibronectin
MMP12	collagen IV, gelatin, vitronectin, laminin
	α1-proteinase inhibitor
MMP19	aggrecan
MMP20	amelogenin
enamalysin	
MMP21	not known
XMMP	
MMP22	not known
CMMP	
MMP23	not known

situations of matrix turnover (Murphy et al 1992). Other mechanisms could include the activation of proMMPs by reactive oxygen species released from inflammatory cells (Saari et al 1990).

Control of MMPs — how are the active MMPs inhibited?

All connective tissues contain members of the TIMP family (TIMPs 1–4) (Murphy & Willenbrock 1995, Denhardt et al 1993). These inhibitory proteins are synthesised by connective tissue cells. TIMPs bind tightly to all known active MMPs with a 1:1 stoichiometry (Gomis-Rüth et al 1997). All four TIMPs have a polypeptide molecular weight of approximately 21 kDa and contain two domains held in a rigid conformation by six disulfide bonds. The TIMPs play an important role in controlling connective tissue breakdown by blocking the action of activated MMPs and preventing activation of the proenzymes. The properties of each

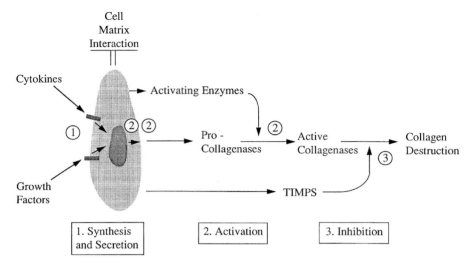

FIG. 3.   Control of MMP activity. MMPs are controlled in different ways. A variety of cytokines and growth factors can up-regulate MMP production and/or TIMP production. These agents often act synergistically and different cell types respond in different ways. Both cell–cell interactions and cell–matrix interactions can modify the response. After leaving the cell in a latent form the enzymes require activation before substrate can be cleaved. Some MMPs can be activated intracellularly and leave the cell in an active form. If the amount of active MMP exceeds the locally available TIMPs then connective tissue breakdown occurs.

member of the TIMP family are summarized in Table 2. All active MMPs can also be inhibited by the serum proteinase inhibitor α_2-macroglobulin, and this large inhibitor is important in blocking MMP activity in serum and other body fluids (Birkedal-Hansen et al 1993).

MMPs and lung disease

A large number of studies indicate that the MMPs are involved in the pathological destruction of joint tissue (Clark et al 1993, Vincenti et al 1994, Hill et al 1994) although their role in lung disease is less clear (Tetley 1997). There is good evidence that neutrophil elastase is involved in the destruction of elastin in lung tissue (see previous chapter). However, both MMP8 and MMP9 are also found in neutrophils (Tschesche et al 1994, Hibbs et al 1985) and can also degrade collagen and elastin, respectively. Macrophages and not neutrophils are the most abundant cell type in both normal and diseased lungs and these cells within diseased lung produce large amounts of MMP9, matrilysin (MMP7) and MMP12 (metalloelastase), all of which can degrade elastin (see Fig. 4). MMP12 is present

TABLE 2   Properties of tissue inhibitors of metalloproteinases

	TIMP-1	TIMP-2	TIMP-3	TIMP-4
MMP inhibition	All	All	All	?All
Mature protein size (Da)	20243	21729	21676	22609
Glycosylation	Yes	No	Yes	No
Localization	Diffusible	Diffusible	ECM bound	?
Gene	Xp11.23. − 11.4	17q2.3–2.5	22q12.1–13.2	?
Transcripts (kb)	0.9	3.5, 1.0	4.5–5.0, (2.8, 2.4)	1.2–1.4
Expression	Inducible	Constitutive	Inducible	?
Major tissue sites	Bone, ovary	Lung, ovaries, brain, testes, heart, placenta	Kidney, brain, lung, heart, ovary	Kidney, placenta, colon, testes, brain, heart, ovary, skeletal muscle
Binding to proMMP	MMP9	MMP2	?	?

at very low levels in normal macrophages but expressed in high levels in the alveolar macrophages of cigarette smokers (Shapiro 1994). Elegant studies by Shapiro's laboratory have shown that in MMP12 knockout mice pulmonary emphysema did not develop when transgenic mice inhaled cigarette smoke (Hautamaki et al 1997). Lung washings have also been shown to contain significant levels of MMP2 and MMP9 (Finlay et al 1997). MMP12 is required for both macrophage accumulation and for tissue destruction seen in emphysema resulting from the long-term inhalation of cigarette smoke and MMPs may also be required to allow macrophages to enter the lung. Transgenic studies have shown that when MMP1 is over-expressed, pulmonary emphysema develops in mice (D'Armiento et al 1992).

There is no doubt that collagen destruction represents a key step in connective tissue turnover, and one that represents an important therapeutic target (Shingleton et al 1996). The loss of collagen as well as elastic tissue has a profound effect on lung function where enlargement of the peripheral air spaces of the lung is accompanied by the destruction of connective tissue in the alveolar walls.

It is interesting to note that MMPs can degrade serpins, the inhibitors of serine proteinases. Many MMPs can destroy the inhibitory activity of α_1-proteinase inhibitor (Table 1). Elastase specifically cleaves TIMP-1 and destroys its inhibitory activity against MMPs. These interactions between serine and metalloproteinases are illustrated in Fig. 5.

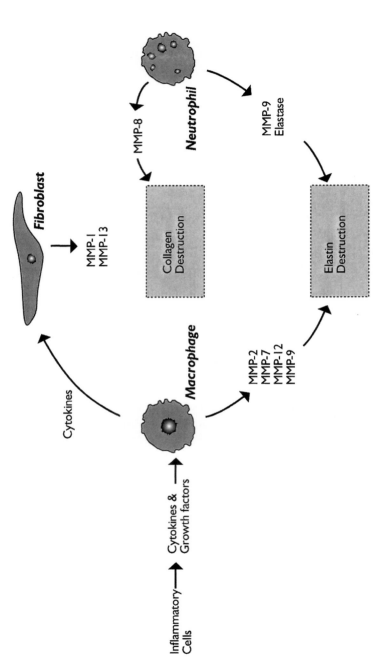

FIG. 4. The role of MMPs in chronic obstructive pulmonary disease. Different cells within inflamed lung tissue produce a variety of cytokines and growth factors that promote the release of MMPs and serine proteinases from cells. Once activated these enzymes can degrade connective tissue collagen if they exceed local TIMP concentrations. Neutrophils contain the serine proteinases elastase and cathepsin G and also the metalloproteinases MMP8 and MMP9, and these enzymes can degrade collagen and elastin, respectively. Macrophages produce both MMP2, MMP9, matrilysin (MMP7) and the metalloelastase MMP12. The relative contribution of each enzyme to the destruction of lung tissue is not yet known.

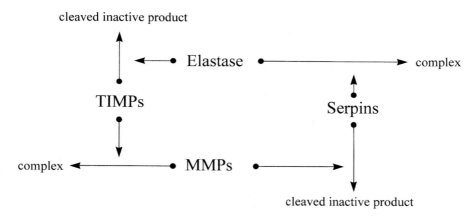

FIG. 5.   Interaction between serine and metalloproteinases. Active MMPs can degrade and inactivate serpins, the inhibitors of serine proteinases whilst elastase can specifically inactivate TIMPs.

In cartilage it is now known that a metalloproteinase but not a MMP is responsible for the cleavage of aggrecan (Sandy et al 1992, Tortorella et al 1999). A membrane-bound metalloproteinase, called TNFα-convertase, is responsible for the release of cell-bound TNFα (Moss et al 1997) in inflammation. These enzymes are members of the ADAMs family (Wolfsberg et al 1995) (A Disintegrin And Metalloproteinase) which are closely related to the snake venom proteinases called reprolysins. It is not yet known if they play a major role in lung matrix turnover either through release of inflammatory mediators or receptors, activation of other metalloproteinases or the degradation of matrix components.

Can synthetic inhibitors of MMPs be used as drugs?

The future prospects for the prevention of connective tissue breakdown using synthetic MMP inhibitors as drugs look promising. Highly specific MMP inhibitors have been made (Porter et al 1994, Beckett et al 1996). Early inhibitors were not stable in biological fluids and were rapidly broken down but current inhibitors remain biologically active after oral ingestion. The recent structural studies of MMPs has allowed the improvement of the existing synthetic inhibitors using molecular modelling at the active site. Some workers support the use of broad spectrum inhibitors as these will test whether MMP inhibition is therapeutically effective before testing specific inhibitors targeted to individual enzymes. If undiscovered MMPs are responsible for connective tissue turnover, then broad spectrum inhibitors could also block their action. If undiscovered enzymes are involved in unrecognized but essential pathways, then undesirable or unexpected side effects could be the result. It will be important to screen potential new compounds in an *in vivo* model as some early compounds have been reported to adversely affect connective tissues peripheral to the joints. Specific inhibitors could avoid such problems but insufficient data are available to pinpoint which enzyme is specific to a single disease. The results of many knockout studies also suggest that where an individual enzyme is deleted then it often has little impact as, presumably, other enzymes that degrade the same substrates can compensate. Some workers are concerned that inhibition of MMPs may lead to excess deposition of matrix within connective tissues thus leading to fibrosis. However it should be possible to treat with MMP inhibitors at a dose that will block excess degradation of matrix but will allow the normal turnover of the connective tissue matrix. Such inhibitors are active in animal models (Karran et al 1995) and initial clinical trials have begun in patients with rheumatoid arthritis.

Acknowledgements

The authors gratefully acknowledge support from the Arthritis Research Campaign and the Wellcome Trust.

References

Barrett AJ 1994 Classification of peptidases. Methods Enzymol 244:1–15
Beckett RP, Davidson AH, Drummond AH, Huxley P, Whittake M 1996 Recent advances in matrix metalloproteinase inhibitor research. Drug Discovery Today 1:16–26
Birkedal-Hansen B, DeCarlo A, Engler JA 1993 Matrix metalloproteinases: a review. Crit Rev Oral Biol Med 4:197–250
Cawston TE 1996 Metalloproteinase inhibitors and the prevention of connective tissue breakdown. Pharmacol Ther 70:163–182
Cawston TE, Ellis AJ, Bigg H, Curry V, Lean E, Ward D 1996 Interleukin-4 blocks the release of collagen fragments from bovine nasal cartilage treated with cytokines. Biochim Biophys Acta 1314:226–232
Cawston TE, Curry VA, Summers CA et al 1998 The role of oncostatin M in animal and human connective tissue collagen turnover and its localisation within the rheumatoid joint. Arthritis Rheum 41:1760–1771
Clark IM, Powell LK, Ramsey S, Hazleman BL, Cawston TE 1993 The measurement of collagenase, tissue inhibitor of metalloproteinases (TIMP) and collagenase-TIMP complex in synovial fluids from patients with osteoarthritis and rheumatoid arthritis. Arthritis Rheum 36:372–379
D'Armiento J, Dalal SS, Okada Y, Berg RA, Chada K 1992 Collagenase expression in the lungs of transgenic mice causes pulmonary emphysema. Cell 71:955–961
Denhardt DT, Feng B, Edwards DR, Cocuzzi ET, Malyankar UM 1993 Tissue inhibitor of metalloproteinases (TIMP, aka EPA): structure, control of expression and biological functions. Pharmacol Ther 59:329–341
Finlay GA, Russell KJ, McMahon KJ et al 1997 Elevated levels of matrix metalloproteinases in bronchoalveolar lavage fluid of emphysematous patients. Thorax 52:502–506

Goldring MB 1993 Degradation of articular cartilage in culture: regulatory factors. In: Woessner JF, Howell DS (eds) Joint cartilage degradation. Marcel Dekker, New York, p 281–345

Gomis-Rüth F-X, Maskos K, Betz M et al 1997 Mechanism of inhibition of the human matrix metalloproteinase stromelysin-1 by TIMP-1. Nature 389:77–81

Hautamaki RD, Kobayashi DK, Senior RM, Shapiro SD 1997 Requirement for macrophage elastase for cigarette smoke-induced emphysema in mice. Science 277:2002–2004

Hibbs MS, Hasty KA, Seyer JM, Kang AH, Mainardi CL 1985 Biochemical and immunological characterisation of the secreted forms of human neutrophil gelatinase. J Biol Chem 260:2493–2500

Hill PA, Murphy G, Docherty AJP et al 1994 The effects of selective inhibitors of matrix metalloproteinases (MMPs) on bone resorption and the identification of MMPs and TIMP-1 in isolated osteoclasts. J Cell Sci 107:3055–3064

Karran EH, Young TJ, Markwell RE, Harper GP 1995 In vivo model of cartilage degradation — effects of a matrix metalloproteinase inhibitor. Ann Rheum Dis 54:662–669

Korzus E, Nagase H, Rydell R, Travis J 1997 The mitogen-activated protein kinase and JAK-STAT signaling pathways are required for an oncostatin M-responsive element-mediated activation of matrix metalloproteinase 1 gene expression. J Biol Chem 272:1188–1196

Moss ML, Jin SLC, Milla ME et al 1997 Cloning of a disintegrin metalloproteinase that processes precursor tumour-necrosis factor-α. Nature 385:733–736 (erratum: 1997 Nature 386:738)

Murphy G, Knäuper V 1997 Relating matrix metalloproteinase structure to function: why the 'hemopexin' domain? Matrix Biol 15:511–518

Murphy G, Willenbrock F 1995 Tissue inhibitors of matrix metalloendopeptidases. Methods Enzymol 248:496–510

Murphy G, Atkinson S, Ward R, Gavrilovic J, Reynolds JJ 1992 The role of plasminogen activators in the regulation of connective tissue metalloproteinases. Ann NY Acad Sci 667:1–12

Pei D, Weiss SJ 1995 Furin-dependent intracellular activation of the human stromelysin-3 zymogen. Nature 375:244–247

Porter JR, Beeley NRA, Boyce BA et al 1994 Potent and selective inhibitors of gelatinase-A 1. Hydroxamic acid derivatives. Bioorg Med Chem Lett 4:2741–2746

Prockop DJ, Kivirikko KI 1995 Collagens; molecular biology, diseases, and potentials for therapy. Annu Rev Biochem 64:403–434

Sandy JD, Flannery CR, Neame PJ, Lohmander LS 1992 The structure of aggrecan fragments in human synovial fluid. Evidence for the involvement in osteoarthritis of a novel proteinase which cleaves the Glu 373-Ala 374 bond of the interglobular domain. J Clin Invest 89:1512–1516

Saari H, Suomalainen K, Lindy O, Konttinen YT, Sorsa T 1990 Activation of latent human neutrophil collagenase by reactive oxygen species and serine proteases. Biochem Biophys Res Commun 171:979–987

Shapiro SD 1994 Elastolytic metalloproteinases produced by human mononuclear phagocytes. Potential roles in destructive lung disease. Am J Respir Crit Care Med 150:S160–S164

Shingleton WD, Hodges DJ, Brick P, Cawston TE 1996 Collagenase: a key enzyme in collagen turnover. Biochem Cell Biol 74:759–775

Springman EB, Angleton EL, Birkedal-Hansen H, van Wart HE 1990 Multiple modes of activation of latent human fibroblast collagenase: evidence for the role of a Cys[73] active-site zinc complex in latency and a 'cysteine switch' mechanism for activation. Proc Natl Acad Sci USA 87:364–368

Tetley TD 1997 Matrix metalloproteinases: a role in emphysema? Thorax 52:495

Tortorella MD, Burn TC, Pratta MA et al 1999 Purification and cloning of aggrecanase-1: a member of the ADAMTS family of proteins. Science 284:1664–1666

Tschesche H, Knäuper V, Kleine T et al 1994 Function and structure of human leucocyte collagenase. J Protein Chem 13:460–461 van Roon JAG, van Roy JLAM, Gmelig-Meyling FHJ, Lafeber FPJG, Bijlsma JWJ 1996 Prevention and reversal of cartilage degradation in rheumatoid arthritis by interleukin-10 and interleukin-4. Arthritis Rheum 39:829–835

Vincenti MP, Clark IM, Brinckherhoff CE 1994 Using inhibitors of metalloproteinases to treat arthritis. Easier said than done? Arthritis Rheum 37:1115–1126

Woessner JF Jr 1991 Matrix metalloproteinases and their inhibitors in connective tissue remodeling. FASEB J 5:2145–2154

Wolfsberg TG, Primåkoff P, Myles DG, White JM 1995 ADAM, a novel family of membrane proteins containing A Disintegrin And Metalloprotease domain: multipotential functions in cell–cell and cell–matrix interactions. J Cell Biol 131:275–278

DISCUSSION

Lomas: Just to add that in the D'Armiento et al (1992) that you referred to, they set out to cause hepatic fibrosis. They tried to target the collagenase to the liver, and the side effect in the transgenic mouse was that this caused emphysema.

Shapiro: I wouldn't call it emphysema as defined by mature airspace destruction and enlargement: these mice had enlarged air spaces which one sees frequently with transgenic mice, particularly if proteases are being released during lung development. There are a lot of questions as to the relevance of this work to emphysema, although clearly when there are big holes in the lung collagen is depleted locally. The role of collagen is important. There is also something special about elastin and the inability to reinitiate the complex elastic fibre assembly following injury. Collagen repair is more efficient, but excess collagen can cause fibrosis, which actually is observed in proximal alveolar structures in emphysema. For that reason, I would be wary about inhibiting collagenase; there are still a lot of unanswered questions.

Rennard: Why do we say that we can't repair elastin fibres?

Shapiro: We need Bob Mecham here to answer that. One can express tropoelastin, which then has to be aligned on a microfibrillar assembly that is composed of at least a handful of other proteins and then cross-linked by lysyl oxidase. I am not saying that this can't be done, but unless all of those components are spatially and temporally regulated, then one might not get a functional elastic fibre.

Rennard: I understand that however if elastin repair were to take place, that this would be a complex process with multiple integrated steps. Clearly elastin fibres can be remodelled during life. You start off as a baby with an elastin network, and it is very unlikely that those elastin structures remain unremodelled or unchanged by the time you reach adulthood. There are data from Phil Stone's group (Stone et al 1997) that newly synthesized tropoelastin can be incorporated into elastin macromolecules *in vitro*. It seems that there should be some capacity for turnover and remodelling of elastin fibres *in vivo*. This is supported by the excretion of desmosine that occurs in all people. We always say that there is no repair of elastin fibres in the lung, but do we really know that?

Campbell: Dr John Pierce has done a lot of thinking about the elastic fibres in the lung, how they get there, and what happens during the development of pulmonary emphysema. We shouldn't have the concept that elastic fibres are randomly distributed in the lung, and they are certainly not randomly distributed in the alveolar septa. What Jack taught me was that there is a continuous fibrous elastic skeleton to the lung. The elastic fibres follow and branch with the airways, and get finer and finer as the airways get smaller. I don't think you need to lose elastic fibres during the development of emphysema, nor do you need to have substantial elastic fibre degradation to produce disruptive effects on the structure of the lung. When there is an elastic fibre under tension in an alveolar septa, remember that this is a very fine fibre. If you injure it enough that it breaks, the ends of that fibre retract and there is no easy opportunity to repair it. Jack Pierce did some very careful studies in the 1960s looking at elastic fibre content in normal lungs compared with emphysematous lungs. He studied lungs with advanced pulmonary emphysema and found that elastin content was the same as in normal lungs (Pierce et al 1961). Thus, as emphysema develops, we are not talking about elastin disappearing from the lung, but rather we are talking about injuring the fibres such that the normal architecture is disrupted. In the normal human lung parenchyma, there is very little evidence that elastic fibres turn over at all. Steve Shapiro has shown by two independent means in the late 1980s that the elastic fibres in your lung parenchyma when you cease postnatal lung growth are essentially the same fibres you have when you die (Shapiro et al 1991). The mean carbon residence time of elastic fibres in the normal lung parenchyma is about 74 years. Elastic fibres do not turn over in the normal lung; therefore, they are relatively permanent structures.

Rennard: That doesn't preclude repair of those structures. The data that are strongly suggestive that repair of these structures can take place are the experiments with β-aminoproprionitrile. Kuhn and colleagues have shown that if you block elastin cross-linking, this can make emphysema worse (Kuhn & Starcher 1980). This suggests that the emphysema is a result not just of the injury, but also of inadequate repair. This means that some repair must be taking place to some structures within the lung. Elastin fibres are candidates for that kind of repair process like other things.

Campbell: I don't have any objection to the notion that new tropoelastin can be laid down on injured elastic fibres. My concern is that if you injure one of the alveolar septal elastic fibres to the extent that it breaks, I don't see how the structure can be restored. I think this is what happened in Dr Charles Kuhn's hamster emphysema model in which he introduced elastase intratracheally. He found that the elastin content in the lungs went down dramatically over the first 24 h (Kuhn et al 1976). The initial fall in lung elastin content was accompanied by a loss of elastic fibres in the alveolar septa, and many alveolar septa had no elastic fibres. Within the next two months, the quantity of elastin in the lungs returned to control values. However, during this time there was a progressive development of structural changes of emphysema. These studies showed that elastin 'repair' and quantitative restoration of elastin could neither restore the original elastic fibres nor prevent the development of emphysema.

Shapiro: Having said how difficult it is to repair an elastic fibre, it is possible in smokers where inflammation and destruction are routine, that there is a minority of smokers who have a defect in repair. It is possible that the inability to repair might be an important one of the genetic factors predisposing to emphysema susceptibility.

Barnes: Are there genetic polymorphisms known in MMPs and TIMPs that could account for disease variability?

Cawston: There is a polymorphism in the *MMP1* gene which Connie Brinckerhoff has looked at in certain cancers (Rutter et al 1998). She finds that the frequency of tumours that invade is increased when this polymorphism is present. It is an insertion of a single nucleotide into the gene which creates an Ets binding site in the promoter region of the *MMP1* gene.

MacNee: What about MMP12 and the lungs?

Shapiro: If any elastase is present uninhibited, it is going to cause lung destruction. MMP12 seems to play a pivotal role in cigarette smoke-induced emphysema in mice. No one has looked for genetic polymorphism to date.

MacNee: What about in human lungs?

Shapiro: It is there, but it probably isn't as prominent as it is in the mouse. Macrophages from mice and humans have about the same elastolytic capacity. In human, MMP9 and MMP12 probably account equally for this activity, while in mice is it largely MMP12

Lomas: In the mouse model, from what I recollect, there aren't many neutrophils. The mice that develop emphysema have lots of macrophages but hardly any neutrophils.

Shapiro: That maybe because we have them in a specific pathogen-free barrier facility. There are also fewer peripheral neutrophils in the mouse than in the human. In smoke-induced emphysema in a mouse, in the periphery in the absence of infection it seems to be more of a macrophage process. If humans were put in a barrier facility we might see the same thing.

Paré: What about the kinetics of the macrophage? We have heard all this talk about how the neutrophil has to go from the vasculature into the interstitium to do the damage. I thought that alveolar macrophages were supposed to stay in the alveolar spaces.

Shapiro: They are present abundantly in the interstitial spaces next to the elastic fibres, and they stay there for a long time.

Hogg: They also divide in the interstitial spaces (Bowden & Adamson 1980, Adamson & Bowden 1980).

Shapiro: When we do bone marrow transplants, we lose host macrophages completely. The macrophages could divide once at most, but they then seem to go away.

Hogg: David Walker has pictures of alveolar macrophages in mitosis in the alveolus. This makes the kinetics that much more difficult to study.

Stockley: They are also long-lived, and are retained for a long time before they die.

Campbell: The party line is that macrophages in the lung can probably divide a limited number of times, but that the supply of mononuclear phagocytes in the lung has to be continuously renewed from the bone marrow. In humans there have been excellent studies of alveolar macrophages, and also Kuppfer cells in the liver, which have very similar kinetics. It takes about three months to turn over either the Kuppfer cell or alveolar macrophage populations. This was learned from bone marrow transplants in humans (Thomas et al 1976, Gale et al 1978).

Stockley: While we are on the subject of monocytes migrating from the periphery into the lung, Ed Campbell, would you summarize your data on monocyte subsets and the phenotype of monocytes that actually migrate?

Campbell: Our information suggests that there is a lot of heterogeneity in peripheral blood monocytes. Dr Caroline Owen has shown that there is a subpopulation of peripheral blood monocytes, comprising about 20% of the total, that has a strikingly neutrophil-like phenotype (Owen et al 1994a). They are the only monocytes in the peripheral blood that have demonstrable chemotactic responses (Owen et al 1994b). They have a neutrophil-like complement of serine proteinases, including leukocyte elastase, cathepsin G and proteinase III. They seem to be the cells that respond to inflammatory stimuli and are poised to migrate outside the vasculature in response to proinflammatory signals. They have very little HLA VR antigen on the surface, so we think these cells, which we have called P monocytes, are poor antigen-presenting cells. However, they are neutrophil-like in may ways, and we think that they have significant potential to cause tissue injury in the proper settings.

Stockley: The reason I raised this issue is because if we are continuously replenishing the lung macrophage population, it may well be that it is this subset of monocytes that are going to be causing essentially the same destructive injury that you get when a neutrophil migrates and using the same enzyme.

Campbell: We probably should emphasize that there is not a black-and-white distinction between young mononuclear phagocytes and neutrophils. As monocytes differentiate, they lose their serine proteinases and develop a complement of metalloproteinases. It is the young circulating monocytes that can have a very neutrophil-like phenotype.

Senior: In thinking about which inflammatory cell may be the culprit in the pathogenesis of emphysema there is a tendency to think of the macrophage as the MMP-harbouring inflammatory cell and the neutrophil as the serine proteinase-harbouring inflammatory cell. As Tim Cawston pointed out, neutrophils have both MMPs and serine proteinases and each of these classes of enzymes have the capacity to degrade inhibitors for the other class of enzymes. We have been studying an immune complex model of bullous pemphigoid in mice. One can prevent the development of blisters either by doing the model in the gelatinase B-deficient mouse or the neutrophil elastase-deficient mouse (Liu et al 2000) It appears that gelatinase B's role in the model is to cleave and inactivate α_1-antitrypsin at the site where the immune complex is fixing complement that will recruit neutrophils and that neutrophil elastase's role is to cleave the hemidesmosome resulting in the blister. The point is that the neutrophil may have a complicated role in inflammatory proteolytic events.

Lomas: What happens when α_1-antitrypsin is cleaved in terms of attracting more neutrophils in?

Senior: From work with Michael Banda we know that cleavage of α_1-antitrypsin by either neutrophil elastase (Banda et al 1988a) or macrophage elastase (Banda et al 1988b) exposes a site adjacent to the cleavage that is chemotactic for neutrophils.

Campbell: Is that a theme that we should be thinking about? Degraded matrix may be a pro-inflammatory stimulus. Steve Shapiro found that in his MMP knockout animals, matrix degradation was important in promoting mononuclear phagocyte influx.

Senior: We have thought for many years that the *in vitro* data showing chemotactic activity of elastin peptides (Senior et al 1984) suggested that elastin-derived and other matrix-derived peptides might be chemotactic *in vivo*. As you say, Steve Shapiro is now obtaining these types of data in the cigarette smoke-exposed mouse.

Campbell: We were talking earlier about how long an inflammatory reaction can be sustained, long after we think that the initial stimulus is gone. I wonder if this is at least one important mechanism by which inflammation might go on and on.

Stockley: But wouldn't the significance of this be that it would also attract cells to the site of the matrix, not to the airway, because they will go to the maximum concentration of the chemoattractant, which if it is degraded matrix, will be the matrix?

Paré: What is the survival advantage of this process? It sounds very destructive.

Shapiro: Why does the macrophage degrade the matrix anyway? It is probably to mop up the debris, but if excessive it can lead to damage.

Rennard: I like the concept that the matrix-derived peptides could be functioning to accelerate the inflammatory response. If little chunks of elastin were being released, the fact that the elastin content of the lung doesn't change much must mean that some new elastin is deposited. If these proteinases can inactivate each other's inhibitors, and therefore lead to lots of proteinases, and these in turn can generate more chemotactic activity to bring in more inflammatory cells, why don't all inflammatory reactions immediately go to complete liquefaction of the tissues? Most inflammatory responses don't do that: they stop at some relatively well controlled level.

Lomas: I have always been impressed by the demonstration that cleaved α_1-antitrypsin and complexes of proteases and their inhibitors are chemotactic for human neutrophils. Our data show that patients with antitrypsin deficiency retain mutant antitrypsin in their livers as chains of polymers. they can also get polymers in their lungs. In our *in vitro* models these polymers are also chemotactic. There is good reason for this. From an immunological point of view, the cleaved antitrypsin, the complex and the polymers all display similar epitopes and are pro-inflammatory. On the anti-inflammatory side is native antitrypisn and antichymotrypsin. I would like to suggest a balance whereby protease inhibitors (antitrypsin and antichymotrypsin) interact with a surface-bound enzymes on neutrophils to switch them off, and this is part of the inflammatory response. On the other side of the balance there are abnormal conformations that recruit neutrophils which can degrade matrix.

Rennard: All of those things are plausible, but I think that there is another aspect of the switch off. The switch off may be programmed into the entire cytokine network from the very beginning. Some cytokines, such as TGFβ, are quite potent anti-inflammatory agents. They are able to block the inflammatory cascade at a number of steps. The cytokine–protease inflammatory cascade is perhaps self-limited by control mechanisms that are initiated from the very beginning. In fact, elastase is a reasonably potent releaser of TGFβ when it is bound to the extracellular matrix. This would be a way for the protease effect to turn off the inflammatory process.

Shapiro: There are probably a lot of inhibitors around. Work from Jay Heinecke and Bill Parks show that oxidants initially will activate metalloproteinases, but then over time they will inactivate them as well. Some of the oxidant defences may therefore be playing a role in turning this system off also.

MacNee: Over the last few years I have followed the elegant studies that have come out of the MMP12 knockout mouse. If you look back at the protease–antiprotease story, I remember the review by Heather Morrison (1987; *The proteinase–antiproteinase theory of emphysema: time for a reappraisal?*), and then we had secretory leukoprotease and MMP12. Is the proof of concept needed from studies in humans on these inhibitors? Is this the way forward? Or do we need more information first?

Shapiro: For MMP12, we have studied this extensively in a mammal. I can't think of anything other than human studies that would give us more information. I think we are ready for human clinical trials. I wouldn't limit this just to MMP12 or even the MMPs: I don't think a single class of protease inhibitors will shut off damage. Cysteine proteases haven't been talked about much, but some are very potent elastases in the right environment. However, even after we inhibit all the proteases, we still have a lung with large holes. We have to understand repair: this is where Don Massaro's work is so elegant.

Stockley: The reason we went to a time for reappraisal was because it was so simple in those days: everyone said it was α_1-antitrypsin, neutrophil elastase, and cigarette smoke was reducing the α_1 function. This was too simplistic to be true. All we were trying to do at that stage was to say that there are other inhibitors of neutrophil elastase that have not been assessed. Back in the 19th century, clotting was fibrinogen being cleaved to fibrin. We now know that there is an enormous cascade of coagulation factors involved in producing that change. This is the message I was trying to get over at that stage. The way that we will have to go forward in emphysema is to take a common pathway mediator and see whether this tells us much about the disease.

Barnes: A very potent endogenous inhibitory mechanism that has not been mentioned is IL-10. IL-10 is extremely effective at switching off MMP expression, particularly MMP9. It up-regulates TIMPs and very potently switches off TNFα and IL-8. It seems to do many of the things needed in chronic obstructive pulmonary disease (COPD) therapy. In asthma, we have reported that alveolar macrophages produce less IL-10 than those from normal people. This correlates with disease severity. There is a polymorphism of the IL-10 promoter which is linked to reduced production of IL-10. This polymorphism is more frequent in severe asthma compared to mild asthma. I think IL-10 is an endogenous mechanism that may not be functioning properly in COPD and this deserves further study. Interestingly, it takes a long time to switch on, so that when macrophages are activated they release inflammatory products rapidly. Switching on IL-10 then takes 24 h, and is the delayed switch off mechanism for the inflammatory genes.

Calverley: What we have actually got are cytokine and chemokine networks that are immense and interconnected. Ultimately, what happens is that some stimulus comes along and pushes this system persistently out of shape. We then end up with tissue destruction. What we are proposing, because we are reductionist scientists, is taking one link out of this by blocking it, and hoping that this will do the job. This may well be the case, but I have a nagging feeling that this problem is too complex to yield to this approach. This is why the mammalian model approach is so valuable. I would really like to know whether we are talking about a continuous distortion or an episodic one: are we seeing loss of lung function over time gradually, or in a step-by-step fashion? Is there evidence that people lose lung function perhaps after episodes that might be associated with greater neutrophil traffic? If someone with α_1-antitrypsin has a pneumonia, does their FEV_1 (forced expiratory volume in one second) fall and never recover?

Stockley: There are data from exacerbations showing step-like drops in some patients with deficiency, but I don't think people have collected enough data prospectively to say that this is a common phenomenon.

MacNee: The original data suggested a decline in lung function for a short period of time following an exacerbation but a return to previous values thereafter.

Calverley: One thing I learned when I got involved in this analysis of rate of FEV_1 decline is that the result depended on what the statistician told you to do with the original data. The curve that is fitted to these data does not look like the raw data.

Rennard: As a principle it is a very difficult thing to do natural history studies on your own species. This is where the mouse is tremendously important. What is the natural history of the emphysema in these mice? Does it develop gradually throughout along a continuous course, or is there a stepwise progression?

Shapiro: Response to cigarette smoke is gradual: it takes a couple of months to start and then it is a fairly linear progression. However, these mice are in a barrier facility and don't get 'exacerbations'. If we were to add infections, perhaps we will observe stepwise progression.

Jeffery: In the mice, are you looking at emphysema developing in lung during its growth phase?

Shapiro: We start the mice smoking at about 10 weeks when the lung is fully developed.

Jeffery: The scenario might be different between a growing lung, a mature lung and a lung undergoing age-related decline.

Senior: As Rob Stockley pointed out, when α_1-antitrypsin deficiency was first identified we thought that we had the answer to the pathogenesis of emphysema in general. We now recognize that we don't, but I think that many people think that at least we understand the pathogenesis of emphysema associated with α_1-antitrypsin deficiency. I am not so sure that we do. Once lung injury begins in antitrypsin deficiency I suspect that the process becomes more complicated than simply neutrophil elastase digesting lung matrix. It is likely that macrophage responses and cytokine networks come into play.

Stockley: I would endorse that absolutely. There are so many questions to be asked of just alpha$_1$-antitrypsin deficiency.

Stolk: In α_1-antitrypsin deficiency-associated emphysema, clearly the trigger is smoking. However, even after they stop smoking there is progression of airflow obstruction of about 60 ml per year. Therefore, it is true that there is still more to learn from this type of patient concerning macrophage responses and cytokine networks.

Stockley: That is fascinating, because we have been looking at the computed tomography (CT) scans in the patients who stop smoking versus the ones that continue to smoke. The CTs are progressing at the same rate so far.

Senior: The complexity of the mechanisms causing lung injury in antitrypsin deficiency may be one of factors limiting the effectiveness of antitrypsin supplementation therapy.

Stockley: We need a controlled trial.

Hogg: This question has been asked many times: why is it that in pneumococcal pneumonia, with all that leukocyte traffic, the architecture is restored when it clears, yet in emphysema there is less traffic but persistent damage?

Stockley: I think that all these neutrophils go along the same pathway. There is just one limited area of destruction, and everything follows this path. But also pneumococci make an anti-elastase.

Jeffery: Is the architecture destroyed in pneumococcal pneumonia? I believe not.

Stockley: Not normally, but patients can develop severe bronchial damage with loss of connective tissue around the bronchus.

Senior: A number of years ago Tom Martin instilled LTB4 into the middle lobe of normal volunteers via a bronchoscope. An acute neutrophilic alveolitis developed, but there was little increase in lavage fluid protein compared to fluid recovered from the lingula that was given only saline (Martin et al 1989). Apparently, large numbers of neutrophils can cross the alveolo-capillary barrier without causing much structural damage.

Hogg: As I mentioned earlier, Wright provided a beautiful diagram and pictures of the elastin network in the lung in 1961 (Wright 1961). It would be really interesting to try to figure out how the neutrophils come in contact with the elastic network and develop a model to predict how damage to elements in the network might change the pressure volume characteristics of the lung tissue.

Campbell: To me, the proinflammatory stimulus is in the *airway* in pneumococcal pneumonia. Neutrophils are just passing through. They don't get activated in the interstitium.

Hogg: Do we know the transit time of neutrophils in interstitial space in any disease?

Campbell: Greg Downey did some studies in rabbits, which showed that the mean transit time of neutrophils from the interstitial space to alveolar airspaces was 60 minutes or less (Downey et al 1993).

Hogg: In the studies on pneumoccocal pneumonia, the neutrophils appeared in the alveoli about an hour after instillation of bacteria (Doerschuk et al 1994). However, we don't know the time that it takes to adhere and migrate out of the vascular space, through the endothelium, interstitium and alveolar epithelium into the airspace. Interestingly, only about 1–2% of the cells delivered to the pneumonia site actually migrated into the airspaces.

MacNee: Repeated LPS instillation in animal lungs produces emphysematous-like lesions. Isn't the problem one of chronicity, with the repeated insult of cigarette smoke that initiates an inflammatory response with the up-regulation of the protective mechanisms.

Rennard: Returning to the issue of smoking cessation in the α_1-antitrypsin patients where the disease appears to progress at the same rate even after the people have quit. In the lung health study this was not found, but the patients were much milder. What is the threshold that the patients cross where smoking cessation no longer makes any difference?

MacNee: That is an important question.

Calverley: This comes back to one of the things we are thinking about: there is a big difference between having a big hit and what actually happens in the natural environment where there is a repeated small stimulus. It is assessing this change that may be more relevant to why you have got the change you have got, rather than the familiar example that we have all used of why doesn't it happen in pneumococcal pneumonia.

References

Adamson IY, Bowden DH 1980 Role of monocytes and interstitial cells in the generation of alveolar macrophages. II. Kinetic studies after carbon loading. Lab Invest 42: 518–524

Bowden DH, Adamson IY 1980 Role of monocytes and interstitial cells in the generation of alveolar macrophages. I. Kinetic studies in normal mice. Lab Invest 42:511–517

Banda MJ, Rice AG, Griffin GL, Senior RM 1988a The inhibitory complex of human alpha1-proteinase inhibitor and human leukocyte elastase is a neutrophil chemoattractant. J Exp Med 167:1608–1615

Banda MJ, Rice AG, Griffin GL, Senior RM 1988b Alpha 1-proteinase inhibitor is a neutrophil chemoattractant after proteolytic inactivation by macrophage elastase. J Biol Chem 263: 4481–4484

D'Armiento J, Dalal SS, Okada Y, Berg RA, Chada K 1992 Collagenase expression in the lungs of transgenic mice causes pulmonary emphysema. Cell 71:955–961

Doerschuk CM, Markos H, Coxson HO, English D, Hogg JC 1994 Quantitation of neutrophil migration in acute bacterial pneumonia in rabbits. J Appl Physiol 87:2593–2599

Downey GP, Worthen GS, Henson PM, Hyde DM 1993 Neutrophil sequestration and migration in localized pulmonary inflammation. Capillary localization and migration across the interalveolar septum. Am Rev Respir Dis 147:168–176

Gale RP, Sparkes RS, Golde DW 1978 Bone marrow origin of hepatic macrophages (Kupffer cells) in humans. Science 201:937–938

Kuhn C III, Starcher BC 1980 The effect of lathyrogens on the evolution of elastase-induced emphysema. Am Rev Respir Dis 122:453–460

Kuhn C III, Yu S, Chraplyvy M, Linder HE, Senior RM 1976 The induction of emphysema with elastase II. Changes in connective tissue. Lab Invest 34:372–380

Liu Z, Shapiro SD, Zhou X, et al 2000 A critical role for neutrophil elastase in experimental bullous pemphigoid. J Clin Invest 105:113–123

Martin TR, Pistorese BP, Chi EY, Goodman RB, Matthay MA 1989 Effects of leukotriene B4 in
    the human lung. Recruitment of neutrophils into the alveolar spaces without a change in
    protein permeability. J Clin Invest 84:1609–1619

Morrison HM 1987 The proteinase-antiproteinase theory of emphysema: time for a reappraisal?
    Clin Sci (Colch) 72:151–158

Owen CA, Campbell MA, Boukedes SS, Stockley RA, Campbell EJ 1994a A discrete
    subpopulation of human monocytes expresses a neutrophil-like proinflammatory (P)
    phenotype. Am J Physiol 267:L775–L785

Owen CA, Campbell MA, Boukedes SS, Campbell EJ 1994b Monocytes recruited to sites of
    inflammation express a distinctive pro-inflammatory (P) phenotype. Am J Physiol
    267:L786–L796

Pierce JA, Hocott JB, Ebert RV 1961. The collagen and elastin content of the lung in
    emphysema. Ann Intern Med 55:210–222

Rutter JL, Mitchell TI, Buttice G et al 1998 A single nucleotide polymorphism in the matrix
    metalloproteinase-1 promoter creates an Ets binding site and augments transcription.
    Cancer Res 58:5321–5325

Senior RM, Griffin GL, Mecham RP, Wrenn DS, Prasad KU, Urry DW 1984 Val-Gly-Val-Ala-
    Pro-Gly, a repeating peptide in elastin, is chemotactic for fibroblasts and monocytes. J Cell
    Biol 99:870–874

Shapiro SD, Pierce JA, Endicott SK, Campbell EJ 1991 Marked longevity of human lung
    parenchymal elastic fibers deduced from prevalence of D-aspartate and nuclear weapons-
    related radiocarbon. J Clin Invest 87:1828–1834

Stone PJ, Morris SM, Thomas KM, Schuhwerk K, Mitchelson A 1997 Repair of elastase-
    digested elastic fibers in acellular matrices by replating with neonatal rat-lung lipid
    interstitial fibroblasts or other elastogenic cell types. Am J Respir Cell Mol Biol 17:289–301

Thomas ED, Ramberg RE, Sale GE, Sparkes RS, Golde DW 1976 Direct evidence for a bone
    marrow origin of the alveolar macrophage in man. Science 192:1016–1018

Wright RR 1961 Elastic tissue in normal and emphysematous lungs. Am J Pathol 39:355–367

Pulmonary alveolus formation: critical period, retinoid regulation and plasticity

Donald Massaro*† and Gloria DeCarlo Massaro*‡

*Lung Biology Laboratory, †Department of Medicine and ‡Department of Pediatrics, Georgetown University School of Medicine, Washington, DC 20007-2197, USA

Abstract. Pulmonary alveoli, the lung's gas-exchange structures, are formed in part by subdivision (septation) of the saccules that constitute the gas-exchange region of the immature lung. Although little is known about the regulation of septation, relatively recent studies show: (1) all-*trans* retinoic acid (RA) treatment of newborn rats increases septation and prevents the inhibition of septation produced by treatment of newborn rats with dexamethasone, a glucocorticosteroid hormone; (2) treatment with RA of adult rats that have elastase-induced emphysema increases lung elastic recoil, induces the formation of alveoli, and increases volume-corrected alveolar surface area; and (3) in tight-skin mice, which have a genetic failure of septation, and in rats in which septation had previously been prevented by treatment with dexamethasone, treatment with RA partially rescues the failed septation. These findings raise the possibility that treatment with RA will induce the formation of alveoli in humans with pulmonary emphysema.

2001 Chronic obstructive pulmonary disease: pathogenesis to treatment. Wiley, Chichester (Novartis Foundation Symposium 234) p 229–241

Pulmonary alveoli, the lung's gas-exchange structures, are formed in part by the subdivision (septation) of the saccules that constitute the gas-exchange region of the immature lung. Although the architectural process of septation is similar among species, its timing is developmentally regulated, occurring in mice (Amy et al 1977) and rats (Burri 1974) from the 4th through 14th postnatal day and in humans during the last month of gestation and the first few postnatal years (Langston et al 1984, Zeltner & Burri 1987). Septation helps establish the link, highly conserved across the entire spectrum of body sizes of mammals, between alveolar dimensions and the organism's oxygen requirements (Tenney & Remmers 1963). When septation ends, alveoli continue to form until adulthood is reached (Blanco et al 1991) suggesting, as proposed earlier (Massaro et al 1985), that the formation of alveoli is regulated differently during and after the period of septation.

The tight link between alveolar dimensions and oxygen need, the latter elegantly designated 'the call for oxygen' (Krogh 1941), is most commonly disrupted in humans by the failure to septate or by alveolar destruction. Failure to septate occurs in very prematurely born infants as part of the syndrome of bronchopulmonary dysplasia (BPD). Recent advances in therapy of prematurely born babies have changed the characteristics of BPD from a syndrome in which damage to conducting airways predominates to one in which arrest or failure of alveolus formation plays an increasing role in mortality (Husain et al 1998, Jobe 1999). The basis for the arrest of alveolus formation is unclear, but prematurely born babies are commonly treated with corticosteroid hormones and high concentrations of oxygen (O'Brodovich & Mellins 1985), which impair septation in animals (Bucher & Roberts 1981, Massaro et al 1985). Alveolar destruction, as occurs in pulmonary emphysema, is the major cause in adults of disruption of the tight link between oxygen need and alveolar dimensions. Unfortunately, spontaneous post hoc septation does not seem to occur in prematurely born infants who fail to septate (Sobonya et al 1982, Margraf et al 1991, Husain et al 1998) nor is there spontaneous regeneration of destroyed alveoli. Furthermore, a paucity of knowledge of the regulation of septation and of the molecular signalling responsible for the formation of alveoli has impaired the development of medical remediation of impaired septation and alveolar destruction. These deficits in knowledge led to studies aimed at understanding the regulation of the formation of alveoli. The following paragraphs describe the results of some of these studies.

Critical period for alveolus formation

The prescient suggestion (Buckingham et al 1968) that glucocorticosteroid hormones might modulate fetal lung development led to great interest in their effect on prenatal lung development, in particular on the development of the lung's surfactant system. The demonstration that in some species septation is an early postnatal rather than a prenatal event raised the possibility that glucocorticosteroid hormones, which are important regulators of the postnatal development of the gastrointestinal tract (Henning 1981), the anlage of the lung, might also regulate the postnatal architectural development of the lung. The time-course of septation in rats (Burri 1974) and in guinea pigs (Collins et al 1986) and the time-course of changes in serum concentration of glucocorticosteroid hormones in these species (Henning 1978, Jones & Roebuck 1980) led us to consider the possibility that septation occurs during a trough in the serum concentration of these hormones. Furthermore, thinning of the alveolar walls in rats, the only species in which direct measurements have been made (Massaro & Massaro 1986), is accelerated as the serum concentration of corticosteroids rise suggesting corticosteroids might induce thinning of the alveolar wall.

We treated rat pups with dexamethasone, a synthetic glucocorticosteroid hormone, daily from postnatal day 4 through 13, the period septation normally occurs (Burri 1974) and killed them on postnatal day 14. This treatment prevented septation (Massaro et al 1985), accelerated alveolar wall thinning, and diminished the volume fraction of fibroblasts in the wall of the alveoli (Massaro & Massaro 1986). Furthermore, rats allowed to live to age 60 days, i.e. 45 days after the last treatment with dexamethasone (Massaro et al 1985), or to age 90 days (Sahebjami & Domino 1989), 76 days after the drug was discontinued, did not exhibit post hoc septation.

Rat pups, whose dams had been acclimatized to living in 13% O_2 prior to becoming pregnant and that were maintained in 13% O_2 during gestation, and that were themselves maintained in 13% O_2 during the period septation occurs, failed to septate (Massaro et al 1990). Furthermore, when removed from 13% O_2 at age 14 days and allowed to breath air until being killed at age 40 days, 13% O_2-exposed rats failed to septate (Blanco et al 1991).

These studies with dexamethasone and 13% O_2 indicate there is a 'critical' period for septation. That is, if septation does not occur at the appropriate time there is not spontaneous post hoc septation. This characteristic of the developing rat lung seems to also be true for prematurely born babies (Husain et al 1998) and is therefore a target for molecular therapies that could induce septation.

Retinoids and the formation of alveoli

At the onset of our studies on the regulation of the formation of alveoli, circa 1982, the absence of substantial information on the molecular regulation of septation led us to a 'guilt by association' approach, e.g. a low serum concentration of corticosteroid hormones during the period of septation suggested corticosteroids might be inhibitors of septation. In a similar vein, we subsequently arrived at the notion that retinoids might regulate septation in a 'guilt by association' among the following: the concentration of cellular retinoic acid-binding protein I peaks in the lung, but not in liver, during the period of septation (Ong & Chytil 1976); the lung's concentration of cellular retinol binding protein I is high during the period of septation (Ong & Chytil 1976); the lung contains large numbers of vitamin A (retinol) storage cells during the time alveoli are formed but not thereafter (Okabe et al 1984, Vaccaro & Brody 1978); prematurely born infants with BPD have a lower concentration of retinol than premature infants without BPD (Shenai et al 1985), and retinoids are active in developmental events in many tissues. These considerations led us to a series of studies that resulted in the following findings:

(1)   All-*trans* retinoic acid (RA) treatment of newborn rats causes a 50% increase in
        the number of alveoli without an increase of lung volume or alveolar surface
        area. The lack of an increase in surface area in the face of additional alveoli
        suggests the action of a regulatory mechanism to prevent unneeded surface
        area (Massaro & Massaro 1996).
(2)   Treatment with all-*trans* RA prevents the low number of alveoli and low body
        mass-specific surface area caused by treatment with dexamethasone (Massaro
        & Massaro 1996).
(3)   Elastase instilled into lungs of adult rats produces changes characteristic of
        human (Thurlbeck 1967) and experimental (Kaplan et al 1973) emphysema:
        increased lung volume reflecting a loss of elastic recoil, larger but fewer
        alveoli and diminished volume-corrected alveolar surface area. Treatment
        with RA reverses these changes (Massaro & Massaro 1997).
(4)   In tight-skin mice (Green et al 1976), which have a genetic failure of septation
        (Martorana et al 1989), and in rats in which septation had been previously
        prevented by treatment with dexamethasone, treatment with RA partially
        rescues the failed septation (Massaro & Massaro 2000).

Subsequently others found:

(1)   Addition of RA to explants of fetal mouse lungs induces septation (Cilley et al
        1997).
(2)   Treatment of prematurely delivered lambs with retinol induces the formation
        of alveoli (Albertine et al 1999).

The molecular basis for the action of RA on the formation of alveoli is unclear but
is under very active investigation.

Plasticity of the gas-exchange region

That the formation of alveoli can be induced in adult animals may be inferred by
studies demonstrating that treating adult rats (Callas & Adkisson 1982) or adult
hamsters (Thompson 1980) with thyroid hormone increases surface area. The more
recent work using retinoids cited in the preceding section has directly
demonstrated the formation of alveoli can be induced in adult animals. However,
there are published studies, we believe too narrowly interpreted (Sahebjami &
Vassallo 1979, Sahebjami & Wirman 1981, Harkema et al 1984, Karlinsky et al
1986, Kerr et al 1985), from which one can infer a plasticity of the gas-exchange
region in response to calorie intake that is broader, more remarkable, and of
biologically more fundamental importance than the response to thyroid hormone
or to retinoids.

Sahebjami and co-workers, using the then state of the art morphometric procedures, demonstrated adult rats subjected to calorie restriction of as much as 80% for a few weeks had markedly increased distance (Lm) between alveolar walls compared to rats allowed food ad libitum (Sahebjami & Vassallo 1979, Sahebjami & Wirman 1981). Furthermore, when corrected for the lower volume of the lungs of the smaller calorie restricted adult rats, surface area was less in calorie restricted rats than in rats fed ad libitum. Sahebjami et al (Sahebjami & Vassallo 1979, Sahebjami & Wirman 1981), and subsequently others who repeated his studies in adult rats (Harkema et al 1984, Kerr et al 1985) and in the more slowly growing adult hamsters (Karlinsky et al 1986), considered the changes in Lm and surface area to represent 'starvation-induced emphysema'. Sahebjami et al (Sahebjami & Vassallo 1979, Sahebjami & Wirman 1981), again using measurements of Lm and surface area, demonstrated that allowing previously calorie-restricted rats free access to food resulted in a decrease of Lm and an increase of surface area. As nearly as we can glean from their papers the people describing these extraordinary charges in Lm and surface area linked them only to the development of emphysema and did not infer that the fall in Lm in response to refeeding represented the induction of the formation of alveoli in adult rats.

We have made different inferences from their findings and have preliminary data to support our inferences. We agree with those who failed to find evidence of lung tissue destruction in calorie-restricted animals (Harkema et al 1984, Karlinsky et al 1986). Our preliminary results from studies in adult mice is that calorie restriction (CR) does not alter lung volume but increases the volume of individual alveoli, decreases the number of alveoli and lowers surface area; refeeding reverses these findings (G. D. Massaro & D. Massaro, unpublished observations). Furthermore, CR results in a substantial decrease, within 72 hours of a two-thirds decrease of calorie intake, in the amount of DNA per lung and this decrease is partly reversed within 72 hours of ad libitum access to food (S. Radaeva & D. Massaro, unpublished observations).

Based on the conservation, across the full range of body mass in mammals, of the link between oxygen intake and alveolar dimensions (Tenney & Remmers 1963), we propose the fall in oxygen consumption which occurs in calorie restricted animals (Gonzales-Pacheco et al 1993), triggers a cascade of events that results in the degradation of alveoli and of loss, most likely by apoptosis, of lung cells. Conversely, we propose, refeeding, which produces an increase in oxygen consumption, in turn stimulates lung cell replication and induces the formation of alveoli.

In our view such remarkable architectural and cellular plasticity in response to calorie intake must have conferred an evolutionary advantage. Preliminarily, we offer the following thoughts. All species tested breathe at a frequency and tidal volume at which the work of breathing is least; deviation to either side of this nadir increases the work of breathing (Otis et al 1950, Agostoni et al 1959). Calorie restriction lowers the frequency of breathing and tidal volume (Keyes et al 1950). We suggest a remodeling of the lung occurs in response to a calorie induced change in breathing pattern that readjusts the work of breathing to a minimal level. Similarly, refeeding increases oxygen consumption, changes the pattern of breathing, and results in the induction of alveoli. Another, not mutually exclusive possibility is that the lung's loss of cells and a cellular matrix, which we propose are less needed because of a lower oxygen consumption, could provide substrate for more needed tissues, e.g. brain during CR.

Linking the changes in architecture and cell number to calorie-induced changes in oxygen consumption does not say anything about the triggering events, which may be mechanical, i.e. rate of breathing and tidal volume, but that must ultimately be molecular. We are in very active pursuit of a cellular and molecular understanding of the caloric regulation of the physiological turnover (loss and induction of formation) of alveoli. We anticipate this information will eventually be used to induce the formation of alveoli for therapeutic purposes.

Acknowledgements

Supported in part by HL 20366, HL 59432, HL 60115, and HL 37666. DM and GDM are Senior Fellows of the Lovelace Respiratory Research Institute, Albuquerque, NM, USA. DM is Cohen Professor of Pulmonary Research, Georgetown University.

References

Agostoni E, Thimm FF, Fenn WO 1959 Comparative features of the mechanics of breathing. J Appl Physiol 14:679–683

Albertine KH, Pierce RA, Starcher BC, MacRitchie AN, Carlton DP, Bland RD 1999 Lung development is impaired in chronic lung disease of prematurity, FASEB J 13:A1154

Amy RW, Bowes D, Burri PH, Haines J, Thurlbeck WM 1977 Postnatal growth of the mouse lung. J Anat 124:131–151

Blanco LN, Massaro D, Massaro GD 1991 Alveolar size, number, and surface area: developmentally dependent response to 13% O_2. Am J Physiol 261:L370–L377

Bucher JR, Roberts RJ 1981 The development of newborn rat lung in hyperoxia: a dose–response study of lung growth, maturation, and changes in antioxidant enzyme activity. Pediatr Res 15:999–1008

Buckingham S, McNary WF, Sommers SC, Rothschild J 1968 Is lung an analog of Moog's developmental intestine? Phosphatases and pulmonary alveolar differentiation in fetal rabbits. Fed Proc 27:328

Burri PH 1974 The postnatal growth of rat lung. 3. Morphology. Anat Rec 180:77–98

Callas G, Adkisson VT 1980 The effects of desiccated thyroid on the rat lung. Anat Rec 197:331–337

Cilley RE, Zgleszewski S, Zhang L, Krummel TM, Chinoy MR 1997 Retinoic acid induces alveolar septation in fetal murine lungs in culture. Am J Respir Crit Care Med 155:A840

Collins MH, Kleinerman J, Moessinger AC, Collins AH, James LS, Blanc WA 1986 Morphometric analysis of the growth of the normal fetal guinea pig lung. Anat Rec 216:381–391

Gonzales-Pacheco DM, Buss WC, Koehler KM, Woodside WF, Alpert SS 1993 Energy restriction reduces metabolic rate in adult male Fisher-344 rats. J Nutr 123:90–97

Green MC, Sweet HO, Bunker LE 1976 Tight-skin, a new mutation of the mouse causing excessive growth of connective tissue and skeleton. Am J Pathol 82:493–512

Harkema JR, Mauderly JL, Gregory RE, Pickrell JA 1984 A comparison of starvation and elastase models of emphysema in the rat. Am Rev Respir Dis 129:584–591

Henning SJ 1978 Plasma concentration of total and free corticosterone during development in the rat. Am J Physiol 235:E451–E456

Henning SJ 1981 Postnatal development: coordination of feeding, digestion, and metabolism. Am J Physiol 241:G199–G114

Husain AN, Siddiqui NH, Stocker JT 1998 Pathology of arrested acinar development in postsurfactant bronchopulmonary dysplasia. Hum Pathol 29:710–717

Jobe AJ 1999 The new BPD: an arrest of lung development. Pediatr Res 46:641–643

Jones CT, Roebuck MM 1980 The development of the pituitary–adrenal axis in the guinea pig. Acta Endocrinol (Copenh) 94:107–116

Kaplan PD, Kuhn C, Pierce JA 1973 The induction of emphysema with elastase. I. The evolution of the lesion. J Lab Clin Med 82:349–356

Karlinsky JB, Goldstein RH, Ojserkis B, Snider GL 1986 Lung mechanisms and connective tissue levels in starvation-induced emphysema in hamsters. Am J Physiol 251:R282–R288

Kerr JS, Riley DJ, Lanza-Jacoby S et al 1985 Nutritional emphysema in the rat. Influence of protein depletion and impaired lung growth. Am Rev Respir Dis 131:644–650

Keyes A, Brozek J, Henshel A, Mickelsen O, Taylor HL 1950 The biology of human starvation. University of Minnesota Press, Minneapolis, MN, p 601–606

Krogh A 1941 The comparative physiology of respiratory mechanisms. University of Pennsylvania Press, Philadelphia, PA, p 3–8

Langston C, Kida K, Reed M, Thurlbeck WM 1984 Human lung growth in late gestation and in the neonate. Am Rev Respir Dis 129:607–613

Margraf LR, Tomashefki JF, Bruce MC, Dahms BB 1991 Morphogenetic analysis of the lung in bronchopulmonary dysplasia. Am Rev Respir Dis 143:391–400

Massaro D, Massaro GD 1986 Dexamethasone accelerates postnatal alveolar wall thinning and alters wall composition. Am J Physiol 251:R218–R224

Massaro D, Teich N, Maxwell S, Massaro GD, Whitney P 1985 Postnatal development of alveoli. Regulation and evidence for a critical period. J Clin Invest 76:1297–1305

Massaro GD, Massaro D 1996 Postnatal treatment with retinoic acid increases the number of pulmonary alveoli in rats. Am J Physiol 270:L305–L310

Massaro GD, Massaro D 1997 Retinoic acid treatment abrogates elastase-induced pulmonary emphysema in rats. Nat Med 3:675–677 (erratum: 1997 Nat Med 3:805)

Massaro GD, Massaro D 2000 Retinoic acid-treatment partially rescues failed septation in rats and mice. Am J Physiol 278:L955–L960

Massaro GD, Olivier J, Dzikowski C, Massaro D 1990 Postnatal development of lung alveoli: suppression by 13% O_2 and a critical period. Am J Physiol 258:L321–L327

Martorana PA, van Even P, Gardi C, Lungarella G 1989 A 16-month study of the development of genetic emphysema in tight-skin mice. Am Rev Resp Dis 139:226–232

O'Brodovich HM, Mellins RB 1985 Bronchopulmonary dysplasia. Unresolved neonatal acute lung injury. Am Rev Respir Dis 132:694–709

Okabe T, Yorifuji H, Yamada E 1984 Isolation and characterization of vitamin A-storing lung cells. Exp Cell Res 154:125–135

Ong D, Chytil F 1976 Changes in the levels of cellular retinol- and retinoic-acid-binding proteins of liver and lung during perinatal development of rat. Proc Natl Acad Sci USA 73:3976–3978

Otis AB, Fenn WO, Rahn H 1950 Mechanics of breathing in man. J Appl Physiol 2:592–607

Sahebjami H, Domino M 1989 Effects of postnatal dexamethasone treatment on development of alveoli in adult rats. Exp Lung Res 15:961–973

Sahebjami H, Vassallo CL 1979 Effects of starvation and refeeding on lung mechanisms and morphometry. Am Rev Respir Dis 119:443–451

Sahebjami H, Wirman JA 1981 Emphysema-like changes in the lungs of starved rats. Am Rev Respir Dis 124:619–624

Shenai JP, Chytil F, Stahlman MT 1985 Vitamin A status of neonates with bronchopulmonary dysplasia. Pediatr Res 19:185–188

Sobonya, RE, Logrinoff MM, Taussig LM, Theriault A 1982 Morphometric analysis of the lung in prolonged bronchopulmonary dysplasia. Pediatr Res 16:969–972

Tenney SM, Remmers JE 1963 Comparative quantitative morphology of the mammalian lung diffusing area. Nature 197:54–56

Thompson ME 1980 Lung growth in response to altered metabolic demand in hamsters: influence of thyroid function and cold exposure. Respir Physiol 40:335–347

Thurlbeck WH 1967 Internal surface area and other measurements in emphysema. Thorax 22:483–496

Vaccaro C, Brody JS 1978 Ultrastructure of developing alveoli. I. The role of the interstitial fibroblast. Anat Rec 192:467–479

Zeltner TB, Burri PH 1987 The postnatal growth and development of human lung. II. Morphology. Respir Physiol 67:269–282

DISCUSSION

Dunnill: In the calorie restriction and normal diet experiments, how old were the animals and what was the interval between the two phases of treatment?

Massaro: They were adult mice. The calorie restriction was for two weeks and the period of re-feeding was for three weeks. We wanted to narrow the period as much as we could, to see whether we could work with a short period for studies that we would like to do on gene expression.

Dunnill: What happens if you calorie-restrict earlier on in development?

Massaro: That is an interesting question. If you calorie-restrict adult animals, they lose weight and they drop their oxygen consumption, which I think is a critical issue here. If you calorie-restrict growing animals (and even humans), their metabolic rate doesn't change, but they grow slower. I don't know what happens in the lungs of growing animals.

Agustí: I am very impressed by the caloric effect. Some patients with COPD, particularly those with emphysema, will lose weight. Emiel Wouters has shown clearly that this loss of weight has prognostic value and can be reversed. I wonder whether these two phenomena are connected.

Massaro: This is why Hamid Sahebjami did these studies in the first place (Sahebjami & Vassallo 1979). But when he did the first studies he didn't know about the Warsaw ghetto. In World War II the German army surrounded

Warsaw to starve the Jews, and there were remarkable Jewish physicians who did studies on patients even though they were starving themselves. They described physical signs such as low diaphragm and narrow mediastinal and cardiac dullness. They had about 350 autopsies and 13% were said to have emphysema. Of these 13%, half were below the age of 40. The other thing is that there are now reports of emphysema in malnourished people with AIDS. I don't know what is happening here.

Lomas: HIV is perhaps not the best model, as HIV can directly affect the lung. In people with a CD4 count of less than 200 there is small airway disease and bullous changes, equally you can get HIV alveolitis.

MacNee: Emiel Wouters, haven't you done computed tomography (CT) scanning in emphysema patients with and without weight loss?

Wouters: That is right. We have done CT scanning mainly in relation to intermediary metabolism. We have to look at two aspects: the effect of emphysema on weight loss and the model of starvation in relation to the lung parenchyma. For the latter, it is probably better to look in patients with anorexia and then to look at the effects of nutritional intervention in this group. But to do this we probably need more sophisticated CT scanning methods.

Massaro: Sahebjami induced emphysema with elastase in rats, and then he calorie-restricted them (Sahebjami & Domino 1989). When he re-fed them their lung morphometry went back to where they were with just elastase emphysema. It is going to be complicated.

Hogg: I wonder if Michael Dunnill has ever examined the lungs of patients that have become cachectic for any reason to see the effect of starvation on the lung parenchyma?

Dunnill: No, but the effects of starvation are very interesting. I had never heard about this work in Warsaw, but I would like to have a reference for this.

Massaro: I found this recently: it is published in one of the nutrition books. There is also one other interesting study, but they didn't do much with the lung: the University of Minnesota had a long-term study of food deprivation in the early 1950s. They food deprived young people for weeks. They then measured respiratory rate. This is the fascinating part, because I have been trying to think about what regulates this plasticity. I think it is fundamental because I think it is linked to oxygen consumption. The lung only does one thing: it provides a gas exchange surface. When the oxygen consumption is changed, as in calorie restriction, tidal volume and respiratory frequency also change. All species breathe at an optimum frequency where the work of breathing is lowest. If you go on either side of that optimum frequency or volume, the work of breathing goes up. Why does the lung remodel? Perhaps it remodels to diminish the work of breathing.

Rennard: There are other examples where caloric intake can change dramatically. People that exercise or people that work in cold environments can have very different caloric intakes and yet maintain their body weight. Presumably their oxygen consumption would be changed in proportion to caloric intake. Are there changes in these peoples' lung functions?

Massaro: Elite swimmers are thought to have more alveoli in their lungs. This is beyond what you would get if you just had better muscles. Ewald Weibel and colleagues did some nice studies in which they put animals in the cold. If you put animals at 4 °C, they get so cold that they don't grow very well. If you put them at 11°C, their metabolic rate goes way up and their food intake increases, but they grow. In this case the lung surface area increases and the animals seem to get more alveoli. The other way this can be done is to alter metabolic rate with thyroid hormone. If rats are treated with thyroid hormone, the surface area of the lung increases. With respect to people with emphysema, the question is, is the response of rats to thyroid hormone a function of the thyroid hormone effect on oxygen consumption, or is it a function on gene expression unrelated to oxygen consumption, in which case could you potentially use thyroid analogues that don't increase metabolic rate but still have these effects?

Silverman: I have one question about the starvation and re-feeding model. How do you know that it is actual loss and regeneration of alveoli as opposed to distension and retraction of alveoli?

Massaro: In food-deprived animals, the lungs get stiffer in air because calorie restriction interferes with surfactant function, but liquid-filled lungs of calorie-restricted animals don't lose recoil. This means that the tissue recoil is normal.

Hogg: Is the growth that goes on in the remaining lung after pneumonectomy a similar sort of septation process?

Massaro: There are really exciting data from Connie Hsia's lab from Southwestern Medical Center at the University of Texas. She and Bob Johnson have been doing pneumonectomies. When they do pneumonectomies on puppies, they get complete regrowth on the pneumonectomy side. That is, if they take out one lung, the other lung forms enough new alveoli to fill up the other side of the chest. The exciting part is that if you look at their data, it suggests that they are forming new terminal conducting airways. This is the first time I have seen this.

Hogg: Thurlbeck did some studies of pneumonectomy in dogs but I can't remember what they found.

Massaro: I can't remember the age of their animals. I think what they found is that the lung over-distended in adult animals. There was one lobe that seemed to distend especially.

Dunnill: I have done this on pneumonectomy specimens for carcinoma, looking at patients who have survived and haven't had secondaries in the remaining lung. All I could find was overdistension. Admittedly I only looked at very few cases.

Massaro: This would fit with the Dallas data, because when they looked at mature dogs that is all they saw.

Rennard: In your published pictures of the RA receptor (RAR)β knockout mice, where there is a change in septation is there also a change in the thickness of the alveolar wall?

Massaro: We have not measured this, but we suspect it is thicker. One of the hallmarks of the thinning of the wall is that you go from having a double capillary system to a single one. Surface area is increased in this way. I wonder whether the knockout has retained the double capillary system, accounting for the thicker wall.

Rennard: Do you know anything about the compliance of those animals' lungs?

Massaro: We haven't measured the recoil of the mice. We ought to do this. Because we measure the volume of an alveolus and we know the volume of a lung, we obviate the problem of the compliance change that you are worried about.

Agustí: This is fascinating work. When you say 'septation', do you mean that new vessels are also formed here? In order to have that new alveolus functioning (gas exchange), you need new vessels.

Massaro: I agree. We haven't measured this, but when we look at the lung, we don't see any alveoli that don't have vessels.

Lomas: What do we know about the RAR in humans, and what do we know about human lung development in comparison to rodents? Do we develop our lungs in the same way?

Massaro: Not much is known about RARs in humans. The people at Vanderbilt University for years have been pushing the treatment of premature babies with retinol, because their blood levels of retinol are low. They were aiming at the lung, but I don't think they were aiming at septation. There is a recent abstract from the NIH about a large clinical trial where they gave retinol to premature babies. They didn't report survival, but the clinical outcome was a little better in that their blood gases improved more rapidly. No one knows what is going on. With regard to human lung development, septation begins in the last month or so of pregnancy, and continues for an unknown period after this. The dogma is that they stop forming alveoli at about age 8.

Dunnill: The septa in the newborn and up until several months of age are thicker. They contain more in the way of capillaries.

Massaro: I think that alveoli continue to form until age 17 or 18, when the chest stops growing. But, if you are prematurely born, and you get oxygen, you seem to stop making alveoli and septation doesn't seem to occur after it has been blocked, i.e. when the blocking agent or condition is removed.

Dunnill: That is true. If you look at the premature children that are in intensive care and die, they have very hypoplastic lungs.

Hogg: There are reports of high resolution CT scans on older children and adults who had bronchopulmonary dysplasia that show considerable residual lung disease (Aquino et al 1999).

Calverley: There is a clinical corollary of that which is slightly alarming. In order to improve short-term survival from infant respiratory distress syndrome, a common and effective intervention is to give corticosteroids during the latter part of pregnancy. If this is happening, then there may well be a loss of septation and alveoli. This has been a practice for 15–20 years, and it would be interesting to know whether any of those children born prematurely were getting emphysematous changes prematurely, because this is another potential cause of chronic obstructive pulmonary disease (COPD) in the future.

Nadel: I want us to explore the mechanisms by which RARs can have their effects. What are the second messengers involved with RARs?

Massaro: They are nuclear receptors that act with co-activators and co-repressors but do not have second messengers.

Nadel: Are there interactions between the retinoid receptors and growth factors? It is clear that growth factors are involved in the growth development and septation of airways.

Massaro: Retinoids affect the dichotomous branching of the fetal lung. For the last couple of years we have been looking at using differential display analysis of gene expression and subtractive hybridization to find out what genes are involved in septation. What is also exciting is that there are gene chips being produced by Affymetrix which will allow us to measure 9000 genes at once.

Nadel: I was going to suggest that you go back and look at the growth factors that are involved in this early division. There are more ways than just looking for thousands of genes. Perhaps selective inhibitors of growth factors could be used in your studies of fetuses or newborns.

Jeffery: In the general population we believe that cigarette smoking induces much of the emphysema that we see. Is there anything known about the effects of cigarette smoke on the expression of RARs?

Massaro: I don't know of any work on this.

Jackson: It is interesting that the retinoid story probably goes further than emphysema, according to Paul Nettesheim's studies. He has shown that retinoids are involved in the regulation of mucin genes, at least *MUC2*, *MUC5AC* and *MUC5B*. He has taken this work further to suggest that RARα is one of the key players.

References

Aquino SL, Schechter MS, Chiles C, Ablin DS, Chipps B, Webb WR 1999 High-resolution inspiratory and expiratory CT in older children and adults with bronchopulmonary dysplasia. Am J Roentgen 173:963–967

Sahebjami H, Domino M 1989 Effects of starvation and refeeding on elastase-induced emphysema. J Appl Physiol 66:2611–2616
Sahebjami H, Vassallo CL 1979 Effects of starvation and refeeding on lung mechanics and morphometry. Am Rev Respir Dis 119:443–451

Systemic effects of chronic obstructive pulmonary disease

Alvar G. N. Agustí

Institut de Medicina Respiratoria, Hospital Universitari Son Dureta, Andrea Doria 55, 07014 Palma de Mallorca, Spain

Abstract. Traditionally, chronic obstructive pulmonary disease (COPD) has been understood as a disease of the lungs, characterized by irreversible airflow limitation due to chronic bronchitis and/or emphysema. The latter are thought mainly to be the consequence of an excessive inflammatory response to tobacco smoking. Recently, several studies have shown that this pulmonary inflammation may also be detected in the systemic circulation, and that this systemic inflammation may have important clinical consequences. Most prominent among them is the loss of skeletal muscle mass that a significant percentage of patients with COPD will show during the course of their disease. This limits considerably their exercise capacity, jeopardizes their health status, and has a negative impact on their prognosis. Importantly, such prognostic value is independent of the degree of airflow obstruction and potentially reversible with appropriate therapy. This chapter summarizes available evidence supporting the concept that COPD is more than a lung disease (the systemic effects of COPD), and speculates on potential cellular mechanisms as future therapeutic targets.

2001 Chronic obstructive pulmonary disease: pathogenesis to treatment. Wiley, Chichester (Novartis Foundation Symposium 234) p 242–254

Chronic obstructive pulmonary disease (COPD) has been traditionally understood as a disease of the lung parenchyma characterized by accelerated decline of lung function. Accordingly, available therapeutic strategies such as bronchodilators, steroids, antibiotics and/or mucolytics target the lungs. These strategies are effective in relieving symptoms but often useless in improving the prognosis of these patients. Interestingly, the only therapeutic measure that has been convincingly shown to improve survival in COPD is long-term oxygen therapy (LOT) which, by definition, subsidizes the lungs in their function of exchanging oxygen. As such, LOT does not treat the lungs; it treats the rest of the body!

Recently, the concept that COPD is more than a lung disease, and that the treatment of the rest of the body is important in the clinical management of the disease, is gaining acceptance (Noguera et al 1998, Schols et al 1993). For instance, it has been shown that patients with COPD often lose muscle mass

TABLE 1   Systemic effects of chronic obstructive pulmonary disease

Systemic inflammation	Oxidant stress
	Increased levels of circulating cytokines and acute phase proteins
	'Activated' inflammatory cells (lymphocytes/neutrophils)
Skeletal muscle dysfunction	Exercise limitation
	Abnormal mass and structure
	Abnormal enzyme activities and bioenergetics
	Excessive apoptosis
Abnormal nutrition and metabolism	Weight loss
	Increased resting energy expenditure
	Abnormal body composition
	Abnormal amino acid metabolism
Other organs	Endothelial dysfunction
	Renal circulation

during the course of their disease and, more importantly, that this phenomenon has prognostic value independently of other more classical prognostic variables, such as the degree of airflow obstruction (Schols et al 1998). Thus, accepting that COPD is more than a lung disease may open new windows of opportunity for therapeutic intervention. However, because the cellular mechanisms involved in this response remain unknown, no new potentially effective pharmacological agents have yet been developed.

This chapter summarizes the available evidence supporting the concept that COPD is associated with systemic inflammation, and that it can cause (directly and indirectly) harmful effects in distant organs outside of the lungs (Table 1), the so-called systemic effects of COPD. It also discusses potential mechanisms underlying these systemic effects, highlights their clinical relevance and speculates on how a better understanding of these effects may open new therapeutic avenues.

Systemic inflammation in COPD

COPD is characterized by an excessive inflammatory process in the lung parenchyma in response to a variety of known (tobacco smoking, allergens, infections, airborne pollutants and minerals) or unknown agents. Recent studies have shown that this excessive pulmonary inflammation may also be detected in the systemic circulation. For instance, some studies have presented compelling evidence of oxidative stress in the systemic blood in patients with COPD, particularly during exacerbations of the disease (Rahman et al 1996). Other authors have described increased plasma levels of several cytokines (interleukin [IL]-6, IL-8, tumour necrosis factor [TNF]α) and acute phase response proteins (C-reactive protein) in patients with stable COPD (Schols et al 1996, Yasuda et al 1998). In our laboratory, we followed the concentration of several cytokines in plasma during the recovery phase of an exacerbation of COPD requiring hospitalization and found that, despite the use of intravenous steroids, the increased plasma level of cytokines did not change significantly (Sauleda et al 1999). This observation suggests, therefore, that the systemic inflammation seen in COPD is not sensitive to steroid treatment. Such a statement would be in keeping with previous studies reporting a lack of response to steroids of several markers determined in patients with COPD either in induced sputum (Keatings et al 1997) or exhaled air (Agustí et al 1999a). Finally, other studies have shown abnormalities in several circulating inflammatory cells. For instance, Noguera et al (1998) showed that circulating neutrophils in patients with stable COPD exhibited abnormal expression of several surface adhesion molecules, most prominently Mac1 (CD11b), and that this situation changed significantly during exacerbations of the disease. Interestingly, these same neutrophils exhibited a down-regulation of one of the G proteins (Gαs) involved in the intracellular signal transduction pathway linked to CD11b (Noguera et al 1998). The origin and significance of these observations is still unclear. It is likely that the inflammatory process going on in the lung parenchyma (COPD) may somehow activate these cells during their transit through the pulmonary circulation. Alternatively, these abnormalities may be the expression of an intrinsic abnormality of neutrophils in COPD that, when exposed to the effects of smoking, contribute to the excessive inflammatory response that ultimately results in the development of the disease. Preliminary results from our laboratory seem to support this latter hypothesis (Agustí et al 1999b).

Skeletal muscle abnormalities in COPD

Exercise limitation is a frequent complaint in COPD. Traditionally, this has been explained on the basis of the increased work of breathing caused by airflow limitation. Killian et al, however, elegantly showed in 1992 that almost half of the patients with COPD stop exercise because of leg fatigue, not because of dyspnea (Killian et al 1992). This was probably the first indication of a skeletal muscle dysfunction in COPD. Since then, numerous investigators have used very different techniques (morphometry [Jakobsson et al 1990], physiology [Sala et al 1999], biochemistry [Jakobsson et al 1990, Maltais et al 1996, Sauleda et al 1998], imaging [Bernard et al 1998, Payen et al 1993, Wuyam et al 1992] and molecular biology [Maltais et al 1999, Sauleda et al 1998]) to get a better understanding of this abnormality. Recently, the European Respiratory Society and the American Thoracic Society have published a joint document summarizing current knowledge in this field (American Thoracic Society, European Respiratory Society,1999). Today, it is accepted that skeletal muscle in patients with COPD is abnormal, both quantitatively and qualitatively, and that this abnormality may limit considerably their exercise capacity and, hence, may impact significantly their health status (formerly, quality of life) (American Thoracic Society, European Respiratory Society 1999).

The mechanisms explaining the skeletal muscle dysfunction of COPD are not fully understood. Because it can be partially improved by rehabilitation (Lacasse et al 1996, Ries et al 1997), it is thought that sedentarism and inactivity, due in turn to the constrains imposed by the diseased lungs, may play a considerable role (American Thoracic Society, European Respiratory Society 1999). However, other factors, such as those related to the systemic inflammation alluded to above, may also be important. For instance, Sauleda et al showed that the activity of cytochrome oxidase (the terminal enzyme in the mitochondrial electron transport chain) was up-regulated in skeletal muscle biopsies obtained from patients with COPD (Sauleda et al 1998). This observation can not be explained by inactivity. Further, these same authors have described very recently that the same abnormality occurs in circulating lymphocytes in these patients (Sauleda et al 2000). Clearly, lymphocytes are not subjected to the influence of sedentarism. Therefore, other mechanisms are probably operative in the pathogenesis of the skeletal muscle dysfunction of COPD. Potential alternatives will be discussed in the next section, in the context of the nutritional abnormalities that can also occur in these patients.

Nutritional abnormalities in COPD

Many patients with COPD will lose weight during the course of their disease (Schols et al 1993). This is particularly true for patients with severe COPD and chronic respiratory failure, but it is not restricted solely to them (Schols et al 1993). This phenomenon appears to be of prognostic value (Schols et al 1998). Importantly, this prognostic value seems independent of the more traditional prognostic factors related to lung function (FEV_1 or arterial PO_2) (Schols et al 1998). Further, it seems to be reversible with appropriate therapy (pharmaco-therapy, nutritional support and physical therapy) (Schols et al 1998). Because of all these reasons, the phenomenon of weight loss in COPD is highly relevant clinically. However, the mechanisms underlying it are far from clear (Sridhar 1995).

The increased metabolic rate often shown by patients with COPD may contribute to weight loss because of unbalanced nutritional needs (Baarends et al

1997). Tissue hypoxia and systemic inflammation have also been quoted as potential factors contributing to weight loss in COPD (Sridhar 1995). Interestingly, Sauleda et al (1998) showed a significant relationship between the degree of arterial hypoxaemia and the activity of cytochrome oxidase in the skeletal muscle of these patients. Given that cytochrome oxidase actually is the enzyme that consumes oxygen in the mitochondria, this observation is likely to explain the increased resting metabolic rate often seen in COPD, particularly in those patients who lose weight (Baarends et al 1997). Alternatively, given that these patients with increased metabolic rate are precisely those who show evidence of more systemic inflammation (Schols et al 1996), it is conceivable that the latter may also contribute to explain the loss of weight seen in COPD. In this context, particular emphasis has been given to TNFα, a cytokine known to induce muscle wasting in several experimental and clinical conditions (Li et al 1998). Today, we know that the concentration of TNFα itself (Di Francia et al 1994) and several of its receptors (Schols et al 1996) is increased in the plasma of those patients with COPD who lose weight, and that monocytes harvested from these same patients produced (*in vitro*) higher levels of such cytokine than those obtained from healthy controls (de Godoy et al 1996). However, we do not know yet the specific cellular mechanisms by which such increased concentrations of TNFα may cause loss of skeletal muscle mass in these patients, or why only a subset of patients present these abnormalities. As shown in Fig. 1, there are several pathways by which TNFα can induce skeletal muscle wasting. For instance, Li et al (1998) have shown that, *in vitro*, TNFα produces a dose-dependent decrease in myosin heavy chain content through the activation of the transcription factor NF-κB (Fig. 1). Alternatively, TNFα can induce apoptosis (programmed cell death), activate other transcription factors (such as AP-1), induce the expression of the inducible form of the nitric oxide synthase (iNOS) or enhance the generation of reactive oxygen species (ROS) by the mitochondria (Fig. 1). In our laboratory, we have recently explored the possibility that enhanced apoptosis contributes to explain the loss of muscle mass that occurs in some patients with COPD. Our preliminary results indicate that those COPD patients with low body mass index present a significantly higher level of skeletal muscle apoptosis than patients with COPD and normal weight or healthy controls (Agustí et al 2000). Interestingly, a recent report has shown abnormal plasma levels of soluble Fas, an inhibitor of apoptosis, in patients with COPD (Yasuda et al 1998).

Other organs

To date, the evidence supporting the presence of systemic effects in patients with COPD is restricted to that discussed above. This is mostly because the access to other (potentially more important) organs, such as the heart, brain, kidney or

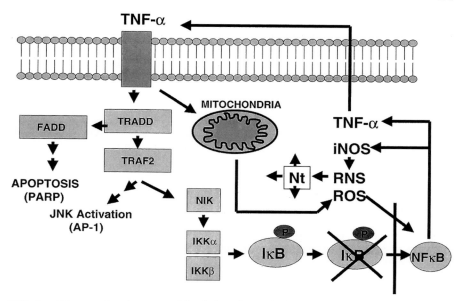

FIG. 1. Diagram showing potential cellular effects of TNFα related to the phenomenon of weight loss seen in some COPD patients. For abbreviations, see text.

liver, is far more difficult than to peripheral blood or skeletal muscle. There are, however, some indications of abnormalities also in some of these 'other organs'. For instance, the kidney appears to present several abnormalities (Campbell et al 1982) likely related to an abnormal endothelial function (Baudouin et al 1992, Howes et al 1996). The latter has been clearly demonstrated in the pulmonary circulation of patients with COPD (Dinh-Xuan et al 1991, Peinado et al 1998) and also in the systemic circulation of smoker subjects (Celermajer et al 1996). The impact of such abnormality in other organs will have to be investigated in the future.

Conclusions and future prospects

The above discussion supports the concept that COPD is *more than* a lung disease (admittedly, it is a *lung* disease) and, importantly, that the treatment of these *extra-pulmonary* effects may be important in the clinical management of the disease. Accordingly, a better understanding of the systemic effects of COPD discussed above may allow the design of new therapeutic strategies that, eventually, will result in a better health status and, hopefully, better prognosis in patients suffering from COPD.

Acknowledgements

The author thanks Drs J. Sauleda, X. Busquets, A. Noguera, C. Miralles, B. Togores, E. Sala, M. Carrera, F. Barbé and S. Batle (Hospital Universitario Son Dureta, Palma de Mallorca) for helpful discussions in the past. Work supported, in part, by ABEMAR, Fondo de Investigación Sanitaria, CICYT and SEPAR.

References

Agustí AGN, Villaverde JM, Togores B 1999a Serial measurements of exhaled nitric oxide during exacerbations of chronic obstructive pulmonary disease. Eur Respir J 14:523–528

Agustí AGN, Noguera A, Batle S et al 1999b Neutrophil function and susceptibility to chronic obstructive pulmonary disease. Eur Respir J 14:358s (abstr)

Agustí AGN, Sauleda J, Batle S et al 2000 Skeletal muscle apoptosis in COPD. Eur Respir J, in press

American Thoracic Society, European Respiratory Society 1999 Skeletal muscle dysfunction in chronic obstructive pulmonary disease. Am J Respir Crit Care Med 159:S1–S40

Baarends EM, Schols AM, Westerterp KR, Wouters EF 1997 Total daily energy expenditure relative to resting energy expenditure in clinically stable patients with COPD. Thorax 52:780–785

Baudouin SV, Bott J, Ward A, Deane C, Moxham J 1992 Short term effect of oxygen on renal haemodynamics in patients with hypoxaemic chronic obstructive airways disease. Thorax 47:550–554

Bernard S, Leblanc P, Whittom F et al 1998 Peripheral muscle weakness in patient with chronic obstructive pulmonary disease. Am J Respir Crit Care Med 158:629–634

Campbell JL, Calverley PMA, Lamb D, Flenley DC 1982 The renal glomerulus in hypoxic cor pulmonale. Thorax 37:607–611

Celermajer DS, Adams MR, Clarkson P et al 1996 Passive smoking and impaired endothelium-dependent arterial dilatation in healthy young adults. N Engl J Med 334:150–154

de Godoy I, Donahoe M, Calhoun WJ, Mancino J, Rogers RM 1996 Elevated TNF-α production by peripheral blood monocytes of weight-losing COPD patients. Am J Respir Crit Care Med 153:633–637

Di Francia M, Barbier D, Mege JL, Orehek J 1994 Tumor necrosis factor-α levels and weight loss in chronic obstructive pulmonary disease. Am J Respir Crit Care Med 150:1453–1455

Dinh-Xuan AT, Higenbottam TW, Clelland CA et al 1991 Impairment of endothelium-dependent pulmonary-artery relaxation in chronic obstructive lung disease. N Engl J Med 324:1539–1547

Howes TQ, Keilty SEJ, Maskrey VL, Deane CR, Baudouin SV, Moxham J 1996 Effect of L-arginine on renal blood flow in normal subjects and patients with hypoxic chronic obstructive pulmonary disease. Thorax 51:516–519

Jakobsson P, Jorfeldt L, Brundin A 1990 Skeletal muscle metabolites and fibre types in patients with advanced chronic obstructive pulmonary disease (COPD), with and without chronic respiratory failure. Eur Respir J 3:192–196

Keatings VM, Jatakanon A, Worsdell YM, Barnes PJ 1997 Effects of inhaled and oral glucocorticoids on inflammatory indices in asthma and COPD. Am J Respir Crit Care Med 155:542–548

Killian KJ, Leblanc P, Martin DH, Summers E, Jones NL, Campbell EJ 1992 Exercise capacity and ventilatory, circulatory, and symptom limitation in patients with chronic airflow limitation. Am Rev Respir Dis 146:935–940

Lacasse Y, Wong E, Guyatt GH, King D, Cook DJ, Goldstein RS 1996 Meta-analysis of respiratory rehabilitation in chronic obstructive pulmonary disease. Lancet 348:1115–1119

Li YP, Schwartz RJ, Waddell ID, Holloway BR, Reid MB 1998 Skeletal muscle myocytes undergo protein loss and reactive oxygen-mediated NF-κB activation in response to tumor necrosis factor α. FASEB J 12:871–880

Maltais F, Simard AA, Simard C, Jobin J, Desgagnés P, LeBlanc P 1996 Oxidative capacity of the skeletal muscle and lactic acid kinetics during exercise in normal subjects and in patients with COPD. Am J Respir Crit Care Med 153:288–293

Maltais F, Sullivan MJ, Leblanc P et al 1999 Altered expression of myosin heavy chain in the vastus lateralis muscle in patients with COPD. Eur Respir J 14:850–854

Noguera A, Busquets X, Sauleda J, Villaverde JM, MacNee W, Agustí AGN 1998 Expression of adhesion molecules and G proteins in circulating neutrophils in chronic obstructive pulmonary disease. Am J Respir Crit Care Med 158:1664–1668

Payen JF, Wuyam B, Levy P et al 1993 Muscular metabolism during oxygen supplementation in patients with chronic hypoxemia. Am Rev Respir Dis 147:592–598

Peinado VI, Barberá JA, Ramirez J et al 1998 Endothelial dysfunction in pulmonary arteries of patients with mild COPD. Am J Physiol 274:L908–L913

Rahman I, Morrison D, Donaldson K, MacNee W 1996 Systemic oxidative stress in asthma, COPD, and smokers. Am J Respir Crit Care Med 154:1055–1060

Ries AL, Carlin BW, Casaburi R et al 1997 Pulmonary rehabilitation: joint ACCP/AACVPR evidence-based guidelines. Chest 112:1363–1396

Sala E, Roca J, Marrades RM et al 1999 Effects of endurance training on skeletal muscle bioenergetics in chronic obstructive pulmonary disease. Am J Respir Crit Care Med 159:1726–1734

Sauleda J, García-Palmer FJ, Wiesner R et al 1998 Cytochrome oxidase activity and mitochondrial gene expression in skeletal muscle of patients with chronic obstructive pulmonary disease. Am J Respir Crit Care Med 157:1413–1417

Sauleda J, Noguera A, Busquets X et al 1999 Systemic inflammation during exacerbations of chronic obstructive pulmonary disease. Lack of effect of steroid treatment. Eur Respir J 14:359s (abstr)

Sauleda J, García-Palmer FJ, Gonzalez G, Palou A, Agustí AGN 2000 The activity of cytochrome oxidase is increased in circulating lymphocytes of patients with chronic obstructive pulmonary disease, asthma and chronic arthritis. Am J Respir Crit Care Med 161:32–35

Schols AM, Soeters PB, Dingemans AM, Mostert R, Frantzen PJ, Wouters EF 1993 Prevalence and characteristics of nutritional depletion in patients with stable COPD eligible for pulmonary rehabilitation. Am Rev Respir Dis 147:1151–1156

Schols AM, Buurman WA, Staal van den Brekel AJ, Dentener MA, Wouters EF 1996 Evidence for a relation between metabolic derangements and increased levels of inflammatory mediators in a subgroup of patients with chronic obstructive pulmonary disease. Thorax 51: 819–824

Schols AM, Slangen J, Volovics L, Wouters EF 1998 Weight loss is a reversible factor in the prognosis of chronic obstructive pulmonary disease. Am J Respir Crit Care Med 157: 1791–1797

Sridhar MK 1995 Why do patients with emphysema lose weight? Lancet 345:1190–1191

Wuyam B, Payen JF, Levy P et al 1992 Metabolism and aerobic capacity of skeletal muscle in chronic respiratory failure related to chronic obstructive pulmonary disease. Eur Respir J 5:157–162

Yasuda N, Gotoh K, Minatoguchi S et al 1998 An increase of soluble Fas, an inhibitor of apoptosis, associated with progression of COPD. Respir Med 92:993–999

DISCUSSION

Stockley: As a great believer in the concepts of Darwinian medicine, I think that abnormalities lead to physiological responses that have stood the test of time. In other words, much of what we are seeing in COPD is the normal physiological response to a low metabolic rate. There is low oxygen uptake because these people are sick from their lungs. This raises an important issue: if we are trying to correct what in actual fact is normal physiology in response to illness, might we be making matters far worse? As an example, take septicaemia: circulating iron content drops, but if you give these patients iron to correct this, it makes the patients sicker. If we are talking about people who are just wasting away because they have bad lungs, and you try to build their muscles up, could this not be in the patients worst interests by increasing metabolic demand?

Agustí: My view is slightly different. To understand why this is happening might be important for providing a more integral care to our patients. Sometimes we forget about the rest of the body. Steroid treatment is a good example of this: now we accept that steroids are not particularly effective in COPD, but until recently this was not the case. Many patients with COPD are actually receiving steroid treatment, although we are only now beginning to realize how this might impact other aspects of the disease.

MacNee: The data from Emiel Wouters' group (Schols et al 1998) show that if you do reverse weight loss by dietary intervention, survival is improved. If you try to increase dietary intake and the patients don't improve their weight, survival is not improved.

Stockley: You can ventilate patients and increase survival. We should be careful about what we are saying: there is quality of life and quantity of life. I agree that rehabilitation actually does improve quality of life because you can re-train people how to operate within the limitations that their lungs have now given them.

Wouters: These patients have impaired lung function but they do not have a lowered oxygen consumption. Alvar Agustí already indicated that resting energy expenditure is increased, and activity-related energy expenditure is very high in COPD patients, because they have a mechanical inefficiency. So what we try to do by muscular intervention is to re-set that oxygen consumption for a given mode. But at this moment I agree that we only have snapshots of the complete disturbances of the metabolic machinery in the muscle cells. Alvar demonstrated elevated cytochrome oxidase in COPD and we have demonstrated high levels of inosine monophosphate even under resting conditions, so the complete energy rich phosphagen metabolism is probably disturbed in COPD. Even the restoration of lung function, for instance by lung transplantation, doesn't restore these muscle abnormalities.

Hogg: I wanted to comment on the role of the systemic response in the pathogenesis of the lung disease. There is good evidence (Weiss et al 1995, Chan-Yeung & Buncio 1984) that the white cell count goes up following chronic exposure to cigarette smoke and acute exposure to air contaminated with particulates (Tan et al 2000). Dr Stephan van Eeden has led a group in Vancouver who showed that if fine particles are put into the lung, they stimulate the bone marrow and shorten the transit times through the mitotic and post-mitotic pool of the marrow (Terashima et al 1996). When you feed alveolar macrophages the same particles *in vitro* and put the supernatants down the lung, it has the same effect on the marrow (Terashima et al 1997). Our data on patients requiring lung resection show the preoperative white cell counts correlate with their smoking history. Examination of the peripheral white blood cell count in smokers also shows high levels of band cells indicating that there is chronic stimulation of the marrow. For these reasons, I think this systemic response to the cigarette smoke might well play a big role in the pathogenesis of the lung lesions.

Stockley: That is slightly different. That is a bad physiological response, but that is because you are insulting the lung to start with. But once you have the lung disease and then you see the metabolic changes that occur, those are because of the lung disease by and large. My initial point really just came down to the muscle side of it: if you are not taking in oxygen very well, your muscles are not going to be able to survive that well.

Agustí: If you measure oxygen uptake in the muscle of COPD patients, it is higher than controls.

Stockley: That is because of a metabolic change within the muscle cell.

Agustí: If you look at several different mitochondrial enzymes, some of them go down and some of them go up. It appears that the mitochondria are uncoupled. The end result of this is that they are inefficient. They cannot provide energy. If you can somehow understand why these patients are undergoing apoptosis in the muscles this would probably make a big impact on their treatment.

Massaro: Have you measured uncoupling of protein 2? This is highly expressed in muscles and lung, and might be involved in what you are talking about.

Agustí: No, but that is a good suggestion. The interesting thing here, linked to the oxidative stress that Bill MacNee was describing in his earlier paper, is that if mitochondrial uncoupling occurs, this might be an important source of reactive oxygen species that in turn might activate gene expression or oxidise proteins.

Massaro: These middle aged or elderly heavy smokers are set up for acute myocardial infarctions. Do you see that?

Wedzicha: We studied plasma fibrinogen in our cohort, both when stable and at exacerbation (Wedzicha et al 2000). The difference between our data and that of Professor Agustí is that our patients are requested to call us as soon as possible at the onset of the symptoms of an exacerbation, so we see them early in the natural history of an exacerbation. When patients were stable the mean plasma fibrinogen level was significantly raised compared to normal subjects at 3.7 g/l and rose further at exacerbation to over 4 g/l. At convalescence at 6 weeks after exacerbation, the fibrinogen level comes back to normal. Thus patients with COPD exacerbations may be at excess risk of acute cardiac ischaemic events around the time of exacerbation. This rise of plasma fibrinogen at exacerbation, which has an infective cause could also explain the proposed effect of respiratory infections in increasing acute coronary events over a period of about 2 weeks after an infection.

Massaro: How many of your patients are you treating for systemic hypertension?

Wedzicha: Not very many; about 10–15%.

MacNee: The co-morbidity in these patients with COPD is very high.

Calverley: There are data from the ISOLDE study about causes of death. In this group of people with an FEV_1 of 35–40% of predicted, the principal cause of death is respiratory, and the second commonest is myocardial infarction.

Rennard: There are also data from the Framingham study (Sorlie et al 1989). The most significant predictor of death for all causes is lung function, and the most common cause of death by all means was cardiac death. Both heart disease and restriction of lung function, without even being severe enough to be classified beyond stage 1 COPD, are obviously related to cigarette smoking. However, if you statistically correct for the presence of cigarette smoking, lung function is still a very strong predictor. If you have COPD, your risk of a cardiac death is much higher than if you are a two-pack per day smoker and your lung function is 80% of predicted. This could be because the lung function is causing cardiac disease or it could support Alvar's suggestion that this is a manifestation of some systemic disorder caused by the underlying cigarette smoking, or it could be due to any of the other things that Alvar suggested. But it certainly supports the concept that cardiac disease is increased in patients who have abnormal lung function.

Lomas: Sten Eriksson, looking at the α_1-antitrypsin-deficient patients, finds a low incidence of ischaemic heart disease and a low incidence of atherosclerosis.

Hogg: In acute coronary thrombosis, it is usually the smaller plaques that rupture to initiate thrombosis. We have as yet unpublished data in rabbits that develop atherosclerosis naturally where we have used histological criteria to define the stability of the plaques. These data suggest that if we expose these animals to particulates found in polluted air, we get both a systemic response of the type I was talking about earlier and destabilization of the atherosclerotic plaques. I think this could be one of the mechanisms linking COPD with heart disease.

MacNee: There are strong relationships between the levels of particulate air pollution and ischaemic cardiac events. We also have data showing that another risk factor for ischaemic heart disease, factor VII in the blood, increases in animal models of particulate exposure.

Nadel: Does starvation cause muscle apoptosis?

Agustí: I don't know.

Nadel: In interpreting some of your data, this may be key. If you lose a lot of weight from a lack of calorific intake without disease, would you get the same apoptosis?

Agustí: Cardiac cachexia probably involves similar mechanisms, and these patients do not have COPD. Many of these chronic conditions probably have a common pathway that might lead to muscle loss by apoptosis.

Wouters: In cardiac patients there is apparently apopotosis in the peripheral muscle cells. In my opinion, you cannot say this is a consequence of starvation because a lot of these patients have a relatively high dietary intake. What is special about COPD patients is that they have a loss of fat-free mass, of muscle mass, which is related to the level of inflammatory markers.

Wedzicha: What happens in asthma with systemic inflammation? When we saw those high fibrinogen levels I mentioned earlier, I hypothesized that this was due the airway inflammation. However, another relationship with fibrinogen was the presence of bronchitic symptoms. Is this the effect of bacterial colonization?

Agustí: In asthma, the activity of cytochrome oxidase in circulating lymphocytes is also much higher than in healthy people and, interestingly, than in patients with COPD (who showed higher values than healthy controls). Interestingly, the same observations occur in patients with chronic arthritis. These observations indicate that these mechanisms may be common to many chronic inflammatory conditions, both of pulmonary and non-pulmonary origin.

Stockley: Isn't this just activation of the lymphocytes?

Agustí: These asthma patients were quite moderate patients, with mild, stable disease. I do not know whether an increase in mitochondrial respiration reflects 'activation'.

Senior: Depression is a common accompaniment to COPD, particularly in advanced stages. I wonder if you could comment about some of the types of measurements that you make as a function of depression.

Agustí: This is an interesting issue. As you know, Almitrine® is a respiratory stimulant that was not very well accepted by the respiratory community because of side effects on peripheral nerve function. However, in the control group not treated with Alimtrine® something like 30% of them showed abnormal peripheral nerve function.

Barnes: I want to return to the issue of TNFα, as this could be an important mediator of many of the systemic effects you have described. TNF used to be infused in patients with renal carcinoma metastases. But patients became ill: they lost weight and became lethargic. Andrew Coats and colleagues at NHLI have shown close relationships between plasma TNF and cachexia in heart failure (Anker et al 1997). This may have very important therapeutic implications, because it is now possible to block TNF using several approaches. In rheumatoid arthritis the symptom improvement with anti-TNF antibody treatment is very rapid and dramatic, and it is very likely a systemic effect, rather than an effect on joints. There may be treatments that will make patients with COPD feel much better and also prevent the wasting of skeletal muscles.

Wouters: It is interesting that part of the problems in COPD are comparable with cardiac cachexia, because there are also changes in the fibre composition of the skeletal muscles, for example. What is so special in COPD is that these changes are related to the impairment in diffusing capacity, at least at this moment. It also seems that the metabolic changes in the muscles differ between chronic bronchitis patients and emphysema patients. It is probably more complicated than in cardiac cachexia.

Rennard: What happens to patients following pneumoreductive surgery with respect to these systemic features?

Agusti: To my knowledge no formal data have been published on this. I talked to Dr J Roca in Barcelona and he told me that his preliminary results suggest that weight improves after pneumonectomy.

References

Anker SD, Chua TP, Ponikowski P et al 1997 Hormonal changes and catabolic/anabolic imbalance in chronic heart failure and their importance for cardiac cachexia. Circulation 96:526–534

Chan-Yeung M, Buncio AD 1984 Leukocyte count, smoking and lung function. Am J Med 76:31–37

Schols AM, Slangen J, Volovics L, Wouters EF 1998 Weight loss is a reversible factor in the prognosis of chronic obstructive pulmonary disease. Am J Respir Crit Care Med 157:1791–1797

Sorlie PD, Kannel WB, O'Connor G 1989 Mortality associated with respiratory function and symptoms in advanced age. The Framingham Study. Am Rev Respir Dis 140:379–384

Tan WC, Qiu D, Liam BL et al 2000 The human bone marrow response to acute air pollution caused by forest fires. Am J Respir Crit Care Med 161:1213–1217

Terashima T, Wiggs B, English D, Hogg JC, van Eeden SF 1996 Polymorphonuclear leukocyte transit times in bone marrow during pneumococcal pneumonia. Am J Physiol 271:L587–L592

Terashima T, Wiggs B, English D, Hogg JC, van Eeden SF 1997 Phagocytosis of small carbon particles (PM10) by alveolar macrophages stimulates the release of polymorphonuclear leukocytes from the marrow. Am J Respir Crit Care Med 155:441–444

Wedzicha JA, Seemungal TAR, MacCallum PK et al 2000 Acute exacerbations of chronic obstructive pulmonary disease are accompanied by elevations of plasma fibrinogen and serum IL-6 levels. Thromb Haemostasis 84:210–215

Weiss ST, Segel MR, Sparrow D, Wager C 1985 Relationship of FEV_1 and peripheral blood leukocyte counts. Am J Epidemiol 142:493–498

Potential novel therapies for chronic obstructive pulmonary disease

Peter J. Barnes

National Heart and Lung Institute, Imperial College School of Medicine, Dovehouse Street, London SW3 6LY, UK

Abstract. While considerable progress has been made in development of drugs for asthma, there have been few advances in the treatment of chronic obstructive pulmonary disease (COPD). New therapeutic approaches to prevent disease progression are urgently needed and these will arise out of better understanding of the disease process at a cell and molecular level. The inflammatory response in COPD differs markedly from that of asthma, with differences in inflammatory cells, mediators and response to therapy. The neutrophilic inflammation is orchestrated by chemotactic factors, such as interleukin (IL)-8, other CXC chemokines and leukotriene B4; receptor blockers (CXCR1, CXCR2, BLT antagonists) or synthesis inhibitors (5′-lipoxygenase inhibitors) might be effective. Tumour necrosis factor (TNF)α may be an important amplifying cytokine and there are several strategies for blocking it (antibodies, soluble receptors, TACE inhibitors). IL-10 is effective in blocking the synthesis of IL-8 and TNFα as well as proteases. Oxidative stress and peroxynitrite may be important in COPD; more effective antioxidants are now in development. The inflammatory response in COPD is essentially steroid-resistant so that alternative anti-inflammatory treatments are needed. Phosphodiesterase 4 inhibitors look promising in early clinical studies. Nuclear factor-κB inhibitors and p38 MAP kinase inhibitors may also be effective. Several protease inhibitors are in development including those for neutrophil elastase, selective matrix metalloproteinase and cathepsin.

2001 Chronic obstructive pulmonary disease: pathogenesis to treatment. Wiley, Chichester (Novartis Foundation Symposium 234) p 255–272

There is a pressing need to develop drugs that control and prevent the progression of chronic obstructive pulmonary disease (COPD). There have been few therapeutic advances in the drug therapy of COPD, in contrast to the enormous advances made in asthma management. There are several possible reasons for the lack of drug development in COPD. COPD has been perceived as 'untreatable' fixed airflow obstruction. Patients with COPD have been treated with anti-asthma therapies, but these drugs may be inappropriate in a disease with a different pathophysiology. Since in most patients COPD is the result of long-term heavy cigarette smoking, it has been felt to be the 'fault' of the patient and therefore less deserving of treatment. Until recently, there has been little research interest in the molecular and cell biology of COPD to identify new therapeutic targets. There are no satisfactory animal models for early drug testing. There are uncertainties about how to test now drugs for COPD, which may require long-term studies in large numbers of patients and a lack of surrogate markers to monitor the short-term efficacy of new treatments.

No drugs that are currently available slow the progression of the disease and new classes of drug are urgently needed. However, some progress is underway and there are several classes of drug that are now in pre-clinical and clinical development (Barnes 1998a,b, 1999). The most important developments in therapy have been in the development of long-acting bronchodilators, including the long-acting β_2-agonists and the long-acting anticholinergic drug tiotropium bromide, which is suitable for once-daily dosing (Disse et al 1999).

Mediator antagonists

Several inflammatory mediators are likely to be involved in COPD as many inflammatory cells and structural cells are activated and there is an on-going inflammatory process, even in patients who have given up smoking. In asthma there are multiple mediators involved (Barnes et al 1998) and blocking the synthesis or receptors of a single mediator has usually been unsuccessful in the development of useful anti-asthma therapies. It is clear that the profile of mediators of COPD differs from that in asthma, so that different drugs are likely to be effective. Since COPD is characterized by a neutrophilic inflammation, attention has largely focused on mediators involved in recruitment and activation of neutrophils or on reactive oxygen species in view of the oxidative stress in COPD (Table 1, Fig. 1)

TABLE 1   Mediator antagonists for COPD

LTB4 antagonists (LY 29311, SC-53228, CP-105,696, SB 201146, BIIL284)
5-LO inhibitors (zileuton, Bay x1005)
Chemokine inhibitors
IL-8 antagonists (SB 225002: CXCR2 antagonist)
MCP antagonists (CCR2 antagonists)
TNF inhibitors (monoclonal antibodies, soluble receptors, TNFα converting enzyme inhibitors)
Antioxidants (stable glutathione analogues, nitrones)
Prostanoid inhibitors (COX-2 inhibitors, thromboxane antagonists, isoprostane receptor antagonists)

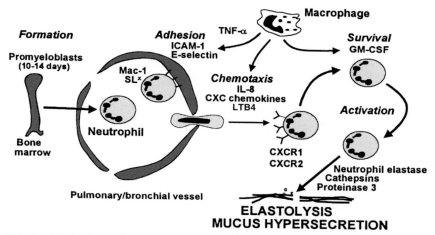

FIG. 1.   Mechanisms of neutrophil inflammation in COPD. Neutrophils formed in the bone marrow from promyeloblasts adhere in the bronchial and pulmonary circulations via adhesion molecules, then traffic into the tissue under the direction of chemotactic factors, such as leukotriene B4 (LTB4) and interleukin 8 (IL-8). They survive in the airway due to growth factors such as granulocyte–macrophage colony-stimulating factor (GM-CSF) and then become activated to release mediators and proteinases.

Leukotriene B4 inhibitors

Leukotriene B4 (LTB4) is a potent chemoattractant of neutrophils and is increased in the sputum of patients with COPD (Hill et al 1999). It is probably derived from alveolar macrophages as well as neutrophil themselves and may be synergistic with interleukin (IL)-8. Selective LTB4 receptor antagonists have now been developed. A potent LTB4 antagonist (LY293111), while ineffective against allergen challenge in asthmatic patients, inhibits the neutrophil recruitment into the airways during the late response, indicating the capacity to inhibit neutrophil chemotaxis in human airways (Evans et al 1996). Several other potent LTB4 receptor (BLT) antagonists are now in development, including SC-53228, CP-105,696, SB 201146 and BIIL284. LTB4 is synthesized by 5′-lipoxygenase (5-LO), of which there are now several inhibitors. 5-LO inhibitors, such as zileuton, are now available in some countries for the treatment of asthma, since they also inhibit the synthesis of cysteinyl leukotrienes, but it is not certain whether they will also be effective in COPD.

Chemokine inhibitors

Several chemokines are involved in neutrophil chemotaxis and mainly belong to the CXC family of chemokines, of which the most prominent member is IL-8. IL-8 levels are markedly elevated in the sputum of patients with COPD (Keatings et al 1996). Blocking antibodies to IL-8 and related chemokines inhibit certain types of neutrophilic inflammation in experimental animals, but may not be suited to long-term therapy in humans, so that there has been a search for small IL-8 receptor antagonists. IL-8 attracts neutrophils via a specific G protein-coupled receptor (CXCR1) and a common receptor shared by other members of the CXC family (CXCR2). A non-peptide inhibitor of CXCR2 (SB225002) has been discovered by screening and blocks the chemotactic response of neutrophils to IL-8 and other CXC chemokines, such as GRO-α which are also likely increased in COPD (White et al 1998). It is not certain whether blocking CXCR2 will be sufficient to block chemotactic activity of IL-8, since CXCR1 is likely also to play a role.

Other chemokines may be involved in COPD. The recruitment of large numbers of activated macrophages (presumably from blood monocytes) may be dependent on CC chemokines, such as monocyte chemotactic proteins (MCP-1–5), which activate CC receptors (CCR2) on macrophages (Luster 1998). Antagonists for CCR2 receptors are now in development.

Inhibitors of tumour necrosis factor-α

Tumour necrosis factor (TNF)α levels are also raised in the sputum of COPD patients (Keatings et al 1996) and TNFα induces IL-8 in airway cells. There is some evidence that the severe wasting in some patients with advanced COPD might be due to circulating TNFα. Humanized monoclonal TNF antibodies, such as infliximab, and soluble TNF receptors, such as etanercept are effective in other chronic inflammatory diseases, such as rheumatoid arthritis and inflammatory bowel disease. There may be problems with long-term administration because of the development of blocking antibodies and the inconvenience of repeated injections. TNFα-converting enzyme, which is required for the release of soluble TNFα, may be a more attractive target as it is possible to discover small molecule TNFα-converting enzyme (TACE) inhibitors, some of which are also matrix metalloproteinase (MMP) inhibitors.

Antioxidants

Oxidative stress is increased in patients with COPD, particularly during exacerbations, and reactive oxygen species contribute to its pathophysiology, as previously discussed (MacNee 2000, this volume). This suggests that antioxidants may be of use in the therapy of COPD. N-acetyl cysteine (NAC) provides cysteine for enhanced production of glutathione (GSH) and has antioxidant effects in vitro and in vivo. More effective antioxidants, including stable glutathione compounds and selenium based drugs, are now in development for clinical use. Spin-trap antioxidants, such as α-phenyl-N-tert-butyl nitrone, are much more potent and inhibit the formation of intracellular reactive oxygen species by forming stable compounds.

New anti-inflammatory treatments

COPD is characterized by chronic inflammation of the respiratory tract, even in ex-smokers. Bronchoalveolar lavage and induced sputum in patients with COPD demonstrate increased numbers of neutrophils and macrophages (Keatings et al 1996). At sites of lung destruction in the lung parenchyma there are increased numbers of macrophages and CD8[+] (cytotoxic) T lymphocytes; similar changes are seen in the airway walls. The mechanisms of the neutrophilic inflammation in COPD is not yet understood, but it is likely that neutrophil chemotactic factors are released into the airways from activated macrophages and possibly from epithelial cells and CD8[+] T lymphocytes. It is important to elucidate more precisely the molecular and cellular mechanisms of COPD in order to identify novel targets for therapy. Our current superficial understanding of COPD suggests that there may be several approaches (Fig. 2).

Because there is chronic inflammation in COPD airways it has been argued that inhaled corticosteroids might prevent the progression of the disease. However, three recent large controlled trials of inhaled corticosteroids have demonstrated no reduction in disease progression (Pauwels et al 1999). This might be predicted by the demonstration that neither inhaled nor oral corticosteroids have any significant effect on neutrophil counts, granule proteins or inflammatory cytokines in induced sputum (Keatings et al 1997) or on proteases or antiproteases in induced sputum (Culpitt et al 1999). The disappointing action of corticosteroids in COPD suggests that novel types of non-steroidal anti-inflammatory treatment may be needed. There are several new approaches to anti-inflammatory treatment in COPD (Table 2).

Phosphodiesterase 4 inhibitors

Inhibition of phosphodiesterases (PDEs) increases cyclic AMP content of neutrophils, resulting in reduced chemotaxis, activation, degranulation and adherence (Nielson et al 1990). The predominant isoenzyme in inflammatory cells is PDE4 (Wang et al 1999) and several PDE4 inhibitors are now in clinical development for asthma. PDE4 inhibitors also inhibit the function of macrophages and CD8[+] T lymphocytes which are also involved in the inflammatory process in COPD (Giembycz et al 1996) (Fig. 3). Many of the first generation PDE4 inhibitors have been limited by side effects, particularly nausea. In second generation PDE4 inhibitors, such as Cilomilast (SB207499), this may be less of a problem and a trial

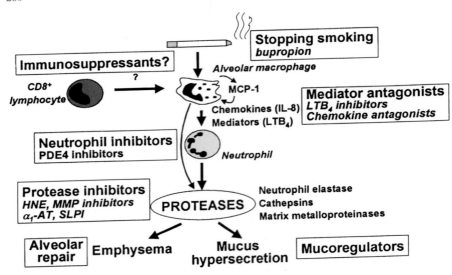

FIG. 2. Targets for COPD therapy based on current understanding of the inflammatory mechanisms. Cigarette smoke (and other irritants) activate macrophages in the respiratory tract that release neutrophil chemotactic factors, including interleukin 8 (IL-8) and leukotriene B4 (LTB4). These cells then release proteases that break down connective tissue in the lung parenchyma, resulting in emphysema, and also stimulate mucus hypersecretion. These enzymes are normally counteracted by protease inhibitors, including α_1-antitrypsin, secretory leukoprotease inhibitor (SLPI) and tissue inhibitor of matrix metalloproteinases (TIMP). Cytotoxic T cells (CD8[+]) may also be involved in the inflammatory cascade.

TABLE 2 New anti-inflammatory drugs for COPD

PDE4 inhibitors (Cilomilast, CP 80633, CDP-840)

NF-κB inhibitors (proteasome inhibitors, IκB kinase inhibitors, IκB-α gene transfer)

Adhesion molecule inhibitors (anti CD11/CD18, anti-ICAM-1, E-selectin inhibitors)

IL-10 and analogues

p38 MAP kinase inhibitors (SB203580, SB 220025, RWJ 67657)

---

of this drug has shown an improvement in lung function and symptoms in patients with moderately severe COPD (Torphy et al 1999).

NF-κB inhibitors

The transcription factor nuclear factor-κB (NF-κB) regulates the expression of IL-8, TNFα and MMPs; its inhibition therefore inhibits neutrophilic inflammation (Barnes & Karin 1997). There are several possible approaches to inhibition of

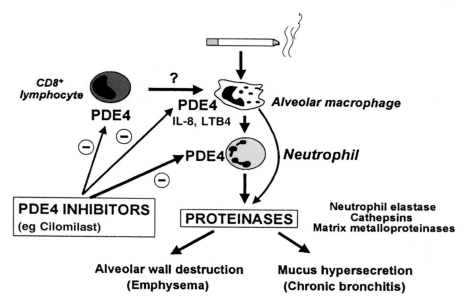

FIG. 3. Effect of phosphodiesterase 4 (PDE4) inhibitors on the inflammatory process in COPD.

NF-κB, including gene transfer of the inhibitor of NF-κB (IκB), a search for inhibitors of IκB kinases (IKK), NF-κB-inducing kinase (NIK) and IκB ubiquitin ligase, which regulate the activity of NF-κB, and the development of drugs that inhibit the degradation of IκB. One concern about this approach is that effective inhibitors of NF-κB may result in immune suppression and impair host defences, since knockout mice which lack NF-κB proteins succumb to septicaemia. However, there are alternative pathways of NF-κB activation that might be more important in inflammatory disease (Nasuhara et al 1999).

Adhesion molecule blockers

Neutrophil recruitment into the lungs and respiratory tract is dependent on adhesion molecules expressed on neutrophils and endothelial cells in the pulmonary and bronchial circulations. Neutrophil adhesion in response to chemotactic factors is characterized by expression of the β2 integrins CD11a/CD18 (LFA-1) and CD11b/CD18 (Mac-1) on the surface of the neutrophil and their interaction with their counterreceptors, including intercellular adhesion molecule 1 (ICAM-1), on endothelial cells. E-selectin on endothelial cells also interacts with sialyl-Lewis[x] on neutrophils. Bronchial biopsies of patients with

COPD have demonstrated increased expression of E-selectin on vessels and ICAM-1 on epithelial cells (Di Stefano et al 1994). Drugs that interfere with these adhesion molecules should therefore inhibit neutrophil inflammation in COPD. Monoclonal antibodies to CD18, ICAM-1 and E-selectin inhibit neutrophil accumulation in animal models of lung inflammation. Analogues of sialyl-Lewis[x] have been developed which block selectins and inhibit granulocyte adhesion (Kogan et al 1998). However, there are concerns about this therapeutic approach for a chronic disease, as an impaired neutrophilic response may increase the susceptibility to infection.

IL-10

IL-10 is a cytokine with a wide spectrum of anti-inflammatory actions. It inhibits the secretion of TNFα and IL-8 from macrophages, but tips the balance in favour of antiproteases, by decreasing the expression of MMPs, while increasing the expression of endogenous tissue inhibitors of matrix metalloproteinases (TIMP) (Lacraz et al 1995). IL-10 is currently in clinical trials for other chronic inflammatory diseases (inflammatory bowel disease, rheumatoid arthritis and psoriasis), including patients with steroid resistance (van Deventer et al 1997). Treatment with daily injections of IL-10 over several weeks has been well tolerated. IL-10 may have therapeutic potential in COPD, especially if a selective activator of IL-10 receptors or signal transduction pathways can be developed.

p38 MAP kinase inhibitors

Mitogen-activate protein (MAP) kinases play a key role in chronic inflammation and several complex enzyme cascades have now been defined. One of these, the p38 MAP kinase pathway, is involved in expression of inflammatory cytokines, including IL-8 and TNFα (Carter et al 1999). Non-peptide inhibitors of p38 MAP kinase, such as SB 203580, SB 220025 and RWJ 67657, have now been developed and these drugs have a broad range of anti-inflammatory effects (Lee et al 1999).

Protease inhibitors

There is compelling evidence for an imbalance between proteases that digest elastin (and other structural proteins) and antiproteases that protect against this in COPD. This suggests that either inhibiting these proteolytic enzymes or increasing antiproteases may be beneficial and theoretically should prevent the progression of airflow obstruction in COPD. Considerable progress has been made in identifying the enzymes involved in elastolytic activity in emphysema and in characterizing the endogenous antiproteases that counteract this activity (this volume: Stockley 2000, Cawston 2000). These include inhibitors of neutrophil elastase, cathepsins and specific (MMPs) with elastase activity (such as MMP-9 and MMP-12). An alternative approach is to give endogenous antiproteases (α_1-antitrypsin, secretory leukoprotease inhibitor, elafin, tissue inhibitor of MMP) either in recombinant form or by viral vector gene delivery.

Mucoregulators

Increased secretion of mucus is found in all patients who smoke heavily, irrespective of airflow obstruction. However, recent epidemiological data suggests that mucus hypersecretion is significantly associated with a more rapid decline in FEV_1 (forced expiratory volume in one second) and increased hospitalization of patients with COPD (Vestbo et al 1996). This suggests that it may be important to develop drugs that inhibit the hypersecretion of mucus, although it is important to find drugs that do not suppress the normal mucus secretion or impair mucociliary clearance. There are several types of mucoregulatory drug in development. These include tachykinin antagonists and inhibitors of sensory neuropeptide release (such as potassium channel openers) since tachykinins are potent mucus secretagogues (Ramnarine et al 1998). Inhibitors of serine proteases may also be effective as these enzymes are very potent at stimulating mucus secretion. In the future specific suppressors of mucin (*MUC*) genes may be developed as there is evidence for increased expression of certain mucin genes, such as *MUC5AC* in COPD.

Alveolar repair

Since a major mechanism of airway obstruction in COPD is loss of elastic recoil due to proteolytic destruction of lung parenchyma, it seems unlikely that this could be reversible by drug therapy, although it might be possible to reduce the rate of progression by preventing the inflammatory and enzymatic disease process. It is even possible that drugs might be developed that will stimulate regrowth of alveoli. Retinoic acid increases the number of alveoli in rats and, remarkably, reverses the histological and physiological changes induced by elastase treatment (Massaro 2000, this volume). It is not certain whether such alveolar proliferation is possible in adult human lungs, however. Retinoic acid activates intracellular retinoic acid receptors, which act as transcription factors to regulate the expression of many genes. The molecular mechanisms involved and whether this can be extrapolated to humans is not yet known. Several retinoic acid receptor subtype agonists have now been developed that may have a greater selectivity for this effect. Hepatocyte growth factor (HGF, also known as scatter factor) has a major effect on the growth of alveoli in fetal lung (Ohmichi et al 1998) and it is possible that in future drugs might be developed that switch on responsiveness to HGF in adult lung or mimic the action of HGF.

Route of delivery

Bronchodilators are currently given as inhalers, metered dose inhalers or dry powder inhalers, that have been optimised to deliver drugs to the respiratory tract in asthma. But in emphysema the inflammatory and destructive process takes place in the lung parenchyma. This implies that if a drug is to be delivered by inhalation that it should have a lower mass median diameter, so that there is preferential deposition in the lung periphery. It may be more appropriate to give therapy parenterally as it will need to reach the lung parenchyma via the pulmonary circulation, but parenteral administration may increase the risk of systemic side effects.

Future directions

New drugs for the treatment of COPD are needed. While preventing and quitting smoking is the obvious preferred approach, this has proved to be very difficult in the majority of patients, and even with bupropion only ~30% of patients are sustained quitters. In addition, it is likely that the inflammatory process initiated by cigarette smoking may continue even when smoking has ceased. Furthermore, approximately 10% of patients with COPD are non-smokers. COPD may be due to other environmental factors (cooking fumes, pollutants, passive smoking, other inhaled toxins) or due to developmental changes in the lungs.

Identification of novel therapeutic targets

It is important to identify the factors that determine why only 10–20% of smokers develop COPD. So far this is little understood, although it is likely that genetic factors are important (Silverman 2000, this volume). Several gene polymorphisms have already been associated with COPD and it will eventually be possible to identify at risk patients and focus more effective therapies on these patients before lung function becomes too impaired (Barnes 1999). Identification of genes that predispose to the development of COPD in smokers may identify novel therapeutic targets. Powerful techniques, including high density DNA arrays (gene chips) are able to identify multiple polymorphisms, differential display may identify the expression of novel genes and proteomics of novel proteins expressed.

Early detection of disease

Since at the moment COPD is irreversible and slowly progressive it will become ever more important to identify early cases before symptoms develop as effective therapies emerge. At present since there are no known drugs that alter the course of disease progression, there is no impetus to diagnose COPD early.

Surrogate markers

Several drugs now in development may be useful in COPD. These include LTB4 antagonists and 5-LO inhibitors, PDE4 inhibitors, new antioxidants, neutrophil elastase and MMP inhibitors. It will be difficult to demonstrate the efficacy of such treatments as determination of the effect of any drug on the rate of decline in lung function will require large studies over at least two years. There is an urgent need to develop surrogate markers, such as analysis of sputum parameters (cells, mediators, enzymes) or exhaled condensates (lipid mediators, reactive oxygen species, cytokines), that may predict the clinical usefulness of such drugs. More research on the basic cellular and molecular mechanisms of COPD and emphysema is urgently needed to aid the logical development of new therapies for this common and important disease for which no effective preventive treatments currently exist.

References

Barnes PJ 1998a Chronic obstructive pulmonary disease: new opportunities for drug development. Trends Pharmacol Sci 19:415–423

Barnes PJ 1998b New therapies for chronic obstructive pulmonary disease. Thorax 53:137–147

Barnes PJ 1999 Genetics and pulmonary medicine. 9. Molecular genetics of chronic obstructive pulmonary disease. Thorax 54:245–252

Barnes PJ, Karin M 1997 Nuclear factor-κB: a pivotal transcription factor in chronic inflammatory diseases. New Engl J Med 336:1066–1071

Barnes PJ, Chung KF, Page CP 1998 Inflammatory mediators of asthma: an update. Pharmacol Rev 50:515–596

Carter AB, Monick MM, Hunninghake GW 1999 Both erk and p38 kinases are necessary for cytokine gene transcription. Am J Respir Cell Mol Biol 20:751–758

Cawston T 2001 Matrix metalloproteinases and inhibitors. In: Chronic obstructive pulmonary disease: pathogenesis to treatment. Wiley, Chichester (Novartis Found Symp 234) p 205–228

Culpitt SV, Nightingale JA, Barnes PJ 1999 Effect of high dose inhaled steroid on cells, cytokines and proteases in induced sputum in chronic obstructive pulmonary disease. Am J Respir Crit Care Med 160:1635–1639

Di Stefano A, Maestrelli P, Roggeri A et al 1994 Upregulation of adhesion molecules in the bronchial mucosa of subjects with chronic obstructive bronchitis. Am J Respir Crit Care Med 149:803–810

Disse B, Speck GA, Rominger KL, Witek TJ JR, Hammer R 1999 Tiotropium (Spiriva): mechanistical considerations and clinical profile in obstructive lung disease. Life Sci 64: 457–464

Evans DJ, Barnes PJ, Spaethe SM, van Alstyne EL, Mitchell MI, O'Connor BJ 1996 The effect of a leukotriene B_4 antagonist LY293111 on allergen-induced responses in asthma. Thorax 51:1178–1184

Giembycz MA, Corrigan CJ, Seybold J, Newton R, Barnes PJ 1996 Identification of cyclic AMP phosphodiesterases 3, 4 and 7 in human $CD4^+$ and $CD8^+$ T-lymphocytes: role in regulating proliferation and the biosynthesis of interleukin-2. Br J Pharmacol 118:1945–1958

Hill AT, Bayley D, Stockley RA 1999 The interrelationship of sputum inflammatory markers in patients with chronic bronchitis. Am J Respir Crit Care Med 160:893–898

Keatings VM, Collins PD, Scott DM, Barnes PJ 1996 Differences in interleukin-8 and tumor necrosis factor-α in induced sputum from patients with chronic obstructive pulmonary disease or asthma. Am J Respir Crit Care Med 153:530–534

Keatings VM, Jatakanon A, Worsdell YM, Barnes PJ 1997 Effects of inhaled and oral glucocorticoids on inflammatory indices in asthma and COPD. Am J Respir Crit Care Med 155:542–548

Kogan TP, Dupré B, Bui H et al 1998 Novel synthetic inhibitors of selectin-mediated cell adhesion: synthesis of 1,6-bis[3-(3-carboxymethylphenyl)-4-(2-α-D-mannopyranosyloxy)-phenyl]hexane (TBC1269). J Med Chem 41:1099–1111

Lacraz S, Nicod LP, Chicheportiche R, Welgus HG, Dayer JM 1995 IL-10 inhibits metalloproteinase and stimulates TIMP-1 production in human mononuclear phagocytes. J Clin Invest 96:2304–2310

Lee JC, Kassis S, Kumar S, Badger A, Adams JL 1999 p38 mitogen-activated protein kinase inhibitors — mechanisms and therapeutic potentials. Pharmacol Ther 82:389–397

Luster AD 1998 Chemokines — chemotactic cytokines that mediate inflammation. New Engl J Med 338:436–445

MacNee W 2001 Oxidants/antioxidants and chronic obstructive pulmonary disease: pathogenesis to therapy. In: Chronic obstructive pulmonary disease: pathogenesis to treatment. Wiley, Chichester (Novartis Found Symp 234) p 169–188

Massaro D 2001 Alveolus formation: critical period, plasticity and retinoid regulation. In: Chronic obstructive pulmonary disease: pathogenesis to treatment. Wiley, Chichester (Novartis Found Symp 234) p 229–243

Nasuhara Y, Adcock IM, Catley M, Barnes PJ, Newton R 1999 Differential IKK activation and IκBa degradation by interleukin-1β and tumor necrosis factor-α in human U937 monocytic cells: evidence for additional regulatory steps in κB-dependent transcription. J Biol Chem 274:19965–19972

Nielson CP, Vestal RE, Sturm RJ, Heaslip R 1990 Effect of selective phosphodiesterase inhibitors on the polymorphonuclear leukocyte respiratory burst. J Allergy Clin Immunol 86:801–808

Ohmichi H, Koshimizu U, Matsumoto K, Nakamura T 1998 Hepatocyte growth factor (HGF) acts as a mesenchyme-derived morphogenic factor during fetal lung development. Development 125:1315–1324

Pauwels RA, Lofdahl CG, Laitinen LA et al 1999 Long-term treatment with inhaled budesonide in persons with mild chronic obstructive pulmonary disease who continue smoking. N Engl J Med 340:1548–1553

Ramnarine SI, Liu YC, Rogers DF 1998 Neuroregulation of mucus secretion by opioid receptors and K(ATP) and BK(Ca) channels in ferret trachea in vitro. Br J Pharmacol 123:1631–1638

Silverman E 2001 Genetics of chronic obstructive pulmonary disease. In: Chronic obstructive pulmonary disease: pathogenesis to treatment. Wiley, Chichester (Novartis Found Symp 234) p 45–65

Stockley RA 2001 Proteases and antiproteases. In: Chronic obstructive pulmonary disease: pathogenesis to treatment. Wiley, Chichester (Novartis Found Symp 234) p 189–204

Torphy TJ, Barnette MS, Underwood DC et al 1999 Ariflo (SB 207499), a second generation phosphodiesterase 4 inhibitor for the treatment of asthma and COPD: from concept to clinic. Pulm Pharmacol Ther 12:131–136

van Deventer SJ, Elson CO, Fedorak RN 1997 Multiple doses of intravenous interleukin 10 in steroid-refractory Crohn's disease. Crohn's Disease Study Group. Gastroenterology 113: 383–389

Vestbo J, Prescott E, Lange P 1996 Association of chronic mucus hypersecretion with FEV1 decline and chronic obstructive pulmonary disease morbidity. Copenhagen City Heart Study Group. Am J Respir Crit Care Med 153:1530–1535

Wang P, Wu P, Ohleth KM, Egan RW, Billah MM 1999 Phosphodiesterase 4B2 is the predominant phosphodiesterase species and undergoes differential regulation of gene expression in human monocytes and neutrophils. Mol Pharmacol 56:170–174

White JR, Lee JM, Young PR et al 1998 Identification of a potent, selective non-peptide CXCR2 antagonist that inhibits interleukin-8-induced neutrophil migration. J Biol Chem 273:10095–10098

DISCUSSION

Jeffery: Why did you leave out the long-acting bronchodilator salmeterol? This seems to have some anti-neutrophil activity, and you mentioned theophylline. These both work through raising cAMP.

Barnes: GlaxoWellcome have now analysed all the salmeterol studies in COPD, and although this was not a primary measure in their studies, they have shown that the number of exacerbations was reduced with salmeterol treatment. This fits in with studies by Rob Wilson at NHLI that salmeterol inhibits bacterial adhesion through a cAMP-dependent mechanism.

Jeffery: We have found from looking at biopsies that salmeterol has down-regulatory effects on neutrophil numbers.

Calverley: There is a clinical trial that is about half-way through, comparing the effects of salmeterol, fluticasone, the combination of both, and placebo on exacerbation rate. Changing the exacerbation rate will be an important endpoint for many other drugs such as LTB4 and PDE4 antagonists.

Rennard: An effect on exacerbations doesn't necessarily mean an anti-inflammatory effect. Friedman, in a careful, retrospective analysis of several studies, demonstrated ipratropium also has an effect on exacerbations (Friedman et 1999). I suppose there could be an anti-inflammatory effect of ipratropium, but I don't think anyone has suggested a mechanism of how that might come to be. Clearly an effect on exacerbations would be clinically useful, but you can't necessarily say that this is due to an anti-inflammatory mechanism.

Nadel: A general comment: over a period of time people have figured out some simple methods of evaluating the efficacy of drugs, and in lung disease: FEV_1. Then after having begun to do this and finding out that there is some relationship between severity of disease morbidity and mortality, FEV_1 has inherited the job of evaluating the drugs. I make the plea that we look at new measures of efficacy that might be better.

Barnes: This is one of the reasons we have invested considerable effort into developing non-invasive surrogate markers of inflammation and oxidative stress in the lungs. If you have a drug that is meant to act on these processes, then this can be measured directly. In COPD it is a particular problem, because we do not have the large changes in lung function that are easily measured.

Stockley: What those surrogates do largely is to give the reassuring phase II step to a pharmaceutical company. This still doesn't get away from the long-term problem of then proving efficacy, where we are still locked into FEV_1 as being our primary efficacy outcome in long term studies. I agree with Jay Nadel: we should try to dispel that myth forever.

Barnes: In recent clinical trials in COPD several parameters, in addition to FEV_1, are measured, including exacerbations and inflammatory markers.

Stockley: But the drug regulation agencies have FEV_1 locked in as their primary outcome measure.

Calverley: They certainly have that locked in. If you want to make a claim about disease modification in COPD, the regulators will look for evidence that you have changed the rate of decline of lung function. This model is tightly related to the intervention we talked about at length earlier in the meeting, namely smoking cessation. However, they also want to see evidence of change in some symptomatic or health status marker. Even the regulators are beginning to recognise that we don't catch the whole story in this disease with FEV_1. This is probably an encouraging trend.

MacNee: Apart from the FEV_1 and a health status measurement, what else are we suggesting? If you suggest computed tomography (CT) scanning, for example, is that without measuring health status or FEV_1?

Rennard: I think we should be suggesting CT scanning people's thighs. We have heard that COPD is a systemic disease. Whether the systemic effects are secondary or how they work isn't clear, but they clearly have impact on patient functionality. Yet we don't know whether any of these drugs affect lean body mass or muscle strength. These would be completely valid clinical endpoints, independent of airflow or change in airflow or natural history of airflow changes. There is quite a milieu of potential systemic effects. If fibrinogen levels are increased and cardiac death is increased in COPD patients, if fibrinogen levels are decreased then cardiac risk may be decreased even if the FEV_1 were unchanged, and this would be a meaningful clinical endpoint. We could come up with a number of potential targets as efficacy endpoints.

Lomas: I agree. I would like to go back to Alvar Agustí's 'DNA' classification, which is simple and snappy, and that is the way we should be going. COPD is not just FEV_1. The CT scans are of greatest value if you are looking at emphysema, and this has been best demonstrated in the α_1-antitrypsin-deficient subgroup of COPD.

MacNee: Why do you think that CT scanning is only useful in this group?

Lomas: It may be useful in other groups, but we don't understand how emphysema develops in COPD, with regard to who gets centrolobular, who gets panlobular. However, we do understand this best in α_1-antitryspin deficiency.

Stockley: The type of emphysema that patients get is not the point, because this can be detected and tracked by high-resolution CT scanning. If that is the target then CT scanning would be immovable as the primary outcome measure.

Lomas: The only caveat is that high-resolution CT scans cannot differentiate between panlobular emphysema and small airways disease.

Hogg: In my opinion, this will never be done by visual assessment but may be accomplished by objective measurements. It is possible to use the basic measurements from the scan to calculate the amount of the lung that is expanded beyond normal. I think this is the way to go because the definition of emphysema is based on expansion beyond the normal.

Nicklin: The apparent efficacy of theophylline in COPD is very interesting and — because of the very positive data for Cilomilast — it is tempting to speculate that it is working via PDE4 inhibition. Do you think that this is the case?

Barnes: The concentrations of theophylline that we are talking about have little inhibitory effect on PDE4. We have found that below the 'therapeutic' concentrations, which have not been shown to cause bronchodilation, theophylline activates an enzyme called histone deacetylase (HDAC)1, which is able to switch off activated inflammatory genes. Interestingly, cigarette smoke has exactly the opposite effect. In A549 cells we have shown a potentiation of steroid effects by theophylline. In asthmatic patients we have found that low doses of theophylline increase HDAC activity in bronchial biopsies and also reduce eosinophil numbers.

Nicklin: Is it acting as a kinase inhibitor at some point?

Barnes: Yes, I think may be working through a MAP kinase pathway.

Poll: The initial data with Cilomilast are pretty impressive. Were you surprised that it appears to work rather quickly to improve FEV_1, and is this related to inflammation?

Barnes: PDE4 inhibitors are complicated by the fact that they are both bronchodilators and anti-inflammatory. It is difficult to know how much of the improvement is due to each mechanism. The arguments in favour of an anti-inflammatory component are the relatively slow onset and the fact that the bronchodilator response to β agonists was the same at the end of the study as it was at the beginning. The bronchodilator response in COPD is very small and would not be expected to give the sort of improvement that was measured.

Hogg: You say that it is important to measure markers of the inflammatory response, but in a sense the small airways test did that. We know that everyone who smokes gets airway inflammation and small airways tests showed abnormalities that correlated with smoking habits. What they didn't do was predict the people who were going to get into long-term trouble. It is also possible that measurements of acute changes in the inflammatory mediators will not help predict those who will develop COPD in the long term.

Barnes: Everything that we have measured in COPD (mediators, cytokines, inflammatory cells, oxidants and proteases) is the same as what we measure in smokers, but the levels are greater.

Hogg: You are really talking about the size of the response therefore, rather than the type of response.

Barnes: COPD appears to be an amplification of the inflammatory response to smoking. What is seen in COPD is an inflammatory response to an irritant, which is exaggerated for reasons that we do not yet understand.

MacNee: We have the opportunity to study inflammatory markers and match these data in smaller groups of patients with the biopsy studies. Although there may be more airway inflammation it doesn't relate necessarily to progression of the disease.

Calverley: This ties up with the regulatory argument, too. We think that endpoints for regulatory studies such as CT change are reasonable. But unless we can go that step further and demonstrate that they are reproducible and robust, we will be no further forward with any of these new markers than we were when we introduced small airways tests that simply showed that people had small airways disease and we could not pick out those who would deteriorate in the future. With respect to the key studies that need doing, we have certainly identified enough possible markers, but we now have to choose the best of these and relate them to the endpoints. This simple and in some cases rather unexciting activity might be the most informative one we could pursue for the next year or two.

Stockley: I don't want to see this meeting close without us getting our balance right. One of the issues I raised earlier is this concept of separating cause and effect. Peter Barnes has just made a valid comment that smokers have a degree of inflammation, and if they have COPD they have more inflammation. The difficulty now is trying to separate that out. Is it that they got their COPD because of that excessive inflammation, or are we just seeing a greater degree of inflammation because COPD has set in? We run the risk of going down blind alleys because we still haven't separated cause and effect.

Barnes: The most promising way forward is to test specific inhibitors in this disease. Animal models are not going to inform us much: we have to do studies in patients with the disease. That is why these drugs that are becoming available are going to be the major step forward in understanding COPD mechanisms. We have learned a lot in asthma by using specific inhibitors.

Stockley: You could do the same in your patients. If you divide them into current and ex-smokers and see if the inflammatory signal changes, that might be more informative.

Barnes: We are planning a study like that now. In some smokers the inflammation will go away and in others it won't.

Rennard: We have done studies on reducing smoking before the development of COPD (Rennard et al 1990), and others have looked at stopping smoking (Sköld et al 1992, 1993a,b, 1996). Clearly, in smokers when they stop or reduce smoking, there can be an associated reduction in inflammation that can be measured by bronchoscopy. So there is a possibility that changing smoking will change inflammation. This doesn't mean that once COPD develops that the inflammatory process won't have some sort of mechanism of self perpetuation. There are a number of important unanswered question.

Stockley: We did a study looking at about 30 patients who continued to smoke and 30 who had given up who were otherwise matched. When we looked at the inflammatory signal, we found a slight reduction in myeloperoxidase, but it was only IL-8 of the cytokines we measured that showed a reduction. We need to do this sort of study and look at TNF, for example, to identify viable targets.

Wedzicha: Going back to what Peter Barnes said about inflammation and exacerbation, I do tend to agree with Steve that exacerbation is not the same thing as inflammation. The only situation in which exacerbations do seem to affect inflammatory markers is if they are associated with viral infections or symptomatic colds. One thing we are going to do is to look at the symptoms associated with exacerbations much more carefully. For example, if sputum is reduced, this may explain why anticholinergics reduce exacerbations.

Barnes: Do you think coughing is a surrogate marker for sputum production?

Stockley: No, it is an inverse marker. The more phlegm you have, the easier it is to cough. Once you start to treat patients, sputum purulence goes down, sputum volume goes down, breathlessness gets better but cough may get worse. And if you see patients — after all, you must have seen one at some stage — and ask them about stopping smoking, they say that if they stop smoking their cough gets worse.

Barnes: We need some more accurate measurement of mucus secretion in COPD.

References

Friedman M, Serby CW, Menjoge SS, Wilson JD, Hilleman DE, Witek TJ Jr 1999 Pharmacoeconomic evaluation of a combination of ipratropium plus salbuterol compared with ipratropium alone and albuterol alone in COPD. Chest 115:635–641

Rennard SI, Daughton D, Fujita J et al 1990 Short-term smoking reduction is associated with reduction in measures of lower respiratory tract inflammation in heavy smokers. Eur Respir J 3:752–759

Sköld CM, Hed J, Eklund A 1992 Smoking cessation rapidly reduces cell recovery in bronchoalveolar lavage, while alveolar macrophage fluorescence remains high. Chest 101:989–995

Sköld CM, Andersson K, Hed J, Eklund A 1993a Short-term *in vivo* exposure to cigarette-smoke increases the fluorescence in rat alveolar macrophages. Eur Respir J 6:1169–1172

Sköld CM, Forslid J, Eklund A, Hed J 1993b Metabolic activity in human alveolar macrophages increases after cessation of smoking. Inflammation 17:345–352

Sköld CM, Blaschke E, Eklund A 1996 Transient increases in albumin and hyaluronan in bronchoalveolar lavage fluid after quitting smoking: possible signs of reparative mechanisms. Respir Med 90:523–529

Closing remarks

William MacNee

ELEGI/Colt Laboratories, Department of Medical and Radiological Sciences, Wilkie Building, The University of Edinburgh, Medical School, Teviot Place, Edinburgh EH8 9AG, UK

We were all reminded at the beginning of this symposium that chronic obstructive pulmonary disease (COPD) is a heterogeneous disease — if it is to be called a disease at all. Jim Hogg's overview of the pathology began by discussing inflammation, then we talked about injury, and then repair. It seems that inflammation is present in every smoker in all compartments of the lungs. We discussed the dynamics of the inflammatory responses which are very complex and events occurring in a micro-environment were clearly very important. A major question that arose from this discussion was whether there is amplification of the inflammatory response in individuals susceptible to the development of COPD. It is clear that we need to know more about the natural history of the pathologies which make up COPD. We now have new techniques to study pathological processes. These include computed tomography (CT) scanning and surrogate markers of inflammation and oxidative stress.

Peter Calverley gave us an overview of current treatments for COPD. Peter's concept of the overlap between pathology, physiology and disease impact is critical: we clearly now need to make other measurements in addition to the FEV_1 (forced expiratory volume in one second) in the assessment of patients with COPD. We mentioned measurements of health status and inspiratory capacity in this regard. We now have much more information on the role of corticosteroids in COPD, although there is still an ongoing debate of their efficacy. If in most patients corticosteroids are ineffective then the study of the mechanism which accounts for this is going to be important. It will also be worth investigating the reason why some COPD patients *do* show a response to corticosteroids. We discussed the stages of the disease which require further study. Studies of the effects of smoking cessation and the possible persistence of inflammation are important experiments in this regard. We also emphasized the importance of characterization of patients with COPD since this is a heterogeneous disease, particularly by biopsy studies and inflammatory markers.

Ed Silverman outlined the interesting area of susceptibility to COPD. Why is it that only 15–20% of smokers get this disease? It has been suggested that a panel of genetic polymorphisms could be used to characterize an individual's susceptibility to COPD. However it is clear that there are still lessons to be learned from the major genetic risk factor for emphysema — α_1-antitrypsin deficiency. The reason why most α_1-antitrypsin-deficient subjects even some who smoke don't develop clinically significant airways obstruction is an important issue worthy of further study and may be due to other genetic factors.

Steve Rennard raised the issue of the dose effects of cigarette smoking which needs to be accounted for in studies of susceptibility. There was an interesting discussion about gender and susceptibility to COPD and there were suggestions that we should be examining genetic factors which may make patients more susceptible to exacerbations of COPD. There was also discussion about the development of animal models and susceptibility, which is a new concept to me. One of the major ideas to come out of this symposium is the need to standardize the technique of CT scanning to quantify emphysema.

Then we came onto the section of the meeting related to mechanisms. One aspect which came over clearly is that mucus hypersecretion is not such an innocent feature as we previously thought. Mucus may have a protective role in large airways, but is problematic in small airways where mucociliary clearance is less developed. Study of the mechanism of hypersecretion should lead to new therapies and Jay Nadel's work on epidermal growth factor receptor (EGFR) blockers is certainly important in this regard. One of the most interesting discussions in the symposium concerned the protease–antiprotease theory of the pathogenesis of emphysema. Rob Stockley talked about serine proteases and then we discussed the matrix metalloproteinases (MMPs). Clearly the dynamics of the protease–antiprotease balance in a microenvironment is quite critical, and may be why the studies done in the past with bronchalveolar lavage (BAL) have not provided any consistent answers. We do need proof of concept with elastase inhibitors. I was fascinated to learn from Tim Cawston about the interaction of MMPs and their inhibitors, which complicates the whole story.

Which mechanisms should we target? An important target in COPD is the exacerbations of the condition. We didn't discuss bacterial colonization and exacerbations, however a high proportion of exacerbations are associated with acute viral infections. Jim Hogg's work on latent adenoviral infection is of interest in susceptibility to COPD but also in enhancing the inflammatory response in the airways. We discussed surrogate markers of inflammation. Exacerbations are a good place to test reproducibility of these markers. There is still controversy over the presence and role of eosinophils in exacerbations and the apparent lack of any increase in neutrophils in induced sputum in exacerbations is intriguing. There now appears to be evidence for the role of corticosteroids in both the prevention of exacerbations and their treatment. We didn't spend time discussing definitions, but it is going to be difficult to compare studies until we have an agreed definition of an exacerbation of COPD. Wisia Wedzicha's data suggests that we should re-examine the relationship between exacerbations and the decline in FEV_1.

In Steve Rennard's paper on injury and repair, we heard that transforming growth factor (TGF)β may be an important factor in the repair processes in small airways. We discussed cytokine networks, integrins and cell–cell interactions involved in this process which may be future targets for treatment. I raised the issue of antioxidant therapy as a treatment which may not only have direct protective effects against the injurious effects of oxidants, but may have more general anti-inflammatory effects through regulation of gene transcription of pro-inflammatory mediators. Peter Jeffery told us of the interesting data on CD8 lymphocytes which appear to be more prominent in smokers who develop COPD. Our impression was that we need to study larger populations of patients in order to cover the heterogeneous population of COPD patients. Peter Jeffery also discussed possible mechanisms which produce the increase in CD8 lymphocytes. A group that are worth studying in this regard are smoking asthmatics. We heard about macrophages from Dr Mantovani who presented some very interesting information on the balance between CCR2 and macrophage chemotactic protein (MCP). MCP may be a future target for therapy. Tim Williams told us about neutrophils, and the time course of the different stimuli producing neutrophil influx into the lungs. We heard that the stimulus is important to the events causing the neutrophil influx. We also discussed work by Jim Hogg's group on the bone marrow release of neutrophils in response to cigarette smoke. This led us to Alvar Agustí's presentation on the new concept that COPD is not only in the lungs but is also a systemic disease. We also heard that emphysema may not be an irreversible condition in Don Massaro's paper on the effect of retinoids on alveolus formation, and the fascinating work on calorie restriction and emphysema. Finally, Peter Barnes told us about the new therapies which are on the horizon particularly new anti-inflammatory treatments.

It has been a fascinating symposium with a particularly intense discussion and I hope will inspire us to further research efforts with a view to new treatments for this disabling condition.

Index of contributors

Subject index